Natural Fullerenes and Related Structures of Elemental Carbon

Developments in Fullerene Science

Volume 7

Series Editor:

Tibor Braun, *Institute of Inorganic and Analytical Chemistry, L. Eötvös University, Budapest, Hungary*

The titles published in this series are listed at the end of this volume.

Natural Fullerenes and Related Structures of Elemental Carbon

by

Frans J.M. Rietmeijer

University of New Mexico,
Albuquerque, NM, U.S.A.

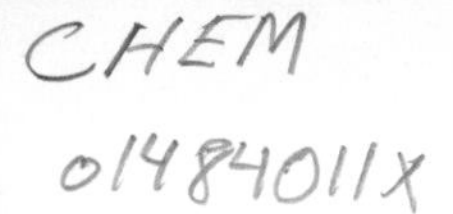

A C.I.P. Catalogue record for this book is available from the Library of Congress.

ISBN-10 1-4020-4134-9 (HB)
ISBN-13 978-1-4020-4134-1 (HB)
ISBN-10 1-4020-4135-7 (e-book)
ISBN-13 978-1-4020-4135-8 (e-book)

Published by Springer,
P.O. Box 17, 3300 AA Dordrecht, The Netherlands.

www.springer.com

Printed on acid-free paper

Printed in the Netherlands.

TABLE OF CONTENTS

Chapter 13
D. Heymann

Chapter 14
L. Becker, R.J. Poreda, J.A. Nuth, F.T. Ferguson, F. Liang and W.E. Billups

Foreword

This is a book on natural fullerenes and related structures of elemental carbon. It may seem ambitious and a glance at the table of contents may confirm this feeling. The original idea for the book, in 2003, belongs to a friend and colleague, Dieter Heymann. In 2004 August, at his suggestion, I was asked if I would be interested in editing this book. With another friend, Alessandra Rotundi, an astronomer by trade, I had just finished a chapter on "Natural carbynes, including chaoite, on Earth, in meteorites, comets, circumstellar and interstellar dust" for a book entitled "Polyynes: Synthesis, Properties, and Application (Cataldo, 2005). So, I was curious to find out if a book on natural fullerenes on Earth, in meteorites, asteroids, comets, and among circumstellar and interstellar dust would reveal systematic patterns of behavior for this particular metastable carbon. I am most pleased with the outcome, which was only possible as result to the hard and dedicated, and above all professional, work by all contributing authors. Naturally occurring fullerenes and related carbons, such as carbynes (linear carbon molecules), are sensitive for their preservation to environmental conditions that should remain perfect throughout geological time following their formation. Natural fullerenes might not be widespread or abundant, which a Google Internet search seems to confirm. A search for 'fullerenes' gives 496,000 hits that reduce to only a fraction when specified 'natural fullerenes' (Table 1). Fullerenes are clearly of interest to astronomers, not in the least because Kroto and coworkers had made this particular connection early on after C_{60} was discovered.

Table 1. Results of a Google WEB search on fullerene(s)

	% Fullerene(s) Hits
natural fullerenes	22
natural + astronomical	4.9
natural + meteorites	3.0
natural + impact	2.6
natural + sediments	0.9

There is correlation between natural fullerenes in 'meteorites' and 'impact' structures (Table 1) that is not surprising since these structures are caused by large meteors hitting the Earth's surface. The entry 'sediments' includes fullerenes in the geological record that occur in bitumen, coal, shungite and fulgurites. Could it mean that natural fullerenes have no geological relevance? The answer is: no! In fact these occurrences will be pivotal in a much better appreciation of the role and the extent of biological activity in the geological record. The major forte of this book is that it puts all known occurrences of natural fullerenes under one cover.

I tremendously enjoyed working with each and every one of the authors on this project. I would like to acknowledge the reviewers whose time and efforts help improving the presentations of the individual chapters. I would like to acknowledge my son Robert who patiently checked the text, which by the very subject matter contains numerous subscripts and superscripts that along with typographical errors were my worst nightmare. His effort reduced the nightmare to a very mild headache. Jim Connolly (UNM) generously shared his computer skills so I could do my job. I thank Aurora Pun who advised on editorial 'savvy'.

At Springer, Emma Roberts and Aaliya Jetha quickly handled each crisis, real or imagined, which made it so much easier for me.

Frans Rietmeijer, Editor

Albuquerque, July 2005

REFERENCE

Cataldo, F., Ed. (2005) *Polyynes: Synthesis, Properties, and Application*, 350p., Taylor & Francis, CRC Press, Taylor & Francis Publishing Group, Boca Raton, Florida, USA.

Overview by the Editor

What possible link could there be between fullerenes in interstellar space, asteroids and in terrestrial rocks? A link is quite plausible for terrestrial rocks associated with meteor craters but what would be the origin(s) of natural fullerenes in bitumen enclosed in pillow lavas, coal, *incl.* shungite, deposits, and in fulgurites? On a planet so richly endowed with life as the Earth, the possibility of biogenic, i.e. algal, activity in the formation of natural fullerenes is exciting and of great interest considering the unique cage structure of fullerenes.

After hydrogen, helium and oxygen, carbon is the fourth most abundant element in the solar system, which is probably also true on a Galactic scale. Carbon ranks only number seventeen in a list of elemental abundances in the Earth's crust. The processes leading from interstellar dust to solar nebula dust and ultimately to asteroids and terrestrial planets are understood with variable degrees of comprehension but the question is what fraction of carbon is in the form of natural C_{60} and lower (e.g. C_{36}) and higher (e.g. C_{1500}) fullerenes. The answer is different for fullerenes in extraterrestrial and terrestrial environments.

In a delightful personal account, Sir Harry in Chapter 1 describes the discovery of C_{60} and its anticipated widespread presence in many astronomical environments. The perfect C_{60} Buckminsterfullerene symmetry (Fig. 1) has an enchanting philosophical beauty that in and by itself would have ensured its longevity as a scientific discovery. It was predicted that C_{60} is abundant in soot such as commonly produced in the laboratory via carbon-vapor condensation and in the soot that provided the bread and butter, and was the curse, of chimney sweeps since times immemorial. It was a bit of a shock when Taylor et al. (1991) were unable to find C_{60} in chimney soot. They realized that this fullerene is degraded in the oxidizing terrestrial atmospheric environment. Regrettable as it might be, this finding would seem to dampen any interest in fullerenes in the geological record. This book shows that it would have been premature.

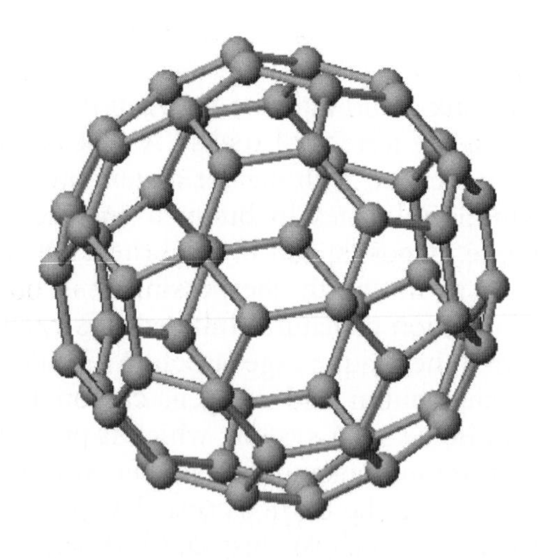

Figure 1. Model of a C_{60} Buckminsterfullerene molecule

The work of Professor Krätschmer, who with his colleagues first showed how fullerene could be extracted from soot and concentrated as fullerite crystals, opened the gateway to explore the science and technology of C_{60}. Chapter 2 explores the complexity and systematic trends in fullerene-production that will be important when searching for natural fullerenes. The laboratory experiments clearly show that carbon vapor condensation is an extremely efficient method for fullerene synthesis, such as the most commonly used arc-discharge method. Such methods were employed in many studies to simulate the physical properties of carbon molecules and carbon solids, amorphous or crystalline, that are expected from high-temperature condensation in astronomical environments that, in reality, also contain hydrogen, and thus (C,H)-molecules. The role of hydrogen in the formation of natural fullerenes, and related structures of elemental carbon, can be explored experimentally by (1) adding hydrogen to a pure carbon vapor in variable C/H ratios, and (2) condensing hydrocarbon plasmas as discussed in Chapter 3. The results of these different approaches show both diversity and similarities among the properties of the condensed carbonaceous molecules and dust.

Astronomers recognize carbon rich (C/O>1) and oxygen-rich (C/O<1) environments of dust condensation (Fig. 2). Silicate-rich dust is formed in the latter, while at C/O>1 carbon molecules, e.g. fullerenes, and dust will form that could be identified by analyzing the diagnostic 217.5-nm carbon extinction feature. The goal of laboratory experiments will be finding a correlation between the observed properties of this feature and the analyzed properties of condensed carbon analogs. On the subject of this famous interstellar extinction feature, Chapter 4 discusses fullerenes in *Diffuse Interstellar Bands*, while simultaneously highlighting the complexities and ingenuity involved in linking laboratory data to astronomical observations.

Figure 2. Dust bands in the Rosetta nebula (ngc2244_2003-02-10_web.jpg) taken by the Hubble Space Telescope

Based on laboratory experiments, Chapter 5 takes the notion that C_{60} is the primary molecule to form via carbon vapor condensation that then may evolve to higher fullerenes. Subsequently, these fullerenes could via solid-state transformations evolve to a number of related elemental carbon nanostructures with spectral properties of the 217.5-nm extinction feature. This chapter also introduces transmission electron microscopy as an enabling tool to visualize the sizes of the hollow fullerenes cages in analog samples of astronomical interest and in carbonaceous chondrite meteorites. The fullerene cages are capable to accommodate a wide range of atoms, including noble gases and metals (see, among many others, Hammond and Kuck, 1992; Kroto and Walton, 1993, and the multitude of peer reviewed research papers). In Chapter 6, this unique fullerene property is accepted as a means to transport noble gases from the atmospheres of carbon-rich stars, encapsulated inside condensed higher fullerenes, to the solar system wherein they accumulated in planetesimals, protoplanets, and carbonaceous asteroids. The fullerene-encapsulated noble gases are considered to be responsible for the noble gas abundances in terrestrial planet atmospheres, *incl.* the Earth's. The very fact that a scenario like this one could even be envisioned, and tested by laboratory experiments and analyses of collected extraterrestrial materials, shows the remarkable revolution of geology to 'earth and planetary sciences' by riding on the coattails of the fantastic journey in Solar System exploration that began with the Voyager space probes. Exploration continues beyond by space-based probes such as the Hubble Space Telescope and the Infrared Space Observatory.

It will be germane to distinguish condensed carbon molecules and clusters from carbon dust. In astronomical environments, the carbon molecule density is orders of magnitude less than in condensation experiments wherein these molecules are a quickly passing stage between vapor and collected dust. It is therefore not surprising that isolated C_{60} molecules appear to be very stable with survival times on an astronomical timescale, but C_{60} condensed in a contained experiment will coagulate to large (higher) fullerenes and, ultimately, to the collectable metastable carbon dust.

We have no direct access to samples of astronomical carbons but we know that astronomical molecules and dust, which were present in the molecular cloud fragment wherein our solar system was formed, are preserved in 4.56 Ga-old comet nuclei and icy asteroids. A fraction of cometary and asteroidal materials reaches the Earth. Picogram debris is collected in the lower stratosphere as interplanetary dust particles that are prime candidates to contain fullerenes. Chapter 7 explains why so far they are elusive although related structures of elemental carbon are present. Much larger debris is recovered at the Earth surface as meteorites, *incl.* carbonaceous chondrite meteorites. Meteorites are 'geologically' processed rocks on airless bodies at great distance from the sun. Environmental deterioration of fullerenes is not a concern until a meteorite has entered the Earth's atmosphere.

A steady stream of meteorites arrives on Earth annually and an occasional big one leaves an impact crater (Gehrels, 1994). Such impacts are highly energetic events wherein target rocks and projectile material are shattered and even evaporated. When carbon is available in either one, or in both, there is the possibility of fullerene formation. In the more gently deposited meteorites, pre-existing fullerenes have survived. Chapter 8 provides a comprehensive overview of fullerenes and related structures of elemental carbon in, mostly carbonaceous, meteorites and their interrelationships. Chapter 9 scrutinizes the evidence for fullerenes associated with major meteorite impact events, e.g. the dinosaur-annihilating Chicxulub event ending the Cretaceous, and the much older Sudbury Impact Structure (Fig. 3).

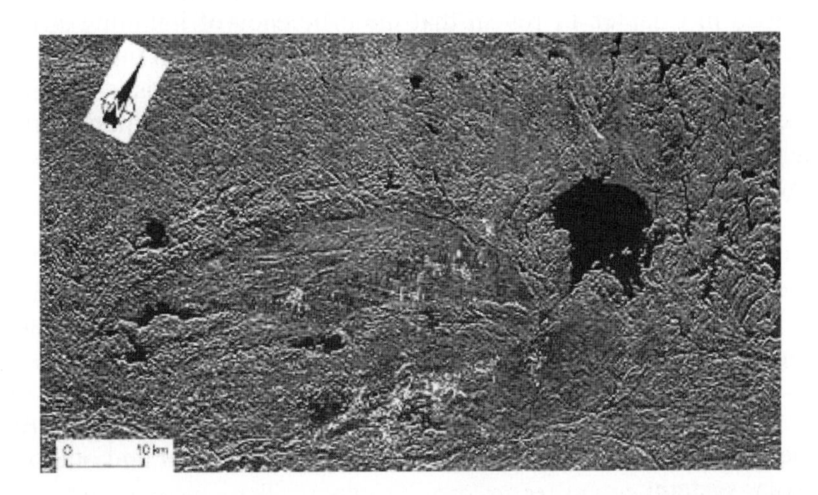

Figure 3. SIR-B radar satellite image showing the elliptical Sudbury Impact Structure in the center zone of this image located northwest of the partially circular lake (black) that is 8-km wide Wanapitei crater that was probably formed 34 Ma ago. The much larger Sudbury structure is a huge impact crater that formed ca. 1.8 Ga ago. It is believed by some that it was originally a circular crater at least 245 km across before tectonic forces more than 900 Ma later deformed the original circular impact structure into its present elliptical shape. Scale bar is 10 km.

Source: http://www.rst.gsfc.nasa.gov (Remote Sensing Education and Outreach Laboratory)

These events could have introduced extraterrestrial fullerenes to the Earth, or produced natural fullerenes during the event itself and its aftermath, for example in global wildfires (Chapters 6, 8 and 9). When natural fullerenes were found to be only associated with these rare events, they would be a mere oddity. It raises an issue of the stability of natural fullerenes. Graphite and cubic diamond are the only thermodynamically stable elemental carbon minerals. All others, *incl.* C_{60} crystals, are metastable solids although they are often treated as carbon allotropes.

For their preservation, metastable fullerenes and other related carbons (e.g. carbynes) would be extremely sensitive to environmental conditions. The record shows that survival of natural fullerenes for billions of years is possible under the precisely right conditions, but what amounts that have survived will be remnants of once more abundant fullerene concentrations. This message comes through in Chapter 10 on C_{60} in bitumen. Its very presence leads to the question of its formation. A lightning strike is an ultra-rapid, high-energy event and, in this regard, it is somewhat akin to conditions during vapor phase condensation. The results of lightning-simulation experiments described in Chapter 11 reveal that the efficiency of lightning strikes for widespread fullerene production is not yet evident but many details are still unknown. A brief description of C_{60} in coal deposits (Chapter 12) suggests that natural fullerenes might be more common than is generally thought. This topic was already the subject of discussion following the first report of C_{60} in shungite (Buseck, 2002). At the time, a biogenic origin of natural fullerenes was seemingly not considered and yet such an origin for terrestrial fullerenes is quite viable as is highlighted in Chapter 13. A 'Deep Hot Biosphere' (Gold, 1999) might yield its own natural fullerenes with encapsulated stable isotopes, such as the noble gases, from deep-seated environments and, thus, be probes of deep crustal processes. A fanciful notion, perhaps, but natural fullerenes could offer new insights in the processes that shaped and continue to shape our planetary environment and others in the solar system.

Aside from the ability of fullerenes to survive in terrestrial rocks and meteorites, there is a matter of fullerene detection in small quantities. It is clear from the combined chapters that at this moment we do have the analytical capabilities to do so. There is always room for improvements as Chapter 14 shows that introduces the new procedure of functionalization in the search for higher fullerenes.

I would like to mention some oddball 'anthropogenic' fullerenes in Chinese ink sticks. Osawa et al. (1997a, 1997b) reported that the amount of C_{60} and C_{70} in the soot used in Chinese ink sticks, a dried mixture of soot produced by slow burning of vegetable oil or pine wood with animal glue, dating from 1620-1772 AD to 1995 AD had decreased from up to 0.1 weight % to 3 ppm due to diffusion of C_{60} molecules to the surface of these sticks and loss via sublimation or oxidation in air. If so, the amount of fullerene measured in a sample could be an original abundance, or a residual fraction, but with no means of quantification. Alternatively, Osawa et al. (1997a, 1997b) suggested that the growth of soot particles (solid-state aging) in these sticks could have increased the ability of older sticks to retain the fullerene molecules. This work suggests that natural fullerenes could be lost initially from soot, but ultimately, fullerene-loss would cease with time, thereby preserving a fraction of the original fullerenes.

Pointing to perhaps the most astonishing fact about natural fullerenes, I note that they are found in rocks ranging from the present day in fulgurites, 65 Ma at Cretaceous-Tertiary Boundary to Ga-old rocks of the Sudbury Impact Structure, the Mítov pillow lavas, and the anthraxolite deposit in Shunga finally the 4.56 Ga-old fullerenes in meteorites. The overall message in the book is that the formation and preservation of fullerene molecules in astronomical environments appears to be easy, but a search for natural terrestrial (planetary) C_{60}, C_{70} and higher fullerenes, and related structures of elemental carbon, is defined by the natural conditions that enabled the preservation of fullerenes, rather than by their mode(s) of formation.

REFERENCES

Buseck, P.R. (2002) Geological fullerenes: review and analysis. *Earth Planet. Sci. Lett.*, 203, 781-792.

Gehrels, T., Ed. (1994) Hazards due to Comets & Asteroids, 1300p., The University of Arizona Press, Tucson & London.

Gold, T. (1999) The Deep Hot Biosphere, 235p., Copernicus, Springer-Verlag New York, Inc., New York, USA.

Hammond, G.S. and Kuck, V.J., Eds. (1992) Fullerenes, Synthesis, Properties, and Chemistry of Large Carbon Clusters. Am. Chem. Soc. Symp. Series, 481, 195p., American Chemical Society, Washington, D.C., USA.

Kroto, H.W. and Walton, D.R.M. (Eds.) (1993) *The Fullerenes; New Horizons for the Chemistry, Physics and Astrophysics of Carbon*, 154p., Cambridge University Press, Cambridge, Great Britain.

Osawa, E., Hirose, Y., Kimura, A., Shibuya, M., Gu, Z. and Li, F.M. (1997a) Fullerenes in Chinese ink. A correction. *Fullerene Sci. Techn.*, 5, 177-194.

Osawa, E., Hirose, Y., Kimura, A., Shibuya, M., Kato, M. and Takezawa, H. (1997b) Seminatural occurence of fullerenes, *Fullerene Sci. Techn.*, 5, 1045-1055.

Taylor, R., Parsons, J.P., Avent, A.G., Rannard, S.P., Dennis, T.J., Hare, J.P., Kroto, H.W. and Walton, D.R.M. (1991) Degradation of C_{60} by light. *Nature*, 351, 277.

Chapter 1

INTROD UCTION: SPACE – PANDORA'S BOX

HAROLD KROTO
Department of Chemisty, University of Essex, Brighton, BN1 9QJ, United Kingdom

1. INTRODUCTION

Space has been the source of inspiration for human beings since time immemorial but until the time of Galileo, Copernicus and Kepler culminating in Newton's gravitational studies it had only a mystical dimension. Since then careful scientific studies became the seed for some of the most important scientific advances. The work of Fraunhoffer on stellar spectra signaled the real beginnings of astrophysics. An important step in our understanding of interstellar material was made by Hartmann who in 1905 detected the presence of atoms in the heralded the start of molecular studies in the interstellar medium (ISM). The mysterious diffuse interstellar bands (DIBs) have also fascinated scientists since 1919 and the UV hump at 217 nm, detected much more recently (but still when it was 2170 Å) has been a similar source of inspiration. The detection of CH, CH^+ and CN in the 1930's and the detection of C_3 and other species in the spectra of comets heralded the start of molecular studies in the ISM and did whet the appetite for further understanding of such matters as the Origin of Life. It was in 1967 Townes and co-workers made the groundbreaking radio studies, which showed that ammonia and water were abundant in the interstellar medium. These experiments opened up a veritable Pandora's box and laboratory microwave spectroscopists (like Takeshi Oka and me) and radioastronomers (such as our Canadian colleagues)

Frans J.M. Rietmeijer (ed.), Natural Fullerenes and Related Structures of Elemental Carbon, 1–6.

formed collaborations to seek and identify as many interstellar molecules as possible.

2. CARBON CHAINS

At about the same time (1974) an undergraduate Anthony Alexander, David Walton and I, were studying long chain carbon molecules by spectroscopic techniques including laboratory microwave spectroscopy. This work led to a collaboration with Takeshi Oka (now at Chicago) and astronomer colleagues Lorne Avery, Norm Broten and John McLeod at the National Research Council in Ottawa Canada to see if we could detect the molecule HC_5N. It seemed a long shot at the time as rough estimates based on the abundances of the other observed molecules suggested that it was likely to be so rare that radioastronomy techniques were unlikely to have sufficient sensitivity. Amazingly the molecule was quite readily detected and much more abundant than expected – at least in some regions of the galaxy. Between 1975 and 1979 we were able to show that the long carbon chain polyynes HC_nN (n = 5, 7 and 9) were relatively abundant in some of the dust clouds that pervade the space between the stars in our galaxy.

3. CARBON STARS

Then the fascinating old carbon-rich red giant IRC+10216 was discovered by infrared astronomy and radio studies indicated that it was surrounded by an expanding shell of dust and gas, which was extremely rich in these carbon chain molecules as well as copious amounts of dust – presumably carbon dust. It was quite close and relatively easy to study in detail.

3.1 Theories of Molecule Production in ISM

At the time the most accepted theories of interstellar molecule production were the Ion-Molecule Reaction theory of Bill Klemperer and Eric Herbst and various Grain Surface Catalysis theories. The laboratory studies of David Smith and Nigel Adams added convincing support for the likelihood that the former approach could explain the existence of many of the molecules detected in the ISM. The latter Grain Surface Catalysis explanation also seemed quite convincing though it was somewhat less amenable to specific laboratory investigation. These approaches were really quite successful in explaining how many of the smaller interstellar molecules might be produced in the dust clouds in which they are observed. My feeling at the time – rightly or wrongly – was that our results on the long carbon chain provided a problem for both these theories and that the IRC+10216 observations suggested at least a partial get-out. To be specific it was not obvious how such relatively large molecules could be built up by the types of sequential bimolecular steps suggested by the Ion-Molecule aficionados and if such highly in-volatile molecules were produced on a grain surface it was not obvious how they might levitate in the cold clouds without the help of David Copperfield. I had for quite some time felt that the production of carbon chains by the high-ish temperature high-ish pressure processes in the atmospheres of carbon stars to be a route worthy of consideration too. After all IRC+10216 was unquestionably pumping out huge amounts of them – did one really need to look any further? The main argument against was that most molecules should be dissociated by the ambient starlight flux. My gut reaction was that perhaps this was true in some cases but why should it always be true. Old carbon rich stars were producing sacks full of these molecules and space was not uniform. Perhaps, in some cases such as that of our favourite cloud TMC1, the long carbon chain molecules had been produced by an old long since dead star and somehow survived the rigours of meandering in space protected in clouds of sooty dust produced at the same time.

4. C_{60} DISCOVERY

These fairly rough ideas were floating around in my mind during the early 1980's by which time a set of particularly interesting papers by Hintenberger and coworkers had really grabbed my imagination. About 1961 they had detected carbon species with as many as 33 atoms by mass spectrometric analysis of the products of a carbon arc discharge. Indeed Otto Hahn had apparently seen carbon species up to about C_{13} prior to 1950. I was visiting Rice University at Easter 1984 when Bob Curl suggested that I pop over to Rick Smalley's lab. Bob was particularly enamoured, as was I, of a beautiful study of the SiC_2 molecule that Rick's group had just completed. On watching Rick bouncing over his beloved 8-foot-high monster TOFMS laser vapourisation cluster beam machine, it struck me that it might provide just the technique to simulate the carbon condensation conditions which occur in the gaseous envelope surrounding IRC+10216. I thought this might give me some compellingly substantial laboratory based ammunition to propagate my idea that our cyanopolyynes might have been produced in stars rather than in the tenuous clouds. I convinced Bob to coerce Rick into a collaboration, which transpired in September of the following year. The fact that we found it relatively easy to produce these species was for me a very satisfying result albeit perhaps a somewhat mundane one for others. Of course the serendipitous detection of C_{60} was the amazing bonus.

In attempts to refine some earlier work on the 217 nm hump in the interstellar UV extinction and pin down the culprit as some sort of carbonaceous clusters, Wolfgang Krätschmer and Don Huffman in 1983 detected an odd feature in roughly this region during laboratory studies of some carbon soot. When our C_{60} study became known they made the brilliant connection that the carrier might be C_{60}. Their beautiful follow-up investigation led ultimately to the extraction of C_{60}. At about the same time (1987-1989) at Sussex some totally unrelated but essentially identical experiments (based of the Hintenberger et al. approach) also revealed interesting results. Lack of funds unfortunately delayed follow-up. Then Krätschmer and Huffman published an extremely interesting paper at a conference on Interstellar Dust in Capri in 1989, which suggested that our Sussex experiments had been on the right track and we quickly resurrected them. A few days after my co-worker, Jonathan Hare, had placed an intriguingly fascinating red benzene solution extracted from our soot on my desk, the beautiful and now world famous manuscript of Krätschmer, Lamb, Fostiropoulos and Huffmam landed on my desk. This paper described how they had extracted a solid material from their soot, which gave a similar wine-red solution in benzene. I cannot really remember the bemused thoughts that went through my mind as I read the words "wine red solution" with our red solution sitting on my desk. By coincidence Nature had asked me to referee this fantastic paper – and the rest is, as they say, history.

5. ASTROPHYSICAL CONNECTION

The amazing aspect of this whole story is how important astrophysical ideas have been at every point. Not only the detection of HC_5N but also SiC_2 result that had prompted Bob Curl to direct me to Rick's Lab. The SiC_2 spectrum had been detected in stellar spectra many years ago and had been studied by Ram Verma at NRC Canada.

Then there was the 217 nm hump as well as the encouraging data from IRC+10216. Both our study, and that of Krätschmer and Huffman were, instigated by the carbon in space connection. Carbon is curious not only is its chemistry amazingly amazing but so is its nuclear chemistry in that there is a fortuitous nuclear resonance that is responsible for not only for the existence of any carbon but any elements at all with atomic number higher than carbon and ultimately our existence.

5.1 RCorBor Stars

RCorBor stars have lost most of their hydrogen and the conditions in their gaseous shells, which consist mainly of He, appear to be so similar to the conditions that occurred in our original cluster beam studies that conventional wisdom suggests that C_{60} must be a significant component of any dust clouds emanating from these types of stars. Direct detection of C_{60} itself by astrophysical spectroscopy will not however be easy. The molecule has no rotational spectrum and the vibrational spectrum is extremely complex broad and difficult to detect for a number of reasons. Similar reservations apply to the detectability of the electronic spectrum too. There is no doubt in my mind that C_{60} is a shady character pervading the dark and dusty alleyways that insinuate themselves among the myriads of stars in our Galaxy. I now think of C_{60} as a love-hate character a bit like the "Third Man". In my mind I recall the famous scene in the film "The Third Man", but instead of the face of Orson Wells and his unforgettably enigmatic expression I see a ball of sixty glistening atoms, suddenly illuminated by light from the upstairs window only to disappear again leaving even more tantalising questions over its whereabouts.

6. EPILOGUE

The hunt for the identification of C_{60} in space is on, and this compendium of papers is bringing us right up to date the present state of play. Chemistry is just over 200 years old as John Dalton read the legendary paper, which provided the first unequivocal evidence of atoms and molecules to members of the Manchester Literary and Philosophical Society on the 20th October 1803. He drew the correct diagrams for those key combustion characters CO and CO_2. Even though we now know that C_{60} is formed in flames it has taken nearly

200 years to discover it. To identify its role in space is taking further painstaking detective work and the studies collected here are piecing together all the evidence needed to unequivocally determine its role. It is yet another aspect of carbon surely of the most enigmatic and amazing element in the periodic table. The more we learn about it the more we still find there is to uncover. Whether my conjecture that the interstellar carbon chains can survive the rigours of space after their formation in stars is right or wrong is still an open question but there is food for thought in the fact that the aim to find supporting circumstantial evidence for this conjecture not only came up with supporting laboratory data but also the discovery of C_{60} *Buckminsterfullerene*. The incentive to understand the myriads of perplexing astrophysical observations is clearly one of the most fruitful forces driving scientific creativity and it invariably results in unexpected revelations about space. Concomitantly such studies have made massive contributions to our general understanding of the Natural and Physical World with impact on the quality of our everyday lives on Earth.

Chapter 2

FORMATION OF FULLERENES

WOLFGANG KRÄTSCHMER
Affiliation Max-Planck-Institut für Kernphysik, Heidelberg, Germany

Abstract: To gain information on fullerene formation, the production methods based on the evaporation of graphite are outlined. In particular, the effects of the buffer gas temperature are considered in connection with experiments performed with a laser furnace arrangement. In addition, fullerene formation by incomplete combustion and by pyrolysis is discussed. The former method is presently applied for large-scale fullerene production and very likely is also the source of most of the fullerenes occurring on earth. The formation processes based on evaporation and on combustion seem to be vastly different indicating that various chemical reaction routes lead to fullerenes.

Key words: Arc discharge; Buckminsterfullerene; buffer gas; C_{60}; C_{70}; carbon chain molecules; carbon molecules; carbon monocyclic ring molecules; carbon vapor condensation; combustion; Cretaceous-Tertiary Boundary (KTB); cryogenic matrices; fulgurite; fullerene; fullerene-generator; fullerene production, graphite particles; growth of molecules; incomplete combustion; laser ablation; laser furnace; polycyclic hydrocarbons; pyrolysis; shungite; smoke particles; soot

1. INTRODUCTION

In the last decade, closed cage fullerene molecules and especially the soccer-ball shaped "Buckminsterfullerene" C_{60} gained much attention. For an overview of the fullerene field there is the reprint collection by Stephens (1993) or the book by Dresselhaus et al. (1996). The fascination with fullerenes is partly coming from the high symmetry of C_{60} and from the fact that fullerenes feature entirely new forms of carbon, namely molecular clusters (Kroto et al., 1985). The superconducting properties of some C_{60}-based compounds stimulated

Frans J.M. Rietmeijer (ed.), Natural Fullerenes and Related Structures of Elemental Carbon, 7–29.

the excitement (Hebard et al., 1991) to such a climax that the magazine *Science* declared C_{60} as the molecule of the year 1991 (*Science*, 1991, 254, issue of 20 December).

Studies of fullerenes were made possible when rather simple and efficient fullerene production methods were discovered. Initially, these were characterized by carbon evaporation (Krätschmer et al., 1990; Lamb and Huffman, 1993; Haufler, 1994) but recently also methods based on incomplete combustion of hydrocarbons were developed (Howard et al., 1991; Murayama et al., 2004). In the former methods, evaporation is usually achieved by an electric arc burning between graphite electrodes. To cool and condense the carbon vapor, the evaporation process takes place in an inert buffer or quenching gas atmosphere. In the case where fundamental studies of fullerene formation are of interest, evaporation is conducted in a laser furnace, as will be described later. In incomplete combustion, hydrocarbons are burned in specially prepared fuel/oxygen mixtures and under controlled conditions. In any case, the soot produced in either of the processes contains fullerenes that can be extracted by non-polar solvents, for example toluene, or by sublimation at 500-800 °C. The efficiency of fullerene production especially for the evaporation methods can be readily tuned to above 10% fullerenes with respect to the soot. In view of the chaotic nature of high-temperature processes and the highly symmetric structures of the formed fullerene such yields are amazingly high. In my opinion these yields have so far not been satisfactorily explained.

2. DIFFERENT PRODUCTION PROCESSES AND FULLERENE YIELDS

Fullerenes are formed from pure carbon or carbon-containing materials by energetic processes because for fullerenes the binding energy per atom is smaller than that of graphite, the most stable form of carbon. Fullerenes are usually produced as a mixture of C_{60}, the smallest stable fullerene, C_{70}, and trace amounts of still larger fullerenes. Fullerenes with hexagon-pentagon structures fulfilling the so-called "isolated pentagon rule" (*IPR*), i.e. displaying no directly adjacent pentagons, are particularly stable. C_{60} is the smallest fullerene to fulfill *IPR*. Apparently, *IPR* defines a limiting surface curvature below which fullerenes become sufficiently chemical inert to survive as intact molecules. Non-*IPR* fullerenes are stable only under special conditions, e.g. in the collision-free environment of molecular beams. Even though the surface curvature of larger fullerenes decreases with size and thus their formation should be favored energetically, C_{60} along with the slightly more stable C_{70} are formed most abundantly. In evaporation processes carried out in an inert atmosphere, the C_{70}/C_{60} ratio will be generally in the 0.1 to 0.2 range that is C_{60} is the most abundant fullerene followed by C_{70} (see, e.g. Ajie et al., 1990). Depending on the flame conditions in the hydrocarbon combustion process, the C_{70}/C_{60} ratio is reported to show large variations including the striking cases in which C_{70} becomes more abundant than C_{60}. In section 3, I will try to relate this to "bottom up", respectively "top down" fullerene assembling processes.

The conclusion to be drawn is that fullerene formation seems not to be a unique process. Apparently there are different pathways leading to fullerenes, some of which are very efficient and commercially attractive while others are much less efficient, and only minute quantities are produced. The latter seems to happen under ambient conditions on earth, that is, natural occurring fullerenes are usually dispersed at rather low-level concentrations. It thus appears to me that – unlike coalmines – profitable fullerene mines are rather unlikely to be opened at some future day, at least on earth. In interstellar space the chance to find lots of fullerenes may be more favorable, especially in the atmospheres of certain stars that contain predominantly carbon and helium. In the condensing outflows of such stars, plenty of fullerenes may reach the interstellar medium. So far, however, no unequivocal evidence for fullerenes in interstellar space is reported.

From carbon containing materials, traces of fullerene seem to form under a variety of virulent conditions, like e.g. lightning strike discharges (Daly et al., 1993) and in laser-beams and ion-beams (Ulmer et al., 1990; Brinkmalm et al., 1992; Gamaly and Chadderton, 1995). The detection of fullerenes in samples of the geologic Cretaceous-Tertiary Boundary (KTB) and Permian-Triassic Boundary (PTB) layers may indicate specific geologic events that were very likely of a catastrophic character, such as the impact of a giant meteor (Heymann et al., 1994; Chijiwa et al., 1999). Fullerenes reported in connection with terrestrial impact craters also point into this direction (Becker et al., 1994).

Some of the issues on fullerene yields will be relevant when exploring the existence of natural fullerenes. In most cases, the yields of fullerenes produced under ambient terrestrial conditions at the (near) surface, or during metamorphic, pressure-temperature processing of rocks, would be very small, not in the least because elements like hydrogen, nitrogen, and oxygen tend to form bonds with carbon and thus would either terminate the molecular growth process leading to fullerenes or destroy existing fullerenes.

3. CARBON MOLECULAR GROWTH

Basically, two limiting formation scenarios are conceivable, viz. (1) a "bottom up" process, in which fullerenes are built up from small precursors such as C atoms and C_2 units, and (2) a "top down" process in which hot, graphitic microparticles cool by decomposing into smaller units that is decomposition into fullerenes or small carbon molecules. The latter decomposition mode may initiate a "bottom up" growth of the fragments, suggesting that "bottom up" and "top down" processes may occur together.

A "bottom up" growth very likely takes place in evaporation – condensation processes either during resistive heating or arc discharge of graphite electrodes. Evidence for this growth process is provided by isotopic scrambling experiments in which electrodes of normal and ^{13}C-enriched graphite were used (Ebbessen et al., 1992; Heath, 1992, and references therein). Fullerene production by laser ablation of graphite seems to be more complicated. In this case the initial ablation process (the term ablation is deliberately chosen since more than just ordinary thermal evaporation is occurring), seems to involve both the "bottom up" and "top down" processes. Their relative contributions seem to depend on the ablating laser energy, i.e. the energy per laser pulse deposited in the graphite target (Moriwaki et al., 1997).

Because of their closed cage, fullerenes certainly are structures which of all possible isomers in configuration space exhibit a minimum of energy and thus should be formed preferentially, provided there are no topological barriers that would inhibit fullerene formation. How fullerene structures can occur and how they can form so efficiently is primarily a matter of molecular growth, which will be considered in the following.

When graphite is heated to about 2500 °C, atomic carbon, C_2 molecules, and linear C_3 molecules account for most of the carbon vapor species. These species are highly reactive and in the quenched vapor will grow readily by polymerization into larger molecules like C_6, C_9, and further. Such polymerization processes can be studied when carbon vapor species are trapped in cryogenic matrices and are allowed to diffuse and react by gently warming the matrices. Optical spectra for carbon vapor species trapped in Argon ice (Krätschmer et al., 1985;

Krätschmer, 1996) are shown in Fig. 1, wherein the growth of linear carbon chain molecules up to C_{15} and larger can be easily recognized.

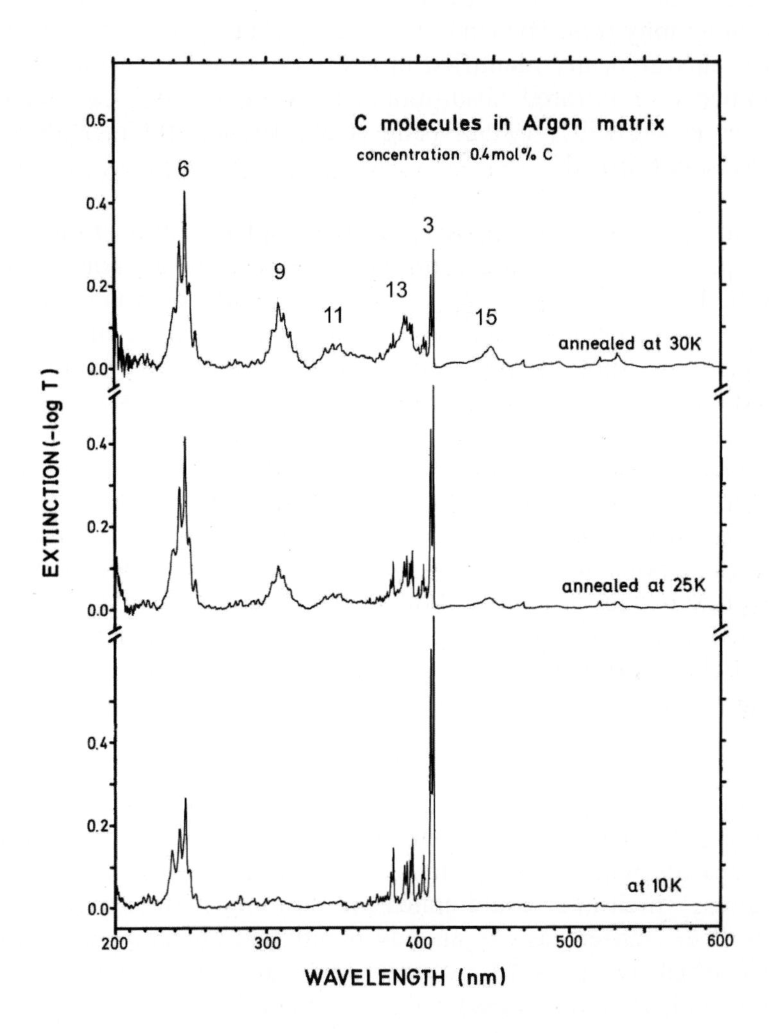

Figure 1. The absorption spectrum of carbon vapor molecules trapped in a matrix of solid Argon at 10 K (bottom). Upon thermal annealing (upper frames) changes in the spectra take place indicating the growth of longer chains at the expense of C_3 that is the most abundant vapor molecule. In the top spectrum, the carbon chain molecule is indicated that produces the particular absorption feature. We followed the assignments of Maier (1997). Data are adopted from Krätschmer (1996)

Such linear chains possess rather intense absorption features centered at known wavelengths (Forney et al., 1996; Maier, 1997; Wyss et al., 1999). Among the larger species formed under these conditions are probably not only linear chains, but also monocyclic rings and other, more complex structures such as were detected by ion chromatography (von Helden et al., 1993a). The spectra of rings were not yet unequivocally identified in matrices. I think that the existing assignments of infrared absorptions to cyclic C_6 and C_8 (Presilla-Marques et al., 1997, 1999; Wang et al., 1997a, 1997b, 2000) still need to be confirmed.

Upon warming such matrices until complete sublimation of the matrix gas, the remaining carbon molecules transform spontaneously from the linear (or monocyclic ring) *sp*-C hybridization into soot-like sp^2-C networks (Wakabayashi et al., 2004). A search for fullerenes in the remaining soot usually remains negative for reasons that will become clear later when the role of the buffer gas is discussed. The transition from linear chains or monocyclic rings to graphitic sp^2-C bonds occurs very likely by the so-called “cycle formation” in which different species combine to form hexagonal, pentagonal, etc. networks. In the context of fullerene built-up, the cycle formation of larger ring molecules seems to open a simple and efficient way to form closed-cage structures. Wakabayashi and Achiba (1992) first proposed this “ring-stacking” process whereby, for example, C_{60} can formally be assembled by stacking the C_{10}, C_{18}, C_{18} and C_{12}, rings and putting a C_2 unit on top (Fig. 2).

Even though on first glance this conjecture may appear naïve, I think that cyclic species may play an important role in assembling carbon cages. The initially formed cages don’t need to be perfect fullerenes at first, but such rings would automatically form closed structures, which later could anneal into a more perfect cages by the action of the buffer gas. As already mentioned our knowledge about cyclic carbon species is still rather limited and the role they play in the carbon condensation process needs to be elucidated.

For a theoretical study of the carbon condensation process by molecular dynamics methods I would like to refer the work by Yamaguchi and Maruyama (1998).

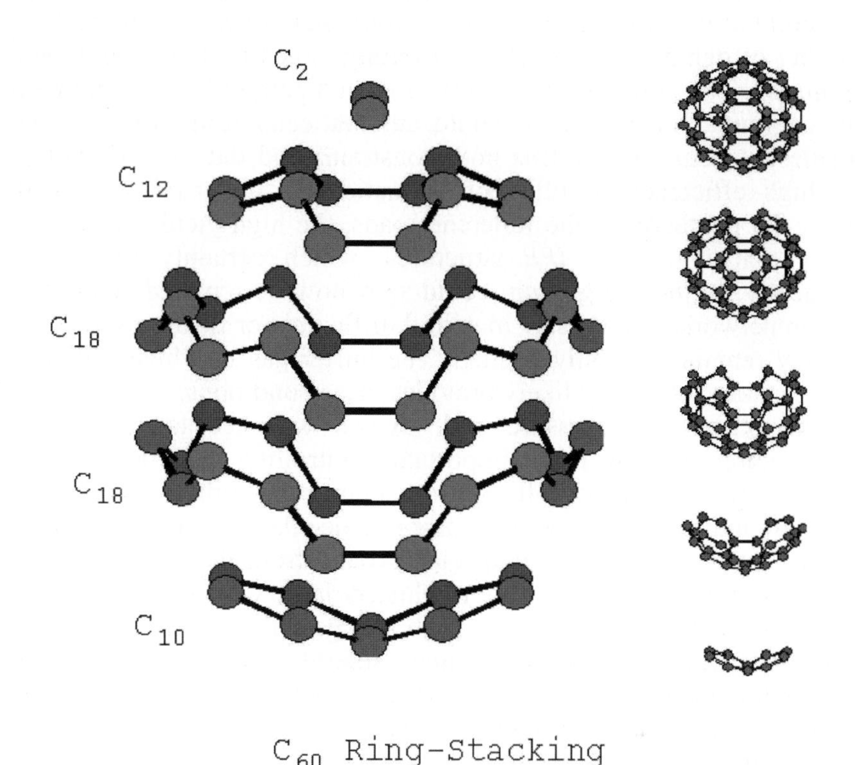

Figure 2. The assembly of C_{60} by monocyclic carbon ring molecules and C_2 according to the "Ring Stacking" mechanism (Wakabayashi and Achiba, 1992). Cross-linking of cyclic species readily leads to closed cage structures

Heath (1992) proposed another fullerene built-up process known as the "fullerene road" that assumed a gradual "bottom up" growth from small molecules to chains and rings and to fullerenes whereby the cages stay closed during growth; hence the name of this process. A very popular fullerene assembling procedure was suggested by Haufler (1994) and Smalley (1992), which is known under the name "pentagon road", and attempts to explain the high fullerene yields.

Here, the cages are assumed to stay open during growth. Again, an sp^2-C network is growing by cycle formation from smaller sp-C precursors. However, growth takes place with the constraints that the curvature of the network fulfils the *isolated-pentagon-rule* and that the number of dangling bonds of the resulting structure tries to assume a minimum. The latter implies that the reaction path follows a minimum energy route. Such process would automatically lead to C_{60} as the smallest structure that fulfils both constrains and thus would explain the high efficiency of fullerene formation. In any case, following either the pentagon or the fullerene roads, the high yield should also imply that faulty non-*IPR* structures, which certainly form very frequently during the growth, could somehow be repaired. Thus, the carbon networks should stay in a kind of flexible or floating state, that is, they remain thermally excited. The buffer gas in which fullerene growth takes place very likely provides these conditions.

The buffer gas pressure and, as we will see later, also its temperature are the most important controlling parameters. The pressure determines the collision rate among the carbon clusters and therefore indirectly the speed of carbon molecular cooling and growth. At low pressures the carbon species expand from their source, grow slowly and may not even grow to clusters large enough for fullerene formation. If the pressure is too high, the growth process will be too fast and overshoot the size range suitable for fullerenes. That nonetheless the pressure dependence of the fullerene yield in the range 10-500 mbar is surprisingly weak, very probably has to do with effect of the buffer-gas temperature that also determines the rate of cluster cooling.

Various other processes relevant for fullerene formation and cage modification were discussed in the literature and for details see von Helden (1993b). I should mention the Stone-Wales mechanism that suggests how the position of carbon atoms of existing fullerene cages can be rearranged (Fowler and Manolopoulos, 1995) and the problem of C_2 capture and release from growing carbon networks.

4. FULLERENE FORMATION BY CARBON VAPOR CONDENSATION

The carbon evaporation and condensation methods applied for fullerene production may also shed some light on the formation processes. In the following, I will describe some of the methods that use graphite as the source material. I will try to draw some conclusions on fullerene formation in such processes.

4.1 Evaporation of Carbon in a Helium Atmosphere

In the decisive experiments of Kroto et al. (1985) leading to the discovery of fullerenes, graphite was evaporated by laser-ablation and in situ mass spectra of the condensing carbon vapor were obtained. The fullerene yield in the soot condensed at the same time as a by-product was apparently very small. In later experiments, fullerenes in soot produced in this manner could be detected by mass spectroscopy (Meijer and Bethune, 1990). When the fullerene discovery was made public, Donald Huffman and the present author resumed work that had started several years before in an attempt to reproduce the most prominent interstellar absorption feature at around 220 nm. This feature was, and still is, believed to originate from strong surface plasmon absorption of graphitic interstellar nanoparticles (see, e.g. Draine 1989) and in our experiments we tried to produce similar graphitic particles by evaporating graphite rods in a helium atmosphere. The apparatus used showed the basic features of a fullerene generator, viz. two graphite electrode rods were kept in mechanical contact by a spring and evaporated by a suitably high current. Initially, the bell jar under which the rods were mounted contained an atmosphere of 1 mbar to 10 mbar He to reproduce interstellar grains.

In the later experiments, we changed the pressure to 100 mbar and to 200 mbar in order to obtain optimum fullerene yields. As long as the helium pressure is kept between 50 mbar and 500 mbar, the pressure dependence of the yield is weak; outside this range the yield decreases. The use of neon, argon, krypton and even nitrogen as buffer gas leads also to fullerenes but the yields in helium are significantly higher. In our initial apparatus, the carbon evaporation rate could be kept steady for several seconds and during that time a remarkable phenomenon could be observed, namely the formation of a "smoke" plume of small particles which appear blue in scattered light. At first, the plume is located concentric to the evaporation site and then it gradually expands. The smoke is a result from the condensation of carbon vapor molecules that are cooled by collisions with the helium atoms.

Thus, a supersaturated carbon vapor is formed leading to spontaneous nucleation of carbon into solid nanoparticles. These particles coagulate to form larger aggregates capable of scattering visible light. The smoke cloud is coupled to the gas by friction and thus follows the developing convection flow of helium through the bell jar. Finally, a layer of soot is deposited on the inside the apparatus. When after some time of evaporation the contact between the graphite electrodes is lost, an arc develops and with the resulting unsteady fluctuations of arc burning, the beautiful picture of the steady stream of smoke is destroyed. The arc formation process however has the advantage that the evaporation of carbon can be maintained without mechanical contact between the graphite rods. The more modern fullerene generators are making use of this advantage where except for igniting the arc there is no mechanical contact required to keep evaporation going. The layout of such a fullerene generator based on a review paper of Haufler (1994) is shown in Fig. 3.

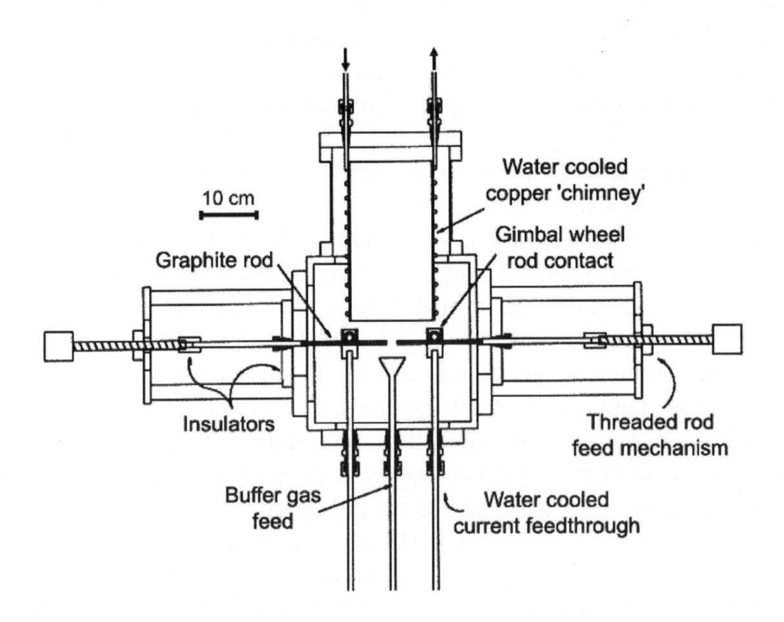

Figure 3. A fullerene arc-generator (after Haufler, 1994). The arc burns between two graphite rods that are clamped by water-cooled electrodes and can be rotated from the outside. The vessel contains helium as the buffer gas. The fullerene containing soot is collected in a water-cooled chimney

When the evaporating graphite rods gradually become consumed it is sufficient to move the rods towards each other keeping the gap between the rods roughly constant. No loss of contact occurs, meaning that the process could, in principle, be carried out continuously until all graphite is consumed. However, there is a practical limit to the time arc-evaporation can be maintained and this has to do with "slag" formation. Slag is a deposit that originates from carbon condensed onto the electrodes and heavily distorts arc burning. But this and most other distorting effects can be overcome (e.g. Dubrovsky et al., 2005).

The requirements for a large-scale fullerene production by the arc method are thus worked out. I may mention that already in the early 1990s the former Hoechst Company produced several kilograms of fullerenes by the arc method but to my knowledge most of the applied conditions remained unpublished. According to my information, CO_2 rather than helium was used as buffer gas.

I might also mention that fullerene generators can be downscaled and simplified to such extent that makes these devices suitable for experiments in school. In youngsters, the bright light of the carbon arc may awaken their interest for natural science.

4.2 The Laser-Furnace Method

As mentioned in the previous section, in the classical laser ablation experiments of Kroto et al. (1985) no significant amounts of fullerenes were formed. In a clever experiment Smalley and co-workers found that the possible reason for the low yield was the temperature of the buffer gas, which was used in their apparatus to cool the forming clusters. The gas is blown into the apparatus at room temperature. In the "laser furnace" apparatus design, laser ablation of the graphite rod takes place in an argon atmosphere that is heated by an oven to temperatures up to about 1200 °C. In detail, argon streams gently through a quartz tube in which the graphite rod is mounted and the entire device is placed inside an oven (Fig. 4).

The laser beam and gas stream usually go into the same direction. Under such conditions, the fullerene yield starts at about 450 °C buffer gas temperature from almost zero to reach a maximum of about 5% at 1200 °C (e.g. Kasuya et al., 1999). Inside the quartz tube, near the exit of the oven with respect to the argon stream, a dark brown deposit of fullerenes condenses and can be washed out by toluene.

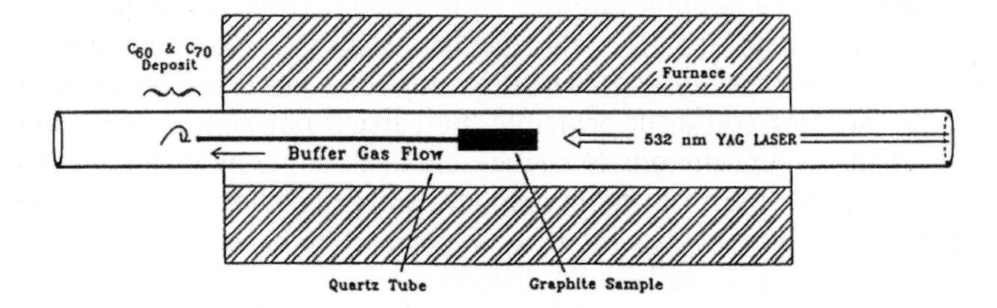

Figure 4. The "laser furnace" (after Smalley, 1992). Laser ablation on the surface of a graphite rod takes place in an argon-filled quartz tube. The tube is inserted in an oven that could heat the argon buffer gas up to 1200 °C. Fullerenes formed in the laser plume will sublime and are transported by the streaming buffer gas to the oven exit where they condense as a brownish film. In this device most production parameters can be controlled

During arc discharge, steady and reproducible conditions are difficult to reach. In the laser furnace, wherein all relevant formation parameters (e.g. ablation laser power, buffer gas temperature and pressure) can be controlled, the reproducibility problem is less severe. Thus, the yield and composition of fullerenes can be monitored as function of these parameters. In short: compared to the arc method, the laser furnace is much more convenient to study the fullerene formation process.

4.3 Fullerene Formation in the Laser Furnace

In this section I report the experiments carried out by Y. Achiba and co-workers at Tokyo Metropolitan University at which place the author had the opportunity to witness the ongoing work. As I already mentioned, the process of laser ablation of graphite is not well understood and appears to differ from a simple thermal evaporation (Moriwaki et al., 1997). Most remarkable is the occurrence of a "laser plume", i.e. a cloud of small hot particles which is ejected at the place of the laser impact and which emits a blackbody-like spectrum.

Another striking difference between arc and laser furnaces is that the laser furnace does not work well with helium but requires argon as buffer gas. The most likely reason being that argon allows a better confinement of the laser plume within the relatively narrow quartz tube. With helium the plume of carbon vapor would expand too much and reach the inner walls of the quartz tube where carbon would condense prematurely.

It is a straightforward task to locate the place within in the tube of the laser furnace where fullerene formation takes place (Kasuya et al., 1999). For this purpose one switches off the ablation laser, shifts the oven along the tube and measures, e.g. by a thermocouple, the temperature at the tip of the graphite rod as function of oven position. Naturally, only the ranges around the oven entrance and exit are of importance since the temperature inside the oven remains constant. Then one switches the ablation laser on and determines the fullerene yield as function of oven position. One obtains bell-shaped distribution functions for the temperature and the fullerene yield (Fig. 5). But – and this is the key result – both distributions are displaced by about 10-30 mm. The conclusion is that fullerene formation takes place at this distance above the tip of the graphite rod, i.e. within the laser plume (Fig. 5, inset). Of course, this result would be expected, but it is nice to know it with certainty.

For studying the time dependence of the laser ablation process in detail, high-speed video and digital cameras were employed. The temperature of the laser plume, its formation and decay could also be studied by measuring the spectral intensity of the emitted light with suitable filters. Figure 6 shows the results obtained for the "cooling down" of the laser plume, i.e. plume temperature as function of time and buffer gas temperature. The other parameters were kept constant. At low buffer gas temperatures where fullerene yields are low, the plume cools down relatively fast. The more the buffer gas temperature is increased, and optimum conditions for fullerene formation are reached, the slower becomes the cooling down process of the plume. Such peculiar behavior can be explained by a release of energy, which heats up the plume and thus is slowing down the cooling process. Y. Achiba has coined the phrase "fullerene fire" to describe the energy release during fullerene formation (Ishigaki et al., 2000).

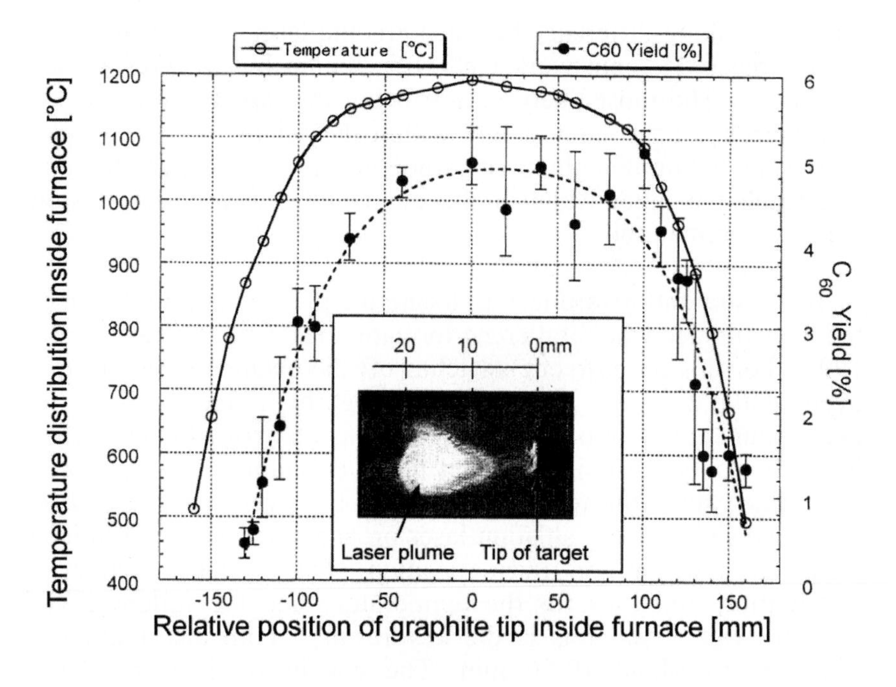

Figure 5. The temperature at the tip of the graphite rod in a laser furnace is shown as function of position along the oven axis (open points). The zero position indicates the oven entrance, respectively exit with respect to the gas flow as shown in Fig. 4. In additional experiments the fullerene yield was measured under the same conditions (solid points). The curves are displaced from each other by about 2 cm, indicating that fullerene formation occurred about 2 cm in front of the graphite tip at the site of the laser plume (inset) (data from Kasuya et al., 1999)

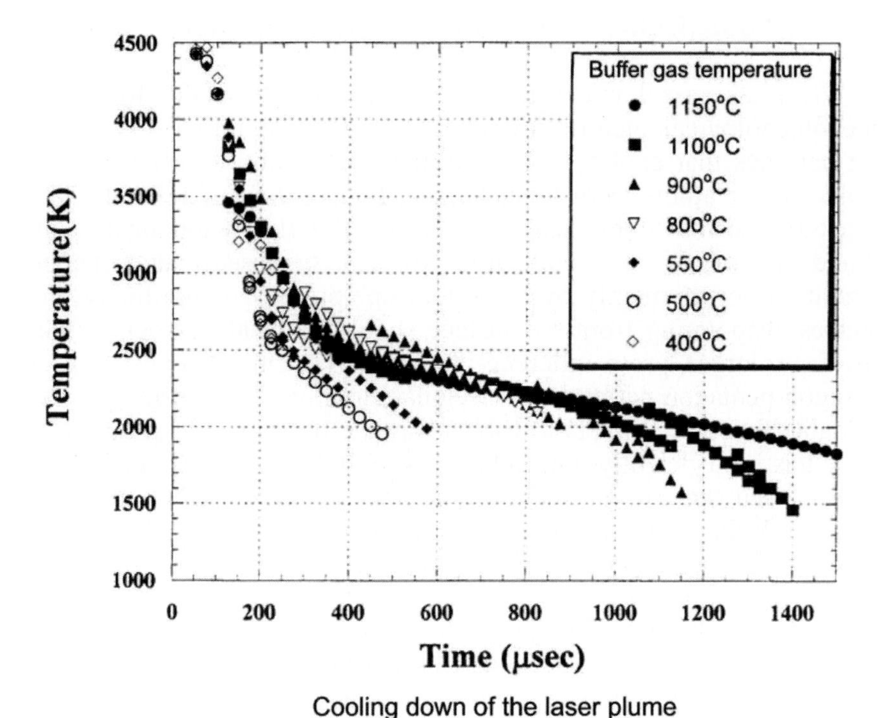

Figure 6. Cooling down curves of the laser plume as derived from time resolved spectroscopic measurements with a high-speed video camera. The plume temperatures are displayed as function of time after laser impact. The spectra are obtained by measuring the intensity of the plume as function of time as seen through different optical filters. Therefore, several laser shots are required to construct the spectrum. The temperature is determined under the assumption that the plume is formed by small particles that emit a continuous blackbody-like spectrum. Notice that under the conditions of fullerene formation, energy is released inside the plume, which slows down the cooling rate (conditions: argon buffer gas at 200 torr; laser power 300 mJ/pulse). The data are taken from Ishigaki et al. (2000)

4.4 Some Conclusions on Fullerene Formation

When taking all experimental evidence together, the following phenomenological picture emerges: laser ablation initially produces hot particles that cool by radiation and evaporation of small carbon molecules or atoms from their surface. Evaporation of fullerene-like structures seems to be less likely since the finally obtained high abundance of C_{60} (as compared to C_{70}) suggests that fullerene formation occurs mainly by a "bottom up" process. Accordingly, the species evaporating from the surface should be small molecules that grow into larger, proto-fullerene-like, units, i.e. clusters consisting of hexagon-pentagon networks of irregular structure. The buffer gas has a double function, viz. (1) to carry away the reaction energy, and (2), if the gas is sufficiently hot, to keep the proto-fullerenes in an excited state so that the carbon network structure could anneal to form the soccer-ball shaped C_{60} molecule or any other regular fullerene. In such a process faulty, energetically unfavorable, networks with structural defects (e.g. adjacent pentagons) will be relaxed by rearrangement of the cages or, if necessary, by the capture or the release of smaller excess carbon molecules, or both. This self-healing feature most likely explains the high fullerene yields in the laser furnace experiments. Since such annealing and relaxation processes are exothermic they should be the source of the observed "fullerene fire".

Returning to the arc process, the buffer gas temperature probably plays a similar key role in determining the fullerene yield, as was the case in the laser furnace experiments. The convection currents driven by the evaporating graphite rods create a steady stream of hot gas wherein irregularly aggregated carbon clusters can anneal to perfect fullerenes. Since evaporation usually takes place in a voluminous vessel, there will be no confinement constrains to use helium as a buffer gas in order to ensure efficient fullerene production. However, the reasons for the beneficial effects of using helium as buffer gas are unclear. Besides chemical inertness and high ionization potential, the exceptionally high heat conductivity of helium is probably a key factor. The temperature of the buffer gas is a controlling parameter and it is thus plausible that the better conducting helium compared to other noble gasses puts it at an advantage.

5. FULLERENES BY INCOMPLETE HYDROCARBON COMBUSTION

Based on in situ mass spectrometric studies, Homann's group made the first report of fullerenes forming in flames as intermediate phases (Gerhardt et al., 1987). When no special care is taken, the fullerenes once formed in flames are burned away and the amount of fullerenes in the resulting remaining soot will be very small. However, employing so-called incomplete combustion conditions, i.e. using sooting flames that burn with a deficit of oxygen, and by selecting suitable fuels (generally benzene), the resultant soot may contain appreciable amounts of fullerenes. Yields varying between 0.2% and 12% with respect to the soot content were reported (Howard et al., 1991).

Recently, the "Frontier Carbon Corporation" in Japan has set up a fullerene production plant for large-scale fullerene production (tons per year). The hope is to stimulate industrial applications of fullerenes and fullerene research at an affordable price (Murayama et al., 2004). Certainly helpful in putting such a plant in operation is the fact that soot and carbon black are already produced by incomplete combustion for industrial purposes on very large scale. For example, soot is used as additive to rubber in car tires to reduce wear. An additional and important advantage of using combustion is the energy requirement. In the combustion process the energy is already contained in the fuel itself whereas the arc process requires an energy supply for making graphite rods and sustaining the electric discharge.

As already mentioned the C_{70}/C_{60} ratio is highly variable and depends on the flame conditions, which implies a fullerene formation process distinctly different from that by evaporation. Ahrens et al. (1994) considered processes that could form fullerenes from flat or slightly curved polycyclic hydrocarbon molecules. One particular scenario suggests that two polycyclic molecules conjugate (Fig. 7) and excess hydrogen will be stripped off from the edges. The resultant cage structures then thermally anneal to fullerenes in the flame. In addition, also monomolecular conversions of polycyclic hydrocarbons into fullerenes were suggested (Homann, 1998).

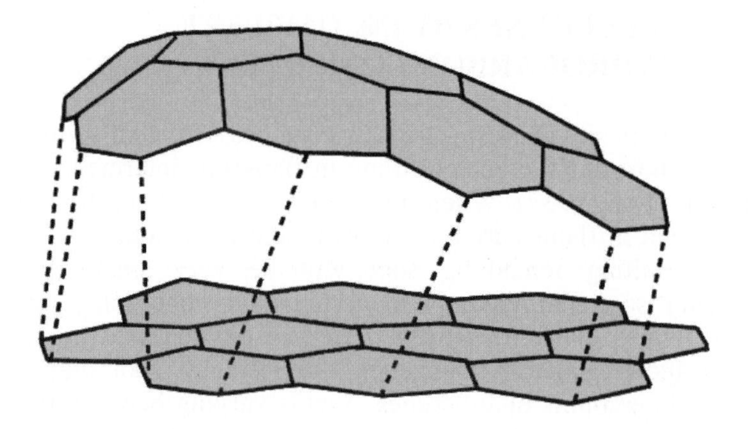

Figure 7. A possible mechanism of fullerene formation in flames according to Ahrens (1994). The usually flat polycyclic hydrocarbons that are produced in abundance in flames are coming together and strip-off their hydrogen atoms at the edges. Newly formed carbon-carbon bonds provide the carbon pentagons required for cage closure. In the figure the upper polycyclic hydrocarbon shows already some curvature due to pentagons in its framework

In this context, it is interesting to note the observation that the most sooting flames do not produce the highest amounts of fullerenes (Howard et al., 1991). Thus, one might infer either there are different origins for soot and fullerenes or, as Homann had suggested, there is a common precursor that converts to soot or fullerenes depending on flame conditions. I think the large abundances of C_{70} suggest that in the combustion process steady "bottom up" growth is accompanied by intense "top down" decay processes. It is certainly reasonable to assume that large, soot-like carbon structures in the heat of the flame may decompose into smaller cages that might ultimately relax into fullerenes. In this case the higher fullerene masses should be preferred when the decay leads to the thermodynamically (compared to C_{60}) more stable C_{70}. However, the still low abundance of the even higher fullerenes beyond C_{70} remains unexplained by a "top down" scenario. In conclusion, fullerene formation in flames appears to be much more complex than in the arc or laser furnace processes. Neither "bottom up" nor "top down" processes can fully account for the observed fullerene abundances. One has to realize that in flames hydrocarbon combustion and carbon condensation are linked and occur simultaneously.

6. FULLERENES BY PYROLYSIS OF SUITABLE PRECURSORS

Organic chemists were intensely searching for methods of fullerene synthesis and preparation that rely on more established methods of organic chemistry and in this sense these might be called "rational". Attempts with solvent-based reactions so far failed since many intermediates in the synthesis process turned out to be insufficiently soluble. As an alternative method, fullerene synthesis by pyrolysis in the gas phase was frequently attempted. Taylor et al. (1993) performed experiments wherein naphthalene was heated in an argon stream and the residual material was investigated for fullerenes. The basic idea was that adjacent hexagons are a repeating motive of a fullerene surface and fullerenes might thus form spontaneously from naphthalene in the gas phase.

Scott et al. (2002) synthesized a precursor compound that should much more facilitate C_{60} formation by gas-phase pyrolysis. But while successful, the reported yields (<1% C_{60}) were amazingly low when compared to that obtained in the less "rational" approaches. The absence of C_{70} in the reaction products validated the synthesis approach since the chosen precursor should not be suitable for the production of any other than C_{60} fullerene.

7. FULLERENE FORMATION UNDER AMBIENT CONDITIONS

The arc process carried out in a nitrogen buffer gas at one atmosphere pressure yields soot containing about 1% fullerenes (Pfänder, 1993). As mentioned above, also CO_2 seems to work for making fullerenes. This is somewhat surprising and raises the question why fullerenes were not discovered much earlier since all technical ingredients were present almost since Faraday's time. By adding more reactive components such as H_2, O_2, or H_2O to the buffer gas, the fullerene yield will drop to a very low level, which can be easily understood, as fullerene built-up is sensitive to chemical distortions. When for example hydrogen or oxygen saturates carbon bonds, the built-up of a carbon network becomes blocked and fullerenes will not form. Thus, laser ablation or ion bombardment of plastics in a vacuum yield only a minute amount of fullerenes barely detectable by mass-spectroscopy. Apparently, impurities of either the carbon source material or the buffer gas, or both, reduce the fullerene yields drastically. In uncontrolled combustion, such as would occur in natural wildfires, the situation seems to be similar and rather low fullerene yields seem to be the rule. For laboratory analyses of such natural "low fullerene samples" a potential fullerene contamination background may pose considerable analytical problems. It is better to perform such studies at "clean" places, i.e. in laboratories where no previous fullerene research was conducted. Alternatively, chemical analyses of samples that might contain fullerenes should adopt rigorous protocols to exclude inadvertent, spurious laboratory contamination by fullerene from extraneous sources within the laboratory.

Even then, reports on natural fullerene findings can be contradictory although for other than laboratory-related reasons. For example, the detection of C_{60} and C_{70} fullerenes in a shungite coal deposit (Buseck et al., 1992) was questioned by Ebbesen et al. (1995) who were unable to confirm the presence of fullerenes in other samples from the same coal deposit, as well as in other carbon-rich samples of similar geological age from Finland, South Africa and Sri Lanka. While the original fullerene detection by Bueseck et al. (1992) is unquestionable, and the high-resolution transmission electron microscope images of the sample consistent with fullerenes, it was their generalization of the importance of fullerenes in this deposit that drew the attention. In a recent review on geological fullerenes, Buseck (2002) concluded that their original report might have been on an unusual sample from within the coal deposit and that perhaps fullerene may have a lightning strike origin.

Similarly, Taylor and Abdul-Sada (2000) challenged the reported occurrence of fullerenes in KTB samples (Heymann et al., 1994) because they were not successful in their own fullerene search in another such sample. Apart from heterogeneity of natural samples, one should consider that fullerenes, when not somehow protected for example within closed cavities, would be relatively easily oxidized (Chibante and Heymann, 1993). I will conclude with the remark that establishing fullerenes as reliable tracers in geochemistry and in other fields of environmental research appears to be a formidable task.

Acknowledgements: I appreciate the helpful comments of A. Rotundi and I thank T. Wakabayashi for stimulating discussions and providing Fig. 2. Dima Strelnikov assisted in the computer graphics. The research was supported by the Deutsche Forschungsgemeinschaft and the Max-Planck-Gesellschaft.

8. REFERENCES

Ahrens, J., Bachmann, M., Baum, Th., Griesheimer, J., Kovacs, R., Weilmünster, P. and Homann K.-H. (1994) Fullerenes and their ions in hydrocarbon flames. *J. Mass. Spectr. and Ion Proc.*, 138, 133-148.

Ajie, H., Alvarez, M.M., Anz, S.J., Beck, R.D., Diederich, F., Fostiropoulos, K., Huffman, D.R., Krätschmer, W., Rubin, Y., Schriver, K.E., Sensharma, D. and Whetten, R.L. (1990) Characterization of the soluble all-carbon molecules C_{60} and C_{70}. *J. Phys. Chem.*, 94, 8630-8633.

Becker, L., Bada, J.L., Winans, R.E., Hunt, J.E., Bunch, T.E. and French, B.M. (1994) Fullerenes in the 1.85-billion-years-old Sudbury Impact Structure. *Science*, 265, 642-645.

Brinkmalm, G., Barofsky, D., Demirev, P., Fenyö, D., Hakansson, P., Johnson, R.E., Reimann, C.T. and Sundquist, B.U.R. (1992) Formation of fullerenes in MeV ion track plasmas. *Chem. Phys. Lett.*, 191, 345-350.

Buseck, P.R. (2002) Geological fullerenes: review and analysis. *Earth Planet. Sci. Lett.*, 203, 781-792.

Buseck, P.R., Tsipurski S.J. and Hettich R. (1992) Fullerenes from the geological environment. *Science*, 257, 215-217.

Chibante, L.P.F. and Heymann, D. (1993) On the geochemistry of fullerenes: stability of C_{60} in ambient air and the role of ozone. *Geochem. Cosmochim. Acta*, 57, 1879-1881.

Chijiwa, T., Arai, T., Sugai, T., Shinohara, H., Kumazawa, M., Takano, M. and Kawakami, S. (1999) Fullerene found in the Permo-Triassic mass extinction period. *Geophys. Res. Lett.*, 26, 767-770.

Daly, T.K., Buseck, P.R., Williams, P. and Lewis, C.F. (1993) Fullerenes from a fulgurite. *Science*, 259, 1599-1601.

Draine, B.T. (1989) On the interpretation of the 217.5 nm feature. *In Interstellar Dust*, L.J. Allamandola and A.G.G.M. Tielens, Eds., Proc. IAU Symp., 135, 313-326, Kluwer Academic Press, Dordrecht, the Netherlands.

Dresselhaus, M.S., Dresselhaus, G. and Eklund, P.C. (1996) Science of Fullerenes and Carbon Nanotubes, 965p., Academic Press, Elsevier.

Dubrovsky, R., Bezmelnitsyn, V. and Sokolov, Y. (2005) Reduction of cathode carbon deposit by buffer gas outflow. *Carbon*, 43, 769-802.

Ebbesen, T.W., Tabuchi, J. and Tanigaki, K. (1992) The mechanistics of fullerene formation. *Chem. Phys. Lett.*, 191, 336-338.

Ebbesen, T.W., Hiura, H., Hedenquist, J.W., de Ronda, C.E.J., Andersen, A., Often, M. and Melezhik, V.A (1995) Origins of fullerenes in rocks; response by P.R. Buseck and S. Tsipursky. *Science*, 268, 1634-1635.

Forney, D., Freivogel, P., Grutter, M. and Maier, J.P. (1996) Electronic absorption spectra of linear carbon chains in neon matrices. IV. C_{2n+1} n=2-7. *J. Chem. Phys.*, 104, 4954-4960.

Fowler, P.W. and Manopoulos, D.E. (1995) *An atlas of fullerenes, International Series of Monographs on Chemistry*, 30, 120-147, Oxford University Press, United Kingdom.

Gamaly, E.G. and Chadderton, L.T. (1995) Fullerene genesis by ion beams. *Proc. R. Soc. Lond. A*, 449, 381-409.

Gerhardt, Ph., Löffler S. and Homann, K.H. (1987) Polyhedral carbon ions in hydrocarbon flames. *Chem. Phys. Lett.*, 137, 306-310.

Haufler, R.E. (1994) Techniques of fullerene production. *In Fullerenes: Recent advances in the chemistry and physics of fullerenes and related materials*, K.M. Kadish and R.S. Ruoff, Eds., *Electrochem. Soc. Proc.*, 94-24, 50-67, The Electrochemical Society, Inc.

Heath, J.R. (1992) Synthesis of C_{60} from small carbon clusters. *In Fullernens-Synthesis, properties, and chemistry of large carbon clusters*, G.S. Hammond and V.J. Kluck, Eds., *Am. Chem. Soc. Symp. Series*, 481, 1-23.

Hebard, A.F., Rosseinsky, M.J., Haddon, R.C., Murphy, D.W., Glarum, S.H., Palstra, T.T.M., Ramirez, A.P. and Kortan, A.R. (1991) Superconductivity at 18K in potassium-doped C_{60}. *Nature*, 350, 600-601.

Heymann, D., Chibante, L.P.F., Brooks, R.R., Wolbach, W.S. and Smalley, R.E. (1994) Fullerenes in the Cretaceous-Tertiary boundary layer. *Science*, 265, 645-647.

Homann, K.-H. (1998) Fullerene and soot formation - New pathways to large particles in flames. *Angew. Chem. Int. Ed. Engl.*, 37, 2435-2451.

Howard, J.B., McKinnon, J.T., Makarovsky, Y., Lafleur, A.L. and Johnson, M.E. (1991) Fullerenes C_{60} and C_{70} in flames. *Nature*, 352, 139-141.

Ishigaki, T., Suzuki, S., Kataura, H., Krätschmer, W. and Achiba, Y. (2000) Characterization of fullerenes and carbon nanoparticles generated with a laser-furnace technique. *Appl. Phys. A*, 70, 121-124.

Kasuya, D., Ishigaki, T., Suganuma, T., Ohtsuka, Y., Suzuki, S., Shiromaru, H., Achiba, Y. and Wakabayashi, T. (1999) HPLC analysis for fullerenes up to C_{96} and the use of the laser furnace technique to study fullerene formation process. *Eur. Phys. J. D*, 9, 355-358.

Krätschmer, W. (1996) Carbon clusters, fullerene cages, and interstellar matter. *In The Chemical Physics of Fullerenes 10 (and 5) years later: The far-reaching Impact of the Discovery of C_{60}*, W. Andreoni, Ed., NATO ASI Series, 27-35, Kluwer Academic Publishers.

Krätschmer, W., Sorg, N. and Huffman, D.R. (1985) Spectroscopy of matrix-isolated carbon cluster molecules between 200 and 850 nm wavelength. *Surface Science*, 156, 814-821.

Krätschmer, W., Lamb, L.D., Fostiropoulos, K. and Huffman, D.R. (1990) Solid C_{60}: a new form of carbon. *Nature*, 347, 354-358.

Kroto, H.W., Heath, J.R., O'Brien, S.C., Curl, R.F. and Smalley, R.E. (1985) C_{60}: Buckminsterfullerene. *Nature*, 318, 162-163.

Lamb, L.D. and Huffman, D.R. (1993) Fullerene Production. *J. Phys. Chem. Solids*, 54, 1635-1643.

Maier, J.P. (1997) Electronic spectroscopy of carbon chains. *Chem. Soc. Rev.*, 1997, 21-28.

Meijer, G. and Bethune, D.S. (1990) Mass spectroscopy confirmation of the presence of C_{60} in laboratory-produced carbon dust. *Chem. Phys. Lett.*, 175, 1-2.

Moriwaki, T., Kobayashi, K., Osaka, M, Ohara, M, Shiromaru, H. and Achiba, Y. (1997) Dual pathway of carbon cluster formation in the laser vaporization. *J. Chem. Phys.*, 107, 8927-8932.

Murayama, H., Tomonoh, S., Alford, J.M. and Karpuk, E. (2004) Fullerene production in tons and more: From science to industry. *Fullerenes, Nanotubes, and Carbon Nanostructures*, 12, 1-9.

Pfänder, N. (1993) Produktion und Trennung der Fullerene; Diplomarbeit (*in German*), Fachhochschule Münster (unpublished).

Presilla-Marquez, J.D., Sheehy, J.A., Mills, J.D., Carrick, P.G. and Larson, C.W. (1997) Vibrational spectra of cyclic C_6 in solid argon. *Chem. Phys. Lett.*, 274, 439-444.

Presilla-Marquez, J.D., Harper, J., Sheehy, J.A., Carrick, P.G. and Larson, C.W. (1999) Vibrational spectra of cyclic C_8 in solid argon. *Chem. Phys. Lett.*, 300, 719-726.

Scott, L.T., Boorum, M.M., McMahon, B.J., Hagen, S., Mack, J., Blank, J., Wegner, H. and deMeijere, A. (2002) A rational chemical synthesis of C_{60}. *Science*, 295, 1500-1503.

Smalley, R.E. (1992) Self-Assembly of the fullerenes. *Acc. Chem. Res.*, 25, 98-105.

Stephens, P.W. (1993) Physics and chemistry of Fullerenes - a reprint collection. *Adv. Series Fullerenes*, 1, 256p., World Scientific Publishing Co.

Taylor, R. and Abdul-Sada, A.K. (2000) There are no fullerenes in the K-T boundary layer. *Fullerene Sci. Techn.*, 8, 47-54.

Taylor, R., Langley, G.H., Kroto, H.W. and Walton, D.R.M. (1993) Formation of C_{60} by pyrolysis of naphtalene. *Nature*, 366, 728-731.

Ulmer, G., Hasselberger, B., Busmann, H.-G. and Campbell, E.E.B. (1990) Excimer laser ablation of polyimide. *Appl. Surf. Sci.*, 46, 272-278.

von Helden, G., Hsu, M.T., Gotts, N. and Bowers, M.T. (1993a) Carbon cluster cations with up to 84 atoms: Structures, formation mechanism, and reactivity. *J. Phys. Chem.*, 97, 8182-8192.

von Helden, G., Gotts, N.G. and Bowers, M.T. (1993b) Experimental evidence for the formation of fullerenes by collisional heating of carbon rings in the gas phase. *Nature*, 363, 60-63.

Wang, S.L., Rittby, C.M. and Graham, W.R.M. (1997a). Detection of cyclic carbon clusters. I. Isotopic study of the ν_4 (e_u) mode of cyclic C_6 in solid Ar. *J. Chem. Phys.*, 107, 6032-6037.

Wang, S.L., Rittby, C.M. and Graham, W.R.M. (1997b) Detection of cyclic carbon clusters. II. Isotopic study of the ν_{12} (e_u) mode of cyclic C_8 in solid Ar. *J. Chem. Phys.*, 107, 7025-7033.

Wang, S.L., Rittby, C.M. and Graham, W.R.M. (2000) On the identification of the vibrational spectrum of cyclic C_8 in solid Ar. *J. Chem. Phys.*, 112, 1457-1461.

Wakabayashi, T. and Achiba, Y. (1992) A model for the C_{60} and C_{70} growth mechanism. *Chem. Phys. Lett.*, 190, 465-468.

Wakabayashi, T., Ong, A.-L., Strelnikov, D. and Krätschmer, W. (2004) Flashing carbon particles on cold surfaces. *J. Phys. Chem. B*, 108, 3686-3690.

Wyss, M., Grutter, M. and Maier, J.P. (1999) Electronic spectra of long odd-number carbon chains C_{17}-C_{21} and C_{13}-C_{21}. *Chem. Phys. Lett.*, 304, 35-38.

Yamaguchi, Y. and Maruyama, S. (1998) A molecular dynamics simulation of the fullerene formation process. *Chem. Phys. Lett.*, 286, 336-342.

Chapter 3

CARBONACEOUS ONION-LIKE PARTICLES: A POSSIBLE COMPONENT OF THE INTERSTELLAR MEDIUM

SETSUKO WADA
Department of Applied Physics and Chemistry, the University of Electro-Communications, Chofugaoka, Chofu, Tokyo 182-8585, Japan

ALAN T. TOKUNAGA
Institute for Astronomy, University of Hawaii, 2680 Woodlawn Dr., Honolulu, Hawaii 96822, USA

Abstract: A carbonaceous material formed from a hydrocarbon plasma called quenched carbonaceous composite (QCC) is shown to have functional groups that approximate the positions of the interstellar 217.5 nm absorption and the infrared emission features at 3.3–11.3 μm. A form of this material, called "dark-QCC" has abundant carbonaceous onion-like particles. We present the results of various experiments involving QCC and conjecture that the carbonaceous onion-like particles in QCC may be a good laboratory analog to the carrier of the interstellar absorption and emission features. A scenario for dust formation from carbonaceous onion-like particles is presented.

Key words: Carbonaceous onion-like particle; circumstellar dust; infrared spectroscopy; infrared emission bands; dust analogs; high-resolution transmission electron microscope (HRTEM) imaging; interstellar dust; interstellar extinction bump; hydrocarbon plasma generation; quenched carbonaceous composite (QCC); UV spectroscopy

1. INTRODUCTION

Two of the outstanding questions regarding the composition of the interstellar medium (ISM) concern the origin of a strong ultraviolet

Frans J.M. Rietmeijer (ed.), Natural Fullerenes and Related Structures of Elemental Carbon, 31–52.

(UV) absorption at 217.5 nm and infrared emission features observed at 3.3, 6.2, 7.7, 8.6, and 11.3 μm that are observed in many sources in and outside of our Galaxy (Fitzpatrick, 2004; Peeters et al., 2004). We will discuss the production of a laboratory analog with spectral properties that suggest that these features could arise from a type of carbonaceous condensate produced from a hydrocarbon plasma. The work described here was originally motivated by the possibility that the 217.5 nm ISM absorption arises from carbonaceous material. A laboratory apparatus was fabricated to produce carbonaceous material that had an absorption band near 220 nm (Sakata et al., 1983, 1994). When the infrared absorption was measured, it was found that this material had absorption bands with assignments similar to that deduced for the ISM infrared emission features.

The 217.5-nm interstellar absorption band dominates the interstellar extinction curve. This absorption band is observed in lines of sight to hot stars and is very common in our galaxy (Fig. 1). There are strong

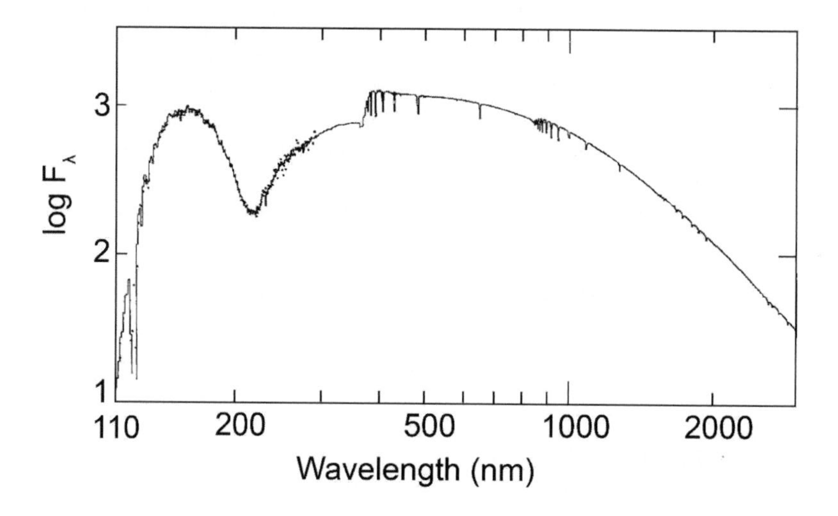

Figure 1. The UV to infrared spectrum of star HD 147701 (adapted from Fitzpatrick, 2004) with a noticeable very strong UV absorption band at 217 nm. The relative flux density F_λ has units of ergs cm^{-2} μm^{-1}

indications that this band involves a type of carbonaceous material (Mennella et al., 1996; Rouleau et al., 1997; Henning et al., 2004).

With regard to the infrared emission features, it is now generally accepted that the 3.3, 6.2, 7.7, 8.6, and 11.3 μm emission bands arise from a material that contains aromatic functional groups (Sellgren, 2001; Peeters et al., 2004) (Fig. 2). These emission features are observed around planetary nebulae, in star formation regions, and in the diffuse ISM. These bands are seen in emission because they arise from very small grains and are easily heated by UV radiation (Sellgren, 2001). Much work has gone into explaining the non-thermal emission mechanism, and a robust conclusion is that the particles giving rise to the emission bands must be less than 5 nm (Draine, 2003; Blakes et al., 2004).

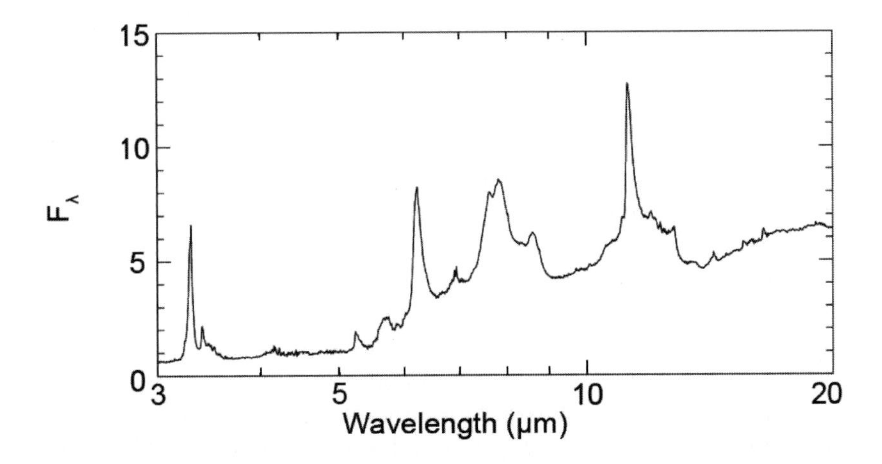

Figure 2. Spectrum of the infrared emission features (*IEF*) in the planetary nebula NGC 7027. Data is adapted from Peeters et al. (2004). The emission lines in the object have been removed so that the infrared emission features may be seen more clearly. Strong emission from the C–H stretch and the C–H out of plane bending in aromatic structures are evident at 3.3 and 11.3 μm. There is also emission at 3.4–3.5 μm that appears to come from side groups on aromatic structures as well. The relative flux density F_λ has units of W m^{-2} μm^{-1}

Precise identification of the emitting material has not been possible to date. Many terms are used to refer to these bands, such as polycyclic aromatic hydrocarbon (PAH) bands, unidentified infrared (*UIR*) emission features, infrared emission features (*IEF*), and others (Peeters et al., 2004). We will refer to these features as "*IEF*" here. Note, however, that these emission bands are commonly referred to in the astronomical literature as "*PAHs*", which is used as a generic term (Hudgins and Allamandola, 2004), even though there has been no spectroscopic detection of any specific PAH molecule in the interstellar medium to date (Clayton et al., 2003).

2. SYNTHESIS OF THE LABORATORY ANALOG AND ITS PROPERTIES

During 1970–1980 many molecules were detected in the ISM and circumstellar environment of stars. To study the formation of molecules, Sakata (1980) built an apparatus for the synthesis of interstellar molecules from simple radicals of plasmic gas. He hypothesized that materials formed from plasmic gases should have some similarities to the materials formed around evolved stars because the materials should be formed from chemically active and simple species.

The plasma was generated using a microwave generator in a small chamber made of quartz glass with a volume of 25 cm^3, which was inserted in a microwave waveguide. Radio frequency generators and microwave generators are often used for formation of plasmic gas. In a plasmic gas formed by a radio frequency generator, the neutral source gas often remains in high abundance. But in Sakata's apparatus, the source gas was decomposed completely into highly excited plasmic gas. The plasmic gas streamed into a vacuum chamber through a nozzle 7 mm in length and 1 mm in diameter. The nozzle collimated the plasmic gas beam. A schematic of the apparatus is shown in Fig. 3. In addition to molecules, solid materials formed in the ejecta from the collimated gas beam. These solids were also expected to be present in the circumstellar environment. This was the start of our study of dust formation in the laboratory.

Methane was initially used as a source of carbon and hydrogen. To make radicals, source gases (CH_4, H_2, etc.) were decomposed in the plasma chamber. By analysis of a quadrupole mass spectrometer, polyynes [$H(C{\equiv}C)_nH$; n = 1–5], and aromatic molecules, benzene, naphthalene, and five-member rings compounds, were detected in the gas beam ejected into vacuum chamber (Sakata, 1980).

A solid material was condensed from the collimated gas beam. A circular spot of a brown-black solid appeared on the quartz-glass substrate placed perpendicular to the beam. A lighter colored brown-black material was deposited around the central spot (substrate B in Fig. 3) (Sakata et al., 1983). This material was named "*quenched carbonaceous composite*" (*QCC*) because it was made from carbon and hydrogen (carbonaceous), was formed through rapidly cooling (quenching) of the plasmic gas, and was a mixture of chemical compounds, i.e. it is a composite solid.

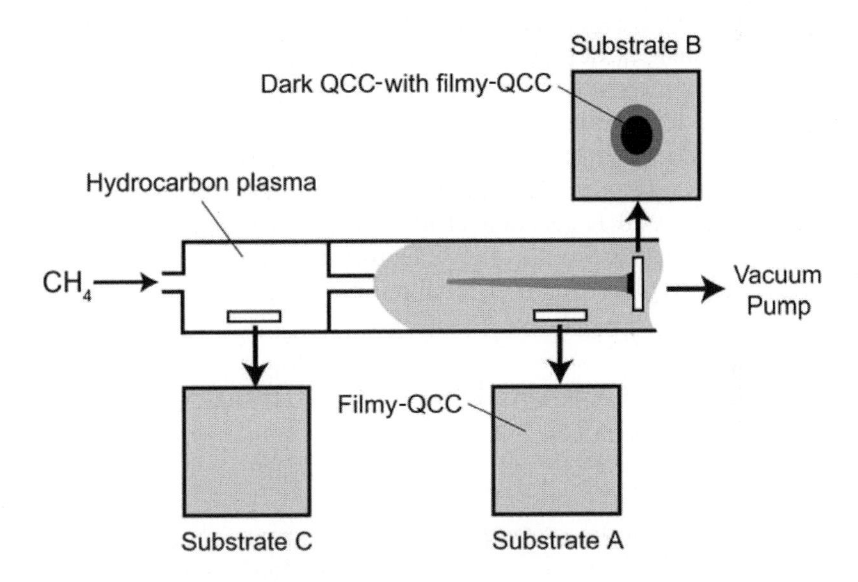

Figure 3. The experimental setup for producing *QCC* samples. The hydrocarbon plasma is produced in a microwave discharge tube. The plasmic gas is injected through a small orifice into a vacuum chamber. *Filmy*-QCC was collected on a quartz substrate A, located on the wall of the vacuum chamber. *Dark*-QCC was collected on substrate B, located in the plasmic beam. A combination of dark and yellow material was collected on substrate C. (for more details, see Sakata et al., 1983, 1994)

A yellow-brown, filmy material was simultaneously deposited on a wall of the apparatus. The material was soluble in organic solvents, acetone, methanol, benzene; it was named "*filmy-QCC*". The color of the material deposited very close to the plasmic chamber was brown, and the color of the material deposited far from the plasmic chamber was yellow. These colors indicate different chemical compositions of the materials.

The solid formed on the substrate in the plasmic beam was named *dark-QCC* (Sakata et al., 1987). The *filmy-QCC* is relatively rich in hydrogen compared with *dark-QCC*. Infrared and mass spectroscope analyses showed that *filmy-QCC* is a mixture of various PAHs and other hydrocarbons. When *filmy-QCC* was heated in a vacuum, it became carbonized, and this material was called "*thermally altered filmy-QCC*" (*TAF-QCC*) (Sakata et al., 1990).

During synthesis of *QCC*, intensive radiation was emitted from the plasmic gas and H_α, H_β, C_2, and CH lines were observed. The emission intensities depended on many factors such as the power of the magnetron generator, duration of the plasma, and gas pressure. Approximate temperatures can be estimated from the emission-line intensities. The temperature of electronic excitation of atomic hydrogen was approximately 6000 K. The vibrational temperature of C_2 was about 3000 K (Sakata et al., 1983). The plasma was a non-equilibrated plasma with the electronic excitation temperature >vibrational temperature >rotational temperature.

3. UV SPECTRA AND STRUCTURE OF *QCC* MATERIALS

3.1 UV Spectra of *QCC*

A *dark-QCC* that formed from a plasmic gas with a starting pressure 4 torr of methane and a microwave input power of ca. 300-350 W showed a broad absorption peak around 220 nm. *Dark-QCC* samples were prepared by washing them with acetone to remove soluble organic molecules. Using different mixtures of hydrogen and methane yielded *QCC* samples with different peak wavelengths (Wada et al., 1998). We obtained *QCC* samples with a maximum absorbance from 200 to 260 nm (Fig. 4). Organic molecular condensates (*filmy-QCC*) formed abundantly together with dark-*QCC*s peaking around 220 nm; *filmy*-QCC was relatively poor in the *dark-*

QCC peaking around 230 nm and absent in the *dark-QCC* peaking at 250 nm. However, any proposed material for the carrier of the 217.5-nm feature must account for the nearly constant wavelength of the peak absorption in the ISM (Rouleau et al., 1997; Henning et al., 2004).

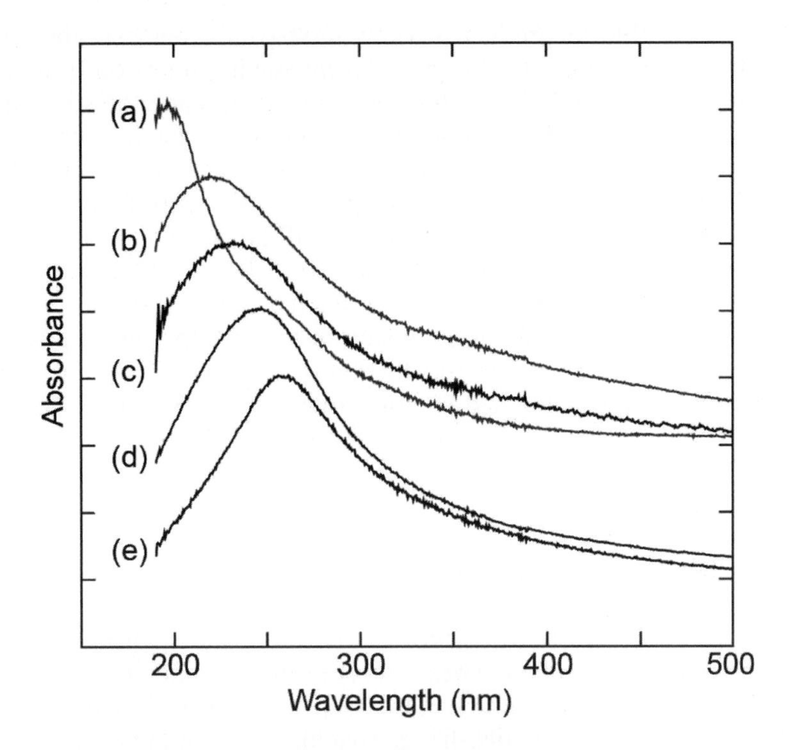

Figure 4. UV spectra of *QCC* as a function of the wavelength of maximum absorbance and gas mixture (a) ca. 200 nm; 100% CH_4, (b) ca. 220 nm; 100% CH_4, (c) ca. 230 nm; 50% H_2, 50% CH_4, (d) ca. 250 nm; 70% H_2, 30% CH_4, (e) ca.260 nm; 100% CH_4.

Samples (b), (c), and (d) were obtained outside of the plasma chamber (substrate B in Fig. 3) and are described by Wada et al. (1998). In addition, we collected solid samples in the plasmic chamber (substrate C in Fig. 3). A yellow solid with maximum absorbance at ca. 200 and a black solid with maximum absorbance at 260 nm were collected and are shown here as curves (a) and (e), respectively

There are some rare classes of stars where the absorption peak is near 250 nm, as in R CrB stars (Hecht et al., 1984; Drilling and Schoenberner, 1989), and thus the ca. 250-nm *QCC* could be a laboratory analog for the absorbing material. The types of carbonaceous material formed in circumstellar environments depend on the chemical abundances and physical conditions, such as density, temperature, and cooling rate of the gas.

3.2 Structure of *QCC* Materials

We were able to produce various *dark-QCC*s with an absorption peak from ca. 200 nm to 260 nm. An interesting question is whether there is a correlation between the peak wavelength and the structure of the *dark-QCC*s. Various kinds of carbon solids have been studied with high-resolution transmission electron microscopy (HRTEM) (Oberlin and Bonmamy, 2001; Inagaki et al., 2004). We studied several samples of deposited *QCC* by HRTEM using a Hitachi H-9000 EM (Sakata et al., 1994; Wada et al., 1998, 1999; Goto et al., 2000). The *QCC* samples were mounted on a carbon thin-film supported by a standard copper, electron microscope, mesh grid. The dark deposit was generally too thick to obtain HRTEM images but good images were obtained in thin areas at the periphery of *QCC* samples (Fig. 5).

3.2.1 Dark-*QCC*

We found a variety of carbon structures in *QCC*. Except for the ca. 200-nm *QCC* sample, all *QCC* deposits have a complex microstructure. In the HRTEM images of the (002) lattice fringes we can see the aromatic graphene sheets stacked in a parallel sequence (Fig. 5d). Lattice fringes have the 0.336-nm spacing for (002) lattice fringes of graphite, but larger lattice-fringe spacings are common for graphitic carbons, even in nicely uniform, parallel ribbons. In many images of *QCC*, the carbon layers are curved and stacked concentrically.

We summarize below the characteristic features seen in the *QCC* images shown in Fig. 5, listed according to increasing wavelength of absorption peak position. For each item we give the color of the material, the composition of the starting gas, the microwave power, and the substrate location (cf. Fig. 3):

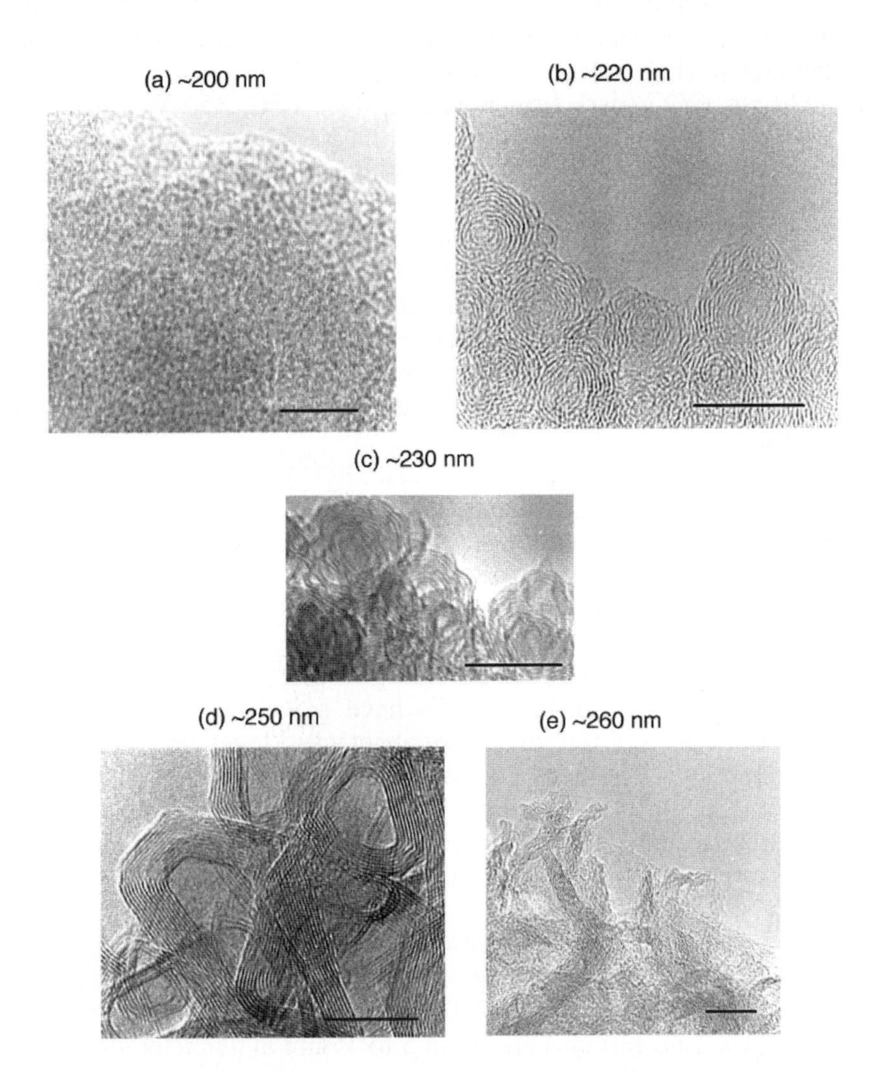

Figure 5. High-Resolution Transmission Electron Micrographs of *QCC* samples corresponding to the UV spectra shown in Fig. 4. The scale bar in each image is 10 nm

1. ~200-nm *QCC* (yellow, 100% CH_4, 270 W, substrate C): Two solids of different color, i.e. yellow and black, were produced in the plasma chamber. The yellow solid appears amorphous in HRTEM image and has an absorption band at about 200 nm. It was not soluble in acetone and is therefore more polymerized than *filmy-QCC* (Fig. 5a).
2. ~220-nm *QCC* (dark brown/black, 100% CH_4, 350 W, substrate B): This dark material is formed in the gas beam ejected from the plasma and contains abundant multi-layer, carbonaceous onion-like particles (Fig. 5b). This material is referred to as *dark-QCC*.
3. ~230-nm *QCC* (dark brown/black, 50% CH_4 + 50% H_2, 270 W, substrate B): The shape of the concentric multi-layer carbon structures is typically polyhedral rather than spherical (Fig. 5c).
4. ~250-nm *QCC* (black, 30% CH_4 + 70% H_2, 270W, substrate B): The linear and bent structures may be polyhedrons (Fig. 5d). The HRTEM appearance is very similar to glassy carbon (Oberlin, 1989).
5. ~260-nm *QCC* (black, 100% CH_4, 300 W, substrate C): The structure is straighter and longer than in the 250-nm *QCC* sample. We also find curled semispherical components attached to peripheral graphitic layers (Fig. 5e).

This series of *QCC* materials shows the growth of the carbon network that occurred between wavelengths 220 nm and 260 nm. The ca. 250-nm and ca. 260-nm *QCC*s have a well-developed carbon network of larger and flat layers that are stacked together. The carbon network in the ca. 250-nm *QCC* has a bent, ribbon-like structure. The number of graphitic layers is nearly the same as the number of layers in the carbonaceous onion-like particles. This suggests that onion-like particles might be a precursor of the ribbon-like structure.

In Fig. 6 we show low-magnification images of another sample of *dark-QCC* along with high-magnification images (insets). The *QCC* sample shows a bump around 220 nm. It is composed of many fine particles. The fine particles are about 5 to 15 nm in diameter and seem to be very similar to the onion-like particles. However, the particles are not perfectly onion-like: *their shells are not always closed.* Some shells are connected to the next shell. The central core of some spherules seems to be vacant (Fig. 6a). This is larger than the size of C_{60}, which is 0.7 nm in diameter. In addition, many shell fragments are attached to their surface.

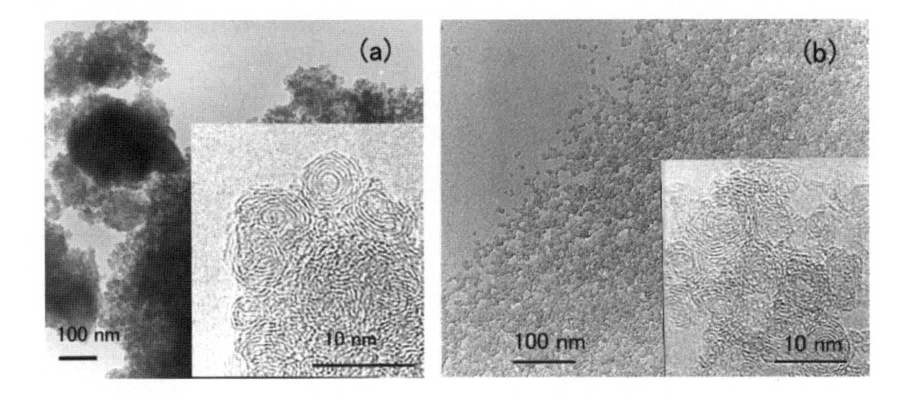

Figure 6. High-Resolution Transmission Electron Micrographs of (a) *Dark-QCC* and (b) *TAF-QCC*

A typical onion-like particle is shown in Fig. 7. The onion-like particle is composed of about 10,000 carbon atoms. The central shell of an onion-like particle is about the same size of C_{450} fullerene. Fullerenes have a closed cage structure because they incorporate five-member rings in their networks. A model of a fragmented shell observed in the *QCC* onion-like particle is presented in Fig. 7.

The concentric spherule on the left side of the HRTEM image of Fig. 7 is composed of more fractured shells than that of the one in the center of this image. It is composed of very small fragments. A structure 0.6 nm in length corresponds to a carbon network with 2–3 carbon rings aligned linearly. PAH molecules with this size are fluorene, coronene, and anthracene, among many others. A structure 1.3 nm in length corresponds to a carbon network with 5–6 carbon rings aligned linearly. Circumanthracene fits this size. However, in the plasma-produced *QCC*, not all of these PAHs may be perfect molecules. It is likely that they lost some peripheral hydrogen atoms. The onion-like particles are an intermediate material between PAHs and carbon onions. They are composed of broken shells, shell fragments, PAH-like components, and possibly fullerene. We view the particles in the HRTEM images as roughly concentric layers with imperfections. The particles contain hydrogen, so they are carbonaceous and not pure carbon onions. We conclude that the onion-like particles with many fragmented shells are the major components of the *QCC*.

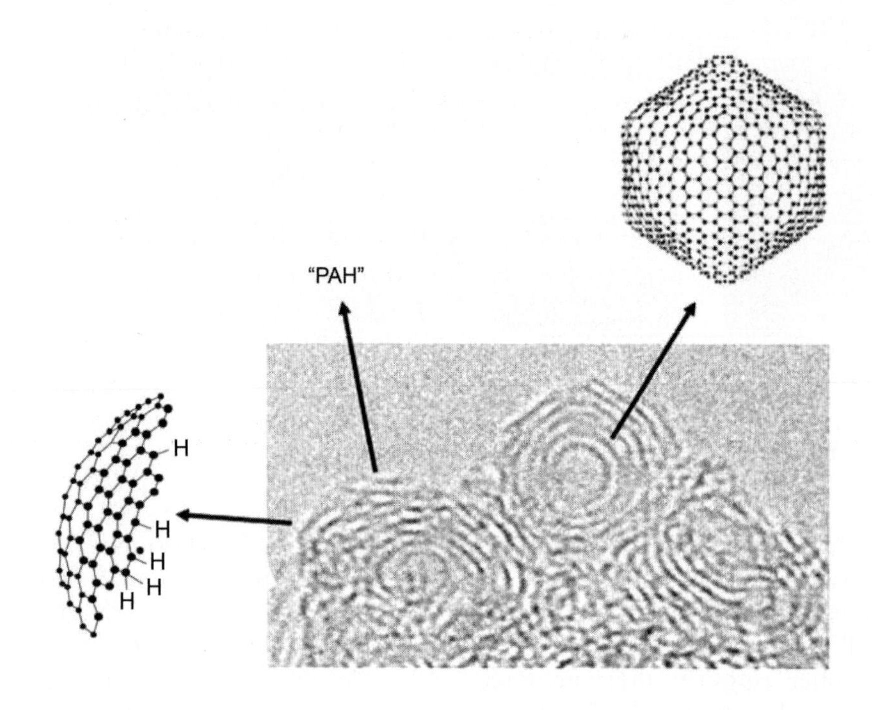

Figure 7. High-resolution transmission electron microscope image of onion-like *QCC* spheres (center) and a C_{960} fullerene model (top right) having the approximate diameter of the second, inner shell of the carbon onion. "PAH" indicates fragments of the carbon onion that are planar or curved structures. A curved fragment of the carbon onion is shown in the sphere on the left. It is only partially hydrogenated, and there are dangling bonds (radicals) as indicated by the "•" symbol

3.2.2 Thermally Altered Filmy-*QCC* (*TAF-QCC*)

PAHs are the main component of the *filmy-QCC*. When the *filmy-QCC* was heated at 350 °C for 20 min inside a small quartz glass vessel, spherules with no discernable structure and a diameter of about

5 to 20 nm appeared in the initially amorphous *filmy-QCC* (Sakata et al., 1994). The sizes of the newly formed spheres are similar to those of the carbonaceous onion-like particles observed in the original *dark-QCC* samples (Fig. 6).

The *filmy-QCC* began to carbonize at temperatures of 500–600 °C. Abundant onion-like particles were found in the carbonized *TAF-QCC*. This *TAF-QCC* shows a 220 nm absorption peak and a 3.3 μm C–H peak (Sakata et al., 1990, 1994). It is unclear whether the onion-like particles were formed in the initially amorphous *filmy-QCC* or formed from the gas that evaporated from the *filmy-QCC* and was trapped in the quartz container.

3.3 Comparison to Other Carbonaceous Materials and the Origin of the 220-nm Absorption Band

In addition to *QCC*, many other types of carbonaceous materials have been proposed as possible carriers of the 217.5-nm absorption. These include processed hydrogenated amorphous carbon (*HAC*) (Duley and Lazarev, 2004; Mennella, 2004), nano-sized carbon grains formed in Ar-H_2 gas mixtures containing carbon vapor (Schnaiter et al., 1998), and coal material (Papoular et al., 1996). Papoular et al. (1996) provide a summary of the properties of these materials. In the case of *HAC*, the 217.5-nm absorption is produced by thermal processing of *HAC* or by UV irradiation (Duley and Lazarev 2004; Mennella, 2004). The 217.5-nm band is also produced by ion bombardment of carbon grains formed in Ar or in H_2 (Mennella et al., 1997).

As already noted, our 220-nm *QCC* samples contain onion-like particles (Figs. 5 and 6) and they are thus not completely amorphous carbon. In this respect, *QCC* is not similar to *HAC*. Another type of carbonaceous laboratory analog that shows onion-like particles is the evaporation of graphite electrodes in the presence of H_2 (see Fig. 4 in Henning et al., 2004).

Soot prepared in Ar-H_2 atmospheres also showed the 220-nm absorption with variations in the wavelength of maximum absorption. Rotundi et al. (1998) studied the correlation between the wavelength of maximum absorption and the structure of particles in the condensed soot samples. They found various forms of carbon particles, *incl.* chain-like aggregates of 7–15 nm spheres, bucky-onions, bucky-tubes, and graphitic carbon ribbons in their samples. They could not find any obvious differences in the abundance of these particles among the soot samples and concluded that the internal structures of chain-like aggregates caused the variations in the wavelength of maximum absorption.

de Heer and Ugarte (1993) investigated the relationship between the peak of the UV absorption and the shape of the carbon onion particles. They made soot samples by arc discharge in a heated tube. By annealing samples at various temperatures, they formed carbon onion particles with different shapes. Their result indicated that the width of the UV absorption band changed with annealing temperature, however, the absorption maximum appeared about 260 nm in all samples, which were dispersed in water.

de Heer and Ugarte (1993) showed that the onion particles annealed at 2250 °C have a more polyhedral shape than the particles of the 220-nm *QCC*. The 220-nm *QCC*, on the other hand, is not stable when heated to such high temperatures. Heating the 220-nm *QCC* at 700 °C for 20 minutes leads to loss of volatile materials (Wada et al., 1999). The peak absorbance of the residual material was reduced by 75% compared to unheated *QCC*, and the peak was shifted to 234 nm. It suggests that onion-like particles in the 220-nm *QCC* are composed of organic or carbonaceous materials that evaporated or are modified at high temperature.

In our experiments, clear differences in the structure were detected in the *QCC*s as the wavelength of the maximum absorption changes. The shift to longer wavelengths of the maximum absorption corresponds to changes in the size and flatness of the carbon network.

From the molecular approach, the absorption band of organic materials at ca. 200 nm results from electronic excitation of π bond electrons (π-π*). It is well known that π electrons are delocalized by the conjugation of π bonds and that the conjugation stabilizes the bonds. As conjugation increases, the absorption peak caused by π electrons is shifted to longer wavelengths. For example, *1,3-butadiene* and *1,3,5-hexatriene* show a peak at 217 nm and 258 nm, respectively.

The wavelength of maximum absorption is at 200 to 260 nm in *QCC*. The 220-nm absorption band can be explained by a

–C=C–C=C–

structure. A wavelength of 260 nm was the maximum found in a well-developed carbon network of *QCC*. Rotundi et al. (1998) also found that by processing soot, the absorption band was shifted up to 260 nm.

Graphite has well-delocalized π electrons but the UV absorption of graphite will shift to shorter wavelengths if the π electrons are localized within a small number of carbon atoms because of the formation of defects. Using amorphous carbon as a starting material, Mennella et al. (1997) performed an experiment to break the C-network by bombardment with helium ions. They found that an absorption peak around 220 nm was produced as a result.

Does a curved structure result in localization of the π electrons? Tomita et al. (2001) made spherical carbon onions by heating particles of nanodiamond to 1700 °C and polyhedral carbon onions by further heating to 1900 °C. By electron energy-loss spectroscopy, they found localized π electrons in the spherical carbon onions. On the other hand, in the polyhedral carbon onions, the π electrons were delocalized. Tomita et al. (2004) calculated the extinction bump of spherical carbon onions with defective structures and found a reasonable fit to the 220-nm interstellar extinction band. However, experiments are needed to confirm these calculations.

From the above discussion, we concluded that the size of conjugation of the π bond is a main factor for the wavelength of maximum absorption. By perturbation of the development of the carbon network, such as the inhibition of the chemical bond formation with hydrogen, defect formation by ion bombardment, or the formation of defective curved structure, the wavelength of maximum absorption is shifted to shorter wavelengths. Although the size of the chemical bonds responsible for the peak wavelength is very small, with the enlargement of the conjugation of the π electrons, the structure of the *QCC* particles is changed as seen in HRTEM images.

Although carbonaceous material with conjugated carbon bonds does have absorption near 220 nm due to π-π* electronic transitions, there is no clear mechanism for achieving the near constancy of the 217.5-nm band observed in the interstellar medium. This is presently an unsolved problem. A major difficulty in producing a good laboratory analog is that we are unable to reproduce isolated particles as in the ISM. There are also other possibilities. From an analysis of interplanetary dust particles, Bradley et al. (2005) presented another idea for the carrier of the interstellar 217.5-nm absorption. They suggested both organic carbon and OH-bearing amorphous silicates could contribute to the absorption.

4. IR SPECTRA

4.1 IR Spectra of *QCC* Materials

The infrared spectra of *QCC* samples are presented in Fig. 8. The spectra of the *dark-QCC* were measured after washing with acetone to remove organic molecules. Samples were directly collected on a KBr or BaF_2 substrate; therefore their thickness is not uniform.

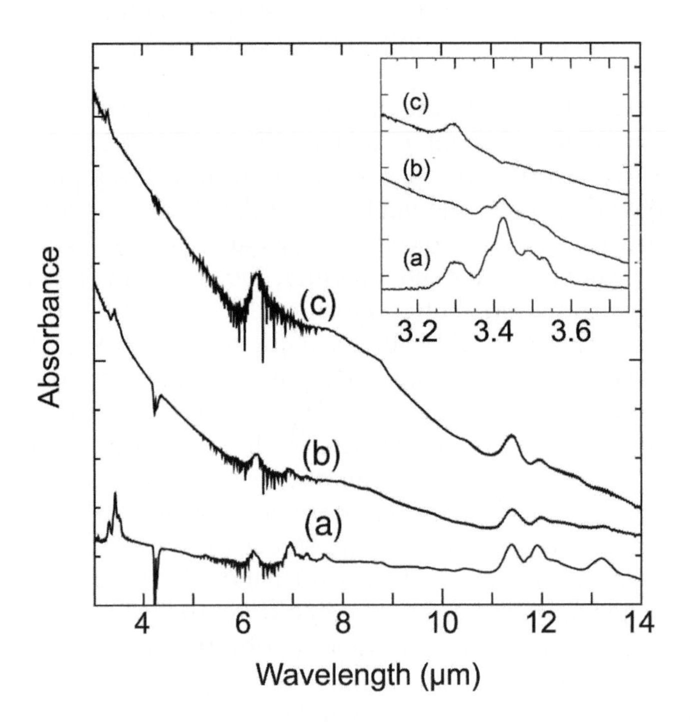

Figure 8. IR spectra of (a) *filmy-QCC*, (b) *dark-QCC*, and (c) *TAF-QCC*. Spectra (a) and (b) are discussed in detail by Wada et al. (2003). The aromatic C–H stretching mode can be seen at 3.3 μm; the C–H out-of-plane bending mode at 11.4-13.2 μm; the skeletal in plane vibration mode of C=C at 6.25 μm; the deformation of CH_2 and the asymmetric deformation of CH_3 at 6.9 μm; and the symmetric deformation of CH_3 at 7.27 μm. We are not sure about the identification of the 7.7 and 8.6 μm bands reported by Sakata et al. (1987)

The *filmy-QCC* shows strong IR bands with weak continuous background absorption (Fig. 8a), while the 220-nm and 250-nm *dark-QCC* material shows strong background absorption (Figs. 8b and 8c). The *TAF-QCC* also shows strong background absorption.

Bands caused by C–H stretching vibration modes appear at 3–4 μm. A band at 3.3 μm (stretching mode) and a band at 11.4 μm (out-of-plane bending mode) are caused by the aromatic C–H bond. They are observed in all samples except the 250-nm *dark-QCC* sample, which is consistent with the HRTEM data showing no major structural difference between *dark-QCC*s and *TAF-QCC*s. Carbonized *TAF-QCC* shows a 3.3 μm band (aromatic C–H). *Dark-QCC* shows weak a 3.3 μm band and strong ca. 3.4 μm and ca. 3.5 μm bands due to the stretching vibration of methylene (CH_2) and methyl (CH_3). *TAF-QCC* is the only material that does not show the ca. 3.4 μm and ca. 3.5 μm bands, since heating had caused the loss of the aliphatic component.

Other bands that arose from duo, trio, and quartet aromatic C–H bonds appear at 11.9–12.0 μm and 13.2–13.3 μm. These bands are strong in the *filmy-QCC* (Fig. 8a), but weak in *TAF-QCC* and *dark-QCC*. A very small hump can be seen at 12.5–12.7 μm in both *QCC* samples (Figs. 8b and 8c).

It is well known that thermal heating destroys aliphatic bonds. We reported a variation of infrared features of the 3–4 μm region during carbonization of *filmy-QCC* at temperatures of 500–600 °C (Sakata et al., 1990; Goto et al., 2000). A band caused by C=C vibration exists at 6.2–6.3 μm in the *QCC* samples. Usually, a single C=C bond shows a band at 5.9–6.1 μm. The peak moves to longer wavelengths when the double bond is weakened by conjugation. Small absorption bands at 7.7 μm and 8.6 μm that correspond to interstellar *IEF* (*infrared emission features*; Fig. 2) are often seen in the broad 8-μm band of *TAF-QCC* and *dark-QCC* (see, Sakata et al., 1987); they are the very weak features seen in Fig. 8. However, *QCC* does not have the strong bands at 7.6/7.9 μm and 8.6 μm in the *IEF*. In addition, there is a difference in the wavelength of the peaks for the out-of-plane bending mode of aromatic C–H, which are at 11.4 μm in *QCC* and 11.25 μm in the *IEF*.

As a whole, the spectra of *TAF-QCC* samples have similarities to the observed interstellar *IEF*, although the relative *QCC* band strengths are different from *IEF*. Similar spectra for heated *HAC* material were obtained (Scott and Duley, 1996), and it indicates that the infrared active functional groups reported by Sakata et al. (1987) have no unique origins in terms of the nature of the carbon phase, the mode of formation, or thermal processing.

4.2 Comparison to Other Carbonaceous Materials

Similar IR absorption bands are observed in *dark-QCC* (Fig. 8), *HAC* heated to 427 °C (Scott and Duley, 1996) and nano-sized carbon grains formed in an H_2 atmosphere (Schnaiter et al., 1998). Thus, these three materials have similar functional groups. In addition, carbonized *TAF-QCC* and *HAC* heated to 527 °C (Scott and Duley, 1996) also show very similar spectra, namely a strong aromatic band at 3.3 μm and a weak aliphatic band at 3.4 μm.

At the level of our current understanding, there appear to be structural differences between these materials even though the infrared active functional groups appear to be similar. Compare, for example, the carbon nanostructures shown in our Figs. 6 and 7 with those shown in Fig. 2 of Scott and Duley (1996) and in Fig. 2 in Schnaiter et al. (1998). In carbonaceous material, atomic groups attached to the peripheral site cause major IR bands. Therefore, similar IR bands appear even if carbonaceous materials would have a different bulk structure. Both the *dark-QCC* and *TAF-QCC* samples are reasonable interstellar dust analogs because they contain nanometer-sized particles (Fig. 7) that have the approximately correct functional groups of the particles that are thought to reside in the ISM.

5. A POSSIBLE SCENARIO FOR THE FORMATION OF THE *IEF* CARRIER

We conjecture that dust processing around evolved stars progresses as follows: Spectral-type M giant and supergiant stars and OH/IR stars are the largest carbon and silicate dust-producing sources in our Galaxy (Gehrz, 1989; Kwok, 2004). Such stars in the last stages of stellar evolution are known as asymptotic giant branch (AGB) stars. Mass loss in this stage of evolution leads to copious production of dust. During the period of maximum mass-loss and dust production, these stars are totally obscured at visible wavelengths, but they are very bright IR sources. Carbonaceous dust, like *dark-QCC*, is formed at the location of high temperature and high gas density in the mass-loss process of AGB stars. On the other hand, organic condensations such as *filmy-QCC* are formed in cooler regions of the circumstellar environment of AGB stars.

The AGB phase ends when mass loss stops and the star collapses into a white dwarf star. The white dwarf star has an extremely high surface temperature and is a source of intense UV radiation. Thus, UV and visible photons irradiate the surrounding dust, and the resulting nebular emission is known as a planetary nebula (PN). As the temperature rises in the PN phase of the central star, organic dusts are heated and thermally altered by UV photons. They are polymerized through dehydrogenation and finally carbonized in a process similar to the production of *TAF-QCC*. Thermal processing is very effective for making isolated aromatic C-H bonds.

Both *filmy-QCC* and *dark-QCC* become mostly carbonaceous onion-like particles containing a small amount of hydrogen through thermal alteration. Both materials exhibit the 220-nm absorption bump. In the photo-dissociation zone of PNs, the carbonaceous onion-like particles are partially destroyed. Shell fragments are gradually broken off from the onion-like particles. The shell fragments absorb UV photons, and are heated stochastically. Some of the fragments are ionized by UV photons. The very small particles (<5 nm) produce the *IEF* but the larger dust, e.g. the carbonaceous onion-like particles (Fig. 7; central object), is not observed because it is not heated stochastically to high temperatures. Finally, the shell fragments and onion-like particles that survived in the shell of a PN are ejected into the ISM.

The *IEF* carrier phases are also widespread in the diffuse ISM of the Galaxy, where they are well mixed with large dust particles (Onaka et al., 1996). According to our hypothesis, the onion-like dust particles are gradually eroded by shock waves, UV irradiation, and bombardment by atoms and ions. These erosion processes create small PAH-like shell fragments in the ISM that become the *IEF* carriers.

When the shell fragments are ionized, strong C=C and –C–C– skeletal emission will be observed because of dipole formation. When onion-like particles absorb UV energy, they show the 217.5-nm extinction bump. Allain et al. (1996) and Vuong and Foing (2000) made estimates of the stability of PAHs in the ISM and concluded that small PAHs are unstable and are easily dehydrogenated, whereas large PAHs are hydrogenated and stable in the ISM.

We note that Jones (1997) and van Winckel (2003) reviewed dust production and processing by UV radiation, and dust ejection into the ISM, and that Tokunaga (1997) and Waters (2004) summarized the rich variety of infrared spectra observed in AGB stars and planetary nebulae. The PAH model was discussed by Hudgins and Allamandola (2004), Scott and Duley (1996) discussed thermally altered HAC, and irradiated HAC was discussed by Mennella et al. (1997, 2002). Papoular et al. (1996) reviewed the coal model. We recommend these papers for other points of view than expressed in our scenario.

6. SUMMARY AND CONCLUSIONS

We synthesized quenched carbonaceous materials from a hydrocarbon plasma. *Dark-QCC* composed of defective onion-like particles showed an absorption feature similar to the interstellar 217.5-nm bump. The absorption maximum shifted to the longer wavelengths as the carbon structure changed into a polyhedral form. The *IEF* in evolved stars is well simulated by thermally-altered *filmy-QCC*. However, the shape and position of the *IEF* in space was not reproduced precisely by the *QCC* spectra. Nevertheless, *QCC* shows functional groups that have many similarities to the observed spectra. We believe *QCC* is a good analog for the carbonaceous dust in space.

Acknowledgements. We thank Alessandra Rotundi for helpful comments on our manuscript. Akira Sakata conceived the method of producing QCC, and he was the driving force behind the development of QCC as a laboratory analog for the carbonaceous materials in the ISM. He passed away in 1995, and we dedicate this paper to him. The authors would like to thank S. Kimura and C. Kaito for their HRTEM analyses at Ritsumeikan University, F. Drabik for assistance with the figures, and L. Good for assistance with the manuscript.

7. REFERENCES

Allain, T., Leach, S. and Sedlmayr, E. (1996) Photodestruction of PAHs in the interstellar medium. I. Photodissociation rates for the loss of an acetylenic group. *Astron. Astrophys.*, 305, 602–615.

Blakes, E.L.O., Bauschlicher, C. and Tielens, A.G.G.M. (2004) Models of the Unidentified Infrared Emission Features. *In Astrophysics of Dust,* A.N. Witt, G.C. Clayton and B.T. Draine, Eds., *ASP Conf. Ser.*, 309, 731–753, Astronomical Society of the Pacific, San Francisco, California, USA.

Bradley, J., Dai, Z.R., Erni, R., Browning, N., Graham, G., Weber, P., Smith, J., Hutcheon, I., Ishii, H., Bajt, S., Floss, C., Stadermann, F. Sandford, S. (2005) An astronomical 2175 Å feature in interplanetary dust particles. *Science*, 307, 244–247.

Clayton, G.C., Gordon, K.D., Salama, F., Allamandola, L.J., Martin, P.G., Snow, T.P., Whittet, D.C.B., Witt, A.N. and Wolff, M.J. (2003) The role of polycyclic aromatic hydrocarbons in ultraviolet extinction. I. Probing small molecular polycyclic aromatic hydrocarbons. *Astrophys. J.*, 592, 947–952.

de Heer, W.A. and Ugarte, D. (1993) Carbon onions produced by heat treatment of carbon soot and their relation to the 217.5 nm interstellar absorption feature. *Chem. Phys. Lett.*, 207, 480–486.

Draine, B.T. (2003) Interstellar dust grains. *Ann. Rev. Astron. Astrophys.*, 41, 241–289.

Drilling, J.S. and Schoenberner, D. (1989) On the nature of newly formed dust around the hydrogen-deficient star V348 Sagittarii. *Astrophys. J.*, 343, L45–L48.

Duley, W.W. and Lazarev, S. (2004) Ultraviolet absorption in amorphous carbons: Polycyclic aromatic hydrocarbons and the 2175 Å extinction feature. *Astrophys. J.*, 612, L33–L35.

Fitzpatrick, E.L. (2004) Interstellar extinction in the Milky Way galaxy. *In Astrophysics of Dust*, A.N. Witt, G.C. Clayton and B.T. Draine, Eds., *ASP Conf. Ser.*, 309, 33–55, Astronomical Society of the Pacific, San Francisco, California, USA.

Gehrz, R.D. (1989) Sources of stardust in the galaxy. *In Interstellar Dust*, L.J. Allamandola and A.G.G.M. Tielens, Eds., *IAU Symp.*, 135, 445–453, Kluwer Academic Publishers.

Goto, M., Maihara, T., Terada, H., Kaito, C., Kimura, S. and Wada, S. (2000) Infrared spectral sequence of quenched carbonaceous composite subjected to thermal annealing. *Astron. Astrophys. Suppl. Ser.*, 141, 149–156.

Hecht, J.H., Holm, A.V., Donn, B. and Wu, C.-C. (1984) The dust around R Coronae Borealis type stars. *Astrophys. J.*, 280, 228–234.

Henning, T., Jäger, C. and Mutschke, H. (2004) Laboratory studies of carbonaceous dust analogs. *In Astrophysics of Dust*, A.N. Witt, G.C. Clayton and B.T. Draine, Eds., *ASP Conf. Ser.*, 309, 603–628, Astronomical Society of the Pacific, San Francisco, California, USA.

Hudgins, D. and Allamandola, L.J. (2004) Polycyclic aromatic hydrocarbons and infrared astrophysics: The state of the PAH model and a possible tracer of nitrogen in carbon-rich dust. *In Astrophysics of Dust,* A.N. Witt, G.C. Clayton and B.T. Draine, Eds., *ASP Conf. Ser.*, 309, 665–688, Astronomical Society of the Pacific, San Francisco, California, USA.

Inagaki, M., Kaneko, K. and Nishizawa, T. (2004) Nanocarbons-recent research in Japan. *Carbon*, 42, 1401–1417.

Jones, A.P. (1997) The lifecycle of interstellar dust. *In From Stardust to Planetesimals*, Y.J. Pendleton and A.G.G.M. Tielens, Eds., *ASP Conf. Ser.*, 122, 97–106, Astronomical Society of the Pacific, San Francisco, California, USA.

Kwok, S. (2004) The synthesis of organic and inorganic compounds in evolved stars, *Nature*, 430, 985–991.

Mennella, V. (2004) Laboratory simulation of processing grains. *In Astrophysics of Dust*, A.N. Witt, G.C. Clayton and B.T. Draine, Eds., *ASP Conf. Ser.*, 309, 629–648, Astronomical Society of the Pacific, San Francisco, California, USA.

Mennella, V., Colangeli, L., Palumbo, P., Rotundi, A., Schutte, W. and Bussoletti, E. (1996) Activation of an ultraviolet resonance in hydrogenated amorphous carbon grains by exposure to ultraviolet radiation. *Astrophys. J.*, 464, L191–L194.

Mennella, V., Baratta, G.A., Colangeli, L., Palumbo, P., Rotundi, A., Bussoletti, E. and Strazzulla, G. (1997) Ultraviolet spectral changes in amorphous carbon grains induced by ion irradiation. *Astrophys. J.*, 481, 545–549.

Mennella, V., Brucato, J.R., Colangeli, L. and Palumbo, P. (2002) C–H bond formation in carbon grains by exposure to atomic hydrogen: The evolution of the carrier of the interstellar 3.4 micron band. *Astrophys. J.*, 569, 531–540.

Oberlin, A. (1989) High-resolution TEM studies of carbonization and graphitization. *In Chemistry and Physics of Carbon*, P.L. Walker, Ed., 22, 1—143, Marcel Dekker, Inc., New York, USA.

Oberlin, A. and Bonmamy, S. (2001) Carbonization and graphitization. *In Graphite and Precursors*, P. Delhaés, Ed., 199–220, Gordon and Breach Science Publishers, Australia.

Onaka, T., Yamamura, I., Tanabe, T., Roellig, T.L. and Yuen, L. (1996) Detection of the mid-infrared unidentified bands in the diffuse galactic emission by IRTS, *Publ. Astron. Soc. Japan*, 48, L59–L63.

Papoular, R., Conard, J., Guillois, O., Nenner, I., Reynaud, C. and Rouzaud, J.-N. (1996) A comparison of solid-state carbonaceous models of cosmic dust, *Astron. Astrophys.*, 315, 222–236.

Peeters, E., Allamandola, L.J., Hudgins, D.M., Hony, S. and Tielens, A.G.G.M. (2004) The unidentified infrared features after ISO. *In Astrophysics of Dust*, A.N. Witt, G.C. Clayton and B.T. Draine, Eds., *ASP Conf. Ser.*, 309, 141–162, Astronomical Society of the Pacific, San Francisco, California, USA.

Rotundi, A., Rietmeijer, F.J.M., Colangeli, L., Mennella, V., Palumbo, P. and Bussoletti, E. (1998) Identification of carbon forms in soot materials of astrophysical interest. *Astron. Astrophys.*, 329, 1087–1096.

Rouleau, F., Henning, T. and Stognienko, R. (1997) Constraints on the properties of the 2175 Å interstellar feature carrier. *Astron. Astrophys.*, 322, 633–645.

Sakata, A. (1980) On the formation of interstellar linear molecules. *In Interstellar Molecules*, B.H. Andrew, Ed., *IAU Symp.*, 87, 325–329, Reidel, Dordrecht, the Netherlands.

Sakata, A., Wada, S., Okutsu, Y., Shintani, H. and Nakada, Y (1983) Does a 2,200 A hump observed in an artificial carbonaceous composite account for UV interstellar extinction? *Nature*, 301, 493–494.

Sakata, A., Wada, S., Onaka, T. and Tokunaga, A.T. (1987) Infrared spectrum of quenched carbonaceous composite (QCC). II – 2 A new identification of the 7.7 and 8.6 micron unidentified infrared emission bands. *Astrophys. J.*, 320, L63–L67.

Sakata, A., Wada, S., Onaka, T. and Tokunaga, A.T. (1990) Quenched carbonaceous composite. III – Comparison to the 3.29 micron interstellar emission feature. *Astrophys. J.*, 353, 543–548.

Sakata, A., Wada, S., Tokunaga, A.T., Narisawa, T., Nakagawa, H. and Ono, H. (1994) Ultraviolet spectra of quenched carbonaceous composite derivatives: Comparison to the '217 nanometer' interstellar absorption feature. *Astrophys. J.*, 430, 311–316.

Schnaiter, M., Mutschke, H., Dorschner, J., Henning, T. and Salama, F. (1998) Matrix-isolated nano-sized carbon grains as an analog for the 217.5 nanometer feature carrier. *Astrophys. J.*, 498, 486–496.

Scott, A. and Duley, W.W. (1996) The decomposition of hydrogenated amorphous carbon: A connection with polycyclic aromatic hydrocarbon molecules, *Astrophys. J.*, 472, L123–125.

Sellgren, K. (2001) Aromatic hydrocarbons, diamonds, and fullerenes in interstellar space: Puzzles to be solved by laboratory and theoretical astrochemistry. *Spectrochim. Acta, A*, 57, 627–642.

Tokunaga, A.T. (1997) A summary of the "UIR" bands. *In Diffuse Infrared Radiation and the IRTS*, H. Okuda, T. Matsumoto and T. Rollig, Eds., *ASP Conf. Ser.*, 124, 149–160, Astronomical Society of the Pacific, San Francisco, California, USA.

Tomita, S., Sakurai, T., Ohta, H., Fujii, M. and Hayashi, S. (2001) Structure and electronic properties of carbon onions. *J. Chem. Phys.*, 114, 7477–7482.

Tomita, S., Fujii, M. and Hayashi, S. (2004) Defective carbon onions in interstellar space as the origin of the optical extinction bump at 217.5 nanometers. *Astrophys. J.*, 609, 220-224.

van Winckel, H. (2003) Post-AGB stars. *Ann. Rev. Astron. Astrophys.*, 41, 391–427.

Vuong, M.H. and Foing, B.H. (2000) Dehydrogenation of polycyclic aromatic hydrocarbons in the diffuse interstellar medium. *Astron. Astrophys.*, 363, L5–L8.

Wada, S., Tokunaga, A.T., Kaito, C. and Kimura, S. (1998) Fitting the unusual UV extinction curve of V348 SGR. *Astron. Astrophys.*, 339, L61–L64.

Wada, S., Kaito, C., Kimura, S., Ono, H. and Tokunaga, A.T. (1999) Carbonaceous onion-like particles as a component of interstellar dust. *Astron. Astrophys.*, 345, 259–264.

Waters, L.B.F.M. (2004) Dust in evolved stars. *In Astrophysics of Dust*, A.N. Witt, G.C. Clayton and B.T. Draine, Eds., *ASP Conf. Ser.*, 309, 229–244, Astronomical Society of the Pacific, San Francisco, California, USA.

Chapter 4

FULLERENES AND RELATED CARBON COMPOUNDS IN INTERSTELLAR ENVIRONMENTS

PASCALE EHRENFREUND
Leiden Institute of Chemistry, Astrobiology Laboratory, 2300 RA Leiden, the Netherlands

NICK COX
Astronomical Institute, University of Amsterdam, 1098 SJ Amsterdam, the Netherlands

BERNARD FOING
ESA, ESTEC/SCI-SR, PO Box 299, 2200 AG Noordwijk, the Netherlands

Abstract: The prediction of the existence of the fullerene C_{60} and its subsequent isolation from soot produced a large interest for fullerenes in the astronomical community. Fullerenes of astronomical origin have only been detected in meteorites and associated with an impact crater on the Long Duration Exposure Facility spacecraft. The discovery of fullerenes led to the hypothesis that they may be stable carbon molecules present in interstellar space. Many authors have suggested the possible presence of C_{60} and carbon onions in astrophysical environments as well as their relation to the diffuse interstellar bands (DIBs) and the UV bump at 220 nm in the interstellar extinction curve. As first evidence for the largest molecule ever detected in space, two diffuse interstellar bands (DIBs) have been identified in the near infrared that are consistent with laboratory measurements of the C_{60} cation. We review the current knowledge on fullerenes in space and present new observations of the interstellar bands at 957.7 nm and 963.2 nm.

Key words: Carbon chemistry; circumstellar environments; diffuse interstellar bands (*DIBs*); fullerenes; interstellar medium (*ISM*); UV spectroscopy

Frans J.M. Rietmeijer (ed.), Natural Fullerenes and Related Structures of Elemental Carbon, 53–69.

1. INTRODUCTION

In the last decade astronomical observations, laboratory simulations and the analyses of extraterrestrial material have enhanced our knowledge regarding the inventory of solid and gaseous carbonaceous matter in the interstellar medium and on small bodies and planetary surfaces (Henning and Salama, 1998; Ehrenfreund and Charnley, 2000; Roush and Cruikshank, 2004). The data have shown that organic chemistry in space is complex and diverse (see Fig. 1). The fact that the same carbonaceous molecules are detected in different galactic space environments suggests that carbon chemistry seems to follow universal pathways (Ehrenfreund et al., 2005).

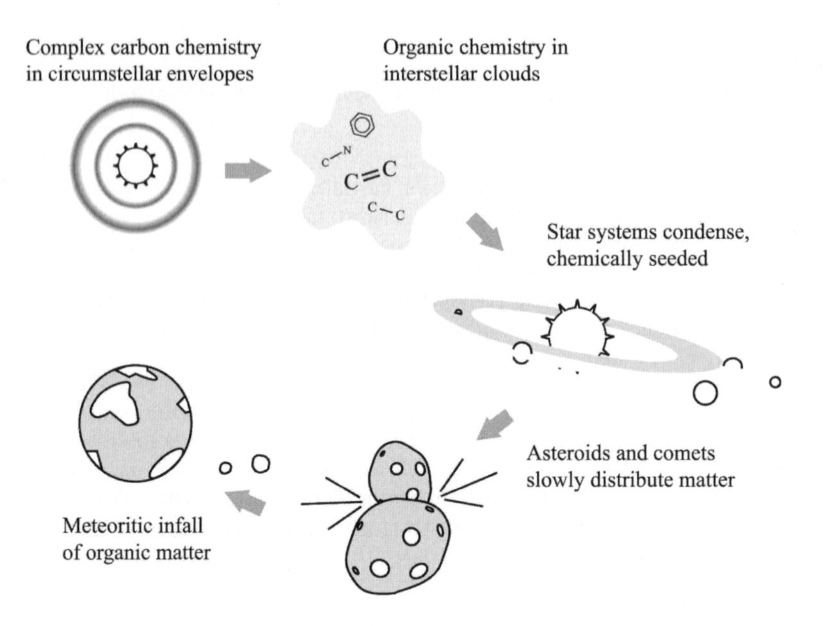

Figure 1. Gas phase and solid-state chemical reactions form a variety of organic molecules in circumstellar and interstellar environments. During the protostellar collapse, interstellar organic molecules in gaseous and solid phases are integrated into protostellar disks from which the planets and smaller solar system bodies form. Remnant planetesimals in the form of comets and asteroids impacted the young planets and delivered organic matter to their surface

Apart from silicates, many carbonaceous constituents were proposed to be present in interstellar dust, including amorphous carbon (AC), hydrogenated amorphous carbon (HAC), diamonds, refractory organic materials, and carbonaceous networks such as coal, soot, graphite, quenched-carbonaceous condensates (QCC), and others (see for a review, Ehrenfreund and Charnley, 2000). All forms of elemental carbon, namely diamonds, graphite and fullerenes are identified in space environments. Diamonds were recently proposed to be the carriers of the 3.4 and 3.5 μm emission bands (Guillois et al., 1999) observed in planetary nebulae. Nanodiamonds are present in primitive meteorites in abundances of up to ca. 1400 ppm (Huss and Lewis, 1995) and are embedded in amorphous carbon of asteroidal interplanetary dust particles (Dai et al., 2002). Graphite is not unambiguously identified in the interstellar medium (Nuth, 1985). It is only present in low abundances in differentiated and undifferentiated stony meteorites (see, Papike, 1998, for detailed meteorite reviews).

In the circumstellar envelopes of carbon-rich, evolved stars complex carbon chemistry occurs that is analogous to carbon soot formation in a candle flame or in industrial smoke stacks. Active acetylene (C_2H_2) chemistry appears to be the starting point for the development of hexagonal aromatic rings of carbon atoms. These aromatic rings probably react further to form large aromatic networks (Frenklach and Feigelson, 1989; Cherchneff et al., 1992). Laboratory simulations in combination with interstellar observations support the idea that the predominant fraction of carbon is incorporated into solid macromolecular carbon (e.g. Pendleton and Allamandola, 2002), amorphous and hydrogenated amorphous carbon (Mennella et al., 1998; Dartois et al., 2004; Duley and Lazarev, 2004), fullerenes (Iglesias-Groth, 2004) or defective carbon onions (Tomita et al., 2004).

Benzene chemistry is the first step to larger polycyclic aromatic hydrocarbons (PAHs) and fullerene-type material. Benzene is thus the key molecule in the pathways of complex carbon compounds in space. Benzene detection was claimed in the Infrared Space Observatory (*ISO*) spectrum of the circumstellar envelope CRL 618 (Cernicharo et al., 2001). We have recently measured the stability of solid, gaseous and ice-embedded benzene against proton bombardment and UV photons. Those data showed that benzene could be available for aromatic carbon chemistry when sufficiently shielded in circumstellar envelopes from protons and UV photons (Ruiterkamp et al., 2005a).

1.1 Fullerenes

Fullerene is the third form of carbon, after diamond and graphite. Kroto et al. (1985) were the first to discuss the polyhedral geometry of the C_{60} fullerene. Krätschmer et al. (1990) were the first to perform the synthesis of macroscopic quantities of C_{60} from condensed soot. The presence of soot material in carbon-rich stars, along with the spontaneous formation and remarkable stability of the fullerene cage, strongly suggest the presence of fullerene compounds in interstellar space (for a review see, Ehrenfreund and Foing, 1997). Rietmeijer et al. (2004) showed that cosmic soot analogs consisted of close packed metastable C_{60} and giant fullerenes.

It was suggested that fullerenes might be formed in small amounts in envelopes of R Coronae Borealis stars (Goeres and Sedlmayr, 1992). Theoretical models show the possible formation of fullerenes in the diffuse interstellar gas via the build-up of C_2 chains up to C_{10} chains from C^+ insertion ion-molecule reactions, and neutral-neutral reactions (Bettens and Herbst, 1997). Petrie and Bohme (2000) have studied ion molecule reactions for fullerene formation.

Fingerprints of the C_{60}^+ ion were discovered in the near-infrared spectra of stars behind diffuse interstellar clouds. The abundance of C_{60}^+ was inferred from optical measurements to be 0.3-0.9% of the cosmic carbon (Foing and Ehrenfreund, 1994, 1997). The search in mid-IR wavelengths for vibrational transitions in emission of C_{60} and C_{60}^+ in the reflection nebula NGC 7023 was negative; only uncertain upper limits could be determined (Moutou et al., 1999). Vibrational spectra of hydrogenated C_{60} were investigated in relation to the unidentified infrared emission bands (Stoldt et al., 2001). Garcia-Lario et al. (1999) attributed the 21-μm, dust feature observed in the C-rich protoplanetary nebula IRAS 16594-4656 to fullerenes, which might be formed during dust fragmentation. However, many different carrier species were proposed for this broad strong emission band (e.g. Hill et al., 1998). Webster (1995) offered an extensive discussion on interstellar hydrogenated fulleranes. C_{60}^+H may be abundant in the interstellar medium (ISM) and endo- and exohedral fullerenes associated with metals (Mg, Al, Si) may be present in abundance in circumstellar and interstellar environments (Kroto and Jura, 1992).

Fullerenes of astronomical (i.e. presolar) origin were detected in meteorites (Becker and Bunch, 1997) and associated with an impact crater on the Long Duration Exposure Facility spacecraft (Radicati di Brozzolo et al., 1994), in low-Earth orbit from February 14, 1980 to April 12, 1984. Using laser-desorption mass spectrometry, in particular higher fullerenes were identified in meteorites (Becker and Bunch, 1997). In this paper we summarize new evidence for fullerenes in space and their relation to the diffuse interstellar bands.

2. FULLERENES AND THE DIFFUSE INTERSTELLAR BANDS

The Diffuse Interstellar Bands (*DIBs*) are absorption bands that are detected in the spectra of more than 100 stars in our galaxy and beyond (Herbig, 1995; Snow, 2001, Ehrenfreund et al., 2002a, Sollerman et al., 2005). They are observed in a spectral range extending from 400 nm to 1.3 μm, but they are most prominent in the regions between 550 and 650 nm. *DIBs* show a large variety in band strengths and bandwidth. Their ubiquitous detection inside and outside our galaxy, and toward supernovae, indicates that the carriers of this spectral feature are (1) chemically stable, (2) abundant and (3) likely based on carbon that, after H and He, is among the most abundant elements in the cosmos. The current consensus based on both, astronomical observations and laboratory studies, is that the *DIB* features are gas phase species, likely in the form of PAHs (Salama et al., 1996) or fullerenes, or both (Ehrenfreund and Foing, 1995).

High-resolution observations continuously reveal new *DIBs* leading to a current number of >200 *DIBs* (Jenniskens and Desert, 1994; Herbig, 1995; O'Tuairisg et al., 2000). Below 440 nm, *DIBs* are difficult to detect in astronomical spectra; this applies also to the spectral regions in the near-IR that are contaminated by numerous telluric lines from the Earth's atmosphere (Foing and Ehrenfreund, 1995). It is the near-infrared (IR) region wherein the transitions of the C_{60} cation, and other large carbon molecules, can be seen. Many bands could be present in the region beyond 1 μm, which is only now becoming accessible for observations in high-resolution mode (Ruiterkamp et al., 2005b). Detection of substructures in some narrow *DIBs* can be interpreted as rotational contours of large molecules, with moments of inertia compatible with C_{60} fullerene compounds (Ehrenfreund and Foing, 1996).

2.1 The C_{60} Cation and Criteria for Spectroscopic Identification

Since C_{60}^+ is expected to be the most stable and dominant fullerene, it was proposed that this polyhedral carbon ion would be an excellent candidate for the *DIB* feature (Léger et al., 1988). It is expected that C_{60} is present as a cation in most of the interstellar regions due to its low ionization potential (7.57 eV). The laboratory spectrum of C_{60} does not show any strong visible bands that match known DIBs (Herbig, 2000; Sassara et al., 2001). Foing and Ehrenfreund (1994, 1997) identified two *DIBs* at 957.7 nm and 963.2 nm (Fig. 2) that are coincident (within 0.1%) with laboratory measurements of C_{60}^+ in a neon matrix (Gasyna et al., 1992; Fulara et al., 1993).

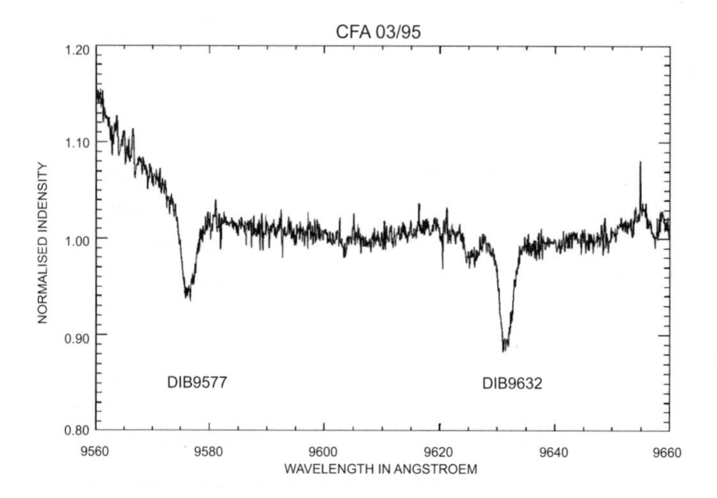

Figure 2. Near infrared spectrum of the star HD183143 observed under very dry conditions with the Canada France Hawaii Telescope. The star spectrum was divided by a spectrum of a reference star of similar spectral type. This allowed us to remove interfering stellar and telluric water lines from the Earth's atmosphere. The two *DIBs* at 957.7 and 963.2 nm (9577 and 9632 Å) are confirmed with the same width of 2.85 Å consistent with the C_{60}^+ assignment and expected rotational contour broadening at temperature 60K ± 10.

Reproduced from Foing and Ehrenfreund (1997; *op. cit.*) by written permission of the copyright holder, the Astronomy and Astrophysics Editor-in-Chief

Bondybey and Miller (1983) discussed that smaller shifts between gas phase and neon-matrix data are expected for IR transitions involving deeper states, compared to excited states that are more strongly affected; a shift of ca. 10 cm^{-1} for C_{60} IR-transitions is therefore plausible. Recently, new near-IR data were acquired from spectra toward highly extincted stars, i.e. dust-rich stellar environments (Galazutdinov et al., 2000). The wavelength difference between the two lines was measured to be 55 Å ± 0.2 (or 60 cm^{-1}); a bandwidth of 2.85 ± 0.1 Å (or 3.1 cm^{-1}) was measured for both bands. It was argued that the two bands appearing in the neon matrix might arise from matrix splitting effects. However, both C_{60} cation bands appear in argon and neon matrices with the same average separation and are probably intrinsic to two major ground state geometries favored by a Jahn-Teller type distortion. According to Bendale et al. (1992), the ground state of C_{60}^{+} departs from the I_h optimized symmetry of neutral C_{60} (two transitions $H_g \rightarrow H_u$ and $G_g \rightarrow H_u$, well separated by 300 cm^{-1}) towards the D_{5d} geometry (five degenerate transitions $E_{1g} \rightarrow A_{1u}$), which is the preferred static equilibrium structure with a Jahn-Teller distortion stabilization energy of 8.1 kcal mol^{-1}.

The correlated strength and width of both *DIBs* suggest that those bands originate from the same carrier. The behavior of these bands in different environments points to an abundant ionized gas phase molecule, surviving most efficiently in regions dominated by hard-UV radiation (Foing and Ehrenfreund, 1997). The width (3.1 cm^{-1}) is compatible with rotational contours of a fullerene molecule (Edwards and Leach, 1993). The changes in band ratio are predicted, as they would correspond to two fundamental levels separated by 60 cm^{-1}, whose populations depend on local excitation temperature conditions. The population ratio can vary significantly in particular in cold lines-of-sight favoring the lower level (the 957.7-nm band). The value of width is compatible with a spherical molecule with rotational constant $\Delta B = 0.0036$ cm^{-1} and the same moment of inertia as C_{60}.

2.2 New Observations of the Near Infrared *DIBs*

The two well-known diffuse interstellar absorption bands at 957.7 nm (9577 Å) (Fig. 3) and 963.2 nm (9632 Å) (Fig. 4) were observed during the nights of 25 to 27 September 2001, with the Ultraviolet and Visual Echelle Spectrograph mounted on the Very Large Telescope pointed towards two target stars, viz. (1) 4U 1907+09 (a high mass X-ray binary star; Cox et al., 2005) and (2) BD -14 5037, both with a substantial amount of dust in their line-of-sight.

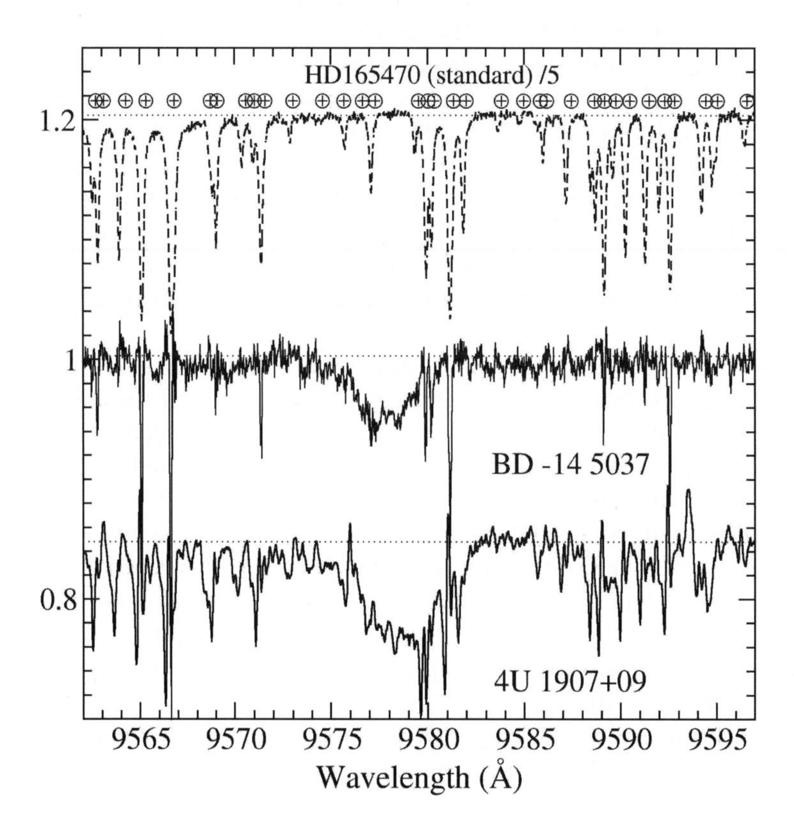

Figure 3. The 957.7 nm (9577 Å) *DIB* range is plotted for the two targets BD-14 5037 [E_{B-V} = 1.57 magnitude (*mag*)] and 4U1907+09 (E_{B-V} = 3.45 *mag*). The spectrum at the top (dotted) is from the standard star HD165470 (scaled to 1/5th intensity). The crossed circles indicate line positions from telluric lines (contaminations from the Earth's atmosphere), and the dotted horizontal lines indicate the approximate spectral continuum

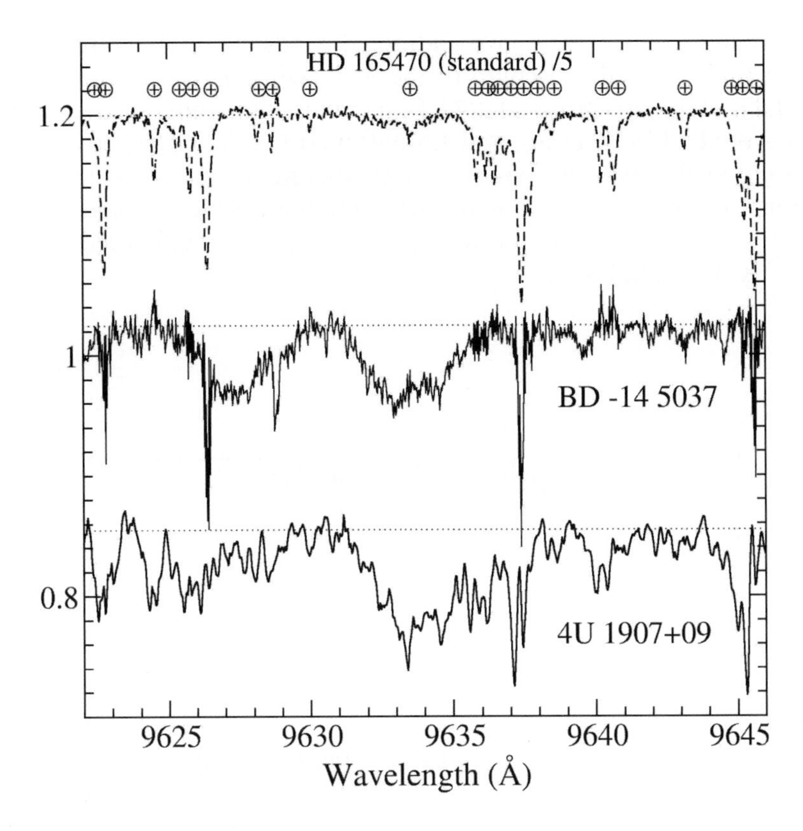

Figure 4. The 963.2 nm (9632 Å) *DIB* range is shown for the two targets, BD-14 5037 (E_{B-V} = 1.57 *mag*) and 4U1907+09 (E_{B-V} = 3.45 *mag*). The spectrum at the top (dotted) is from the standard star HD165470 (scaled to 1/5th intensity). The crossed circles indicate telluric line positions; dotted horizontal lines represent the approximate continuum levels. The absorption feature near 962.7 nm (9627 Å) is a residual stellar He I line

This dust preferentially scatters short wavelengths (bluer light) and causes an apparent reddening of light from distant stars that can be measured in magnitudes (*mag*). The reddening $E_{(B-V)}$ for the lines of sight of the stars 4U 1907+09 and BD -14 5037 was measured to be 3.45 and 1.57 *mag*, respectively. The two bands at 957.7 nm and 963.2 nm are tentatively attributed to the C_{60} cation (e.g. Foing and

Ehrenfreund, 1994, 1997). Two additional bands of C_{60}^+ that are much weaker (only 15% the strength of the 957.7-nm *DIB*) were identified in a laboratory spectrum at 942.1 nm and 936.6 nm, respectively (Fulara et al., 1993). Foing and Ehrenfreund (1997) predicted a C_{60}^+ transition at 941-942 nm based on the laboratory measurements in a neon matrix as well as estimated shifts from the gas phase value.

The region around 942 nm where we searched for additional bands of C_{60}^+ is shown in Fig. 5.

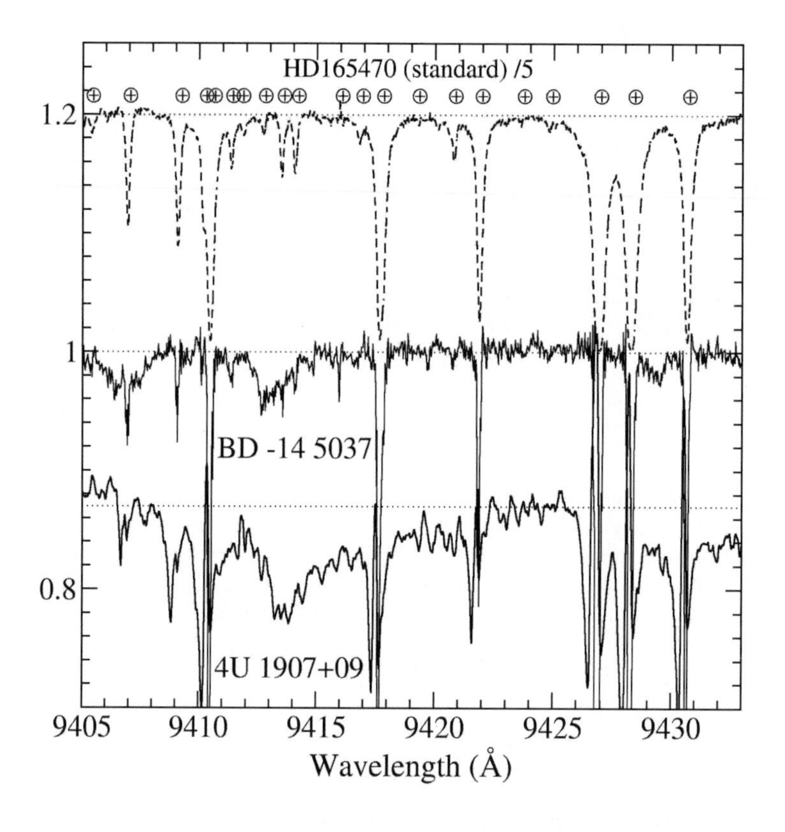

Figure 5. A new tentative band at 941.3 nm (9413 Å) may be a third transition of C_{60}^+ (the feature shows a consistent velocity shifts and a wing asymmetry specific for a possible *DIB*). High resolution, high signal-to-noise spectroscopic observations of targets and standard stars (of exact spectral type) are required to confirm this preliminary result. The standard star and both targets are plotted as in Figs. 3 and 4 but for the 9405–9435 Å spectral range

Telluric lines are corrected for by use of the telluric standard star HD165470, with the relevant spectral range plotted at the top (scaled by 1/5th) of each figure to indicate the strength and positions of the telluric lines.

The correlation between the 957.7 and 963.2 nm (9577 and 9632 Å) diffuse interstellar absorption bands was measured in different interstellar environments, which supports a strong link between the carriers of these two *DIB* features. Galazutdinov et al. (2000) showed that the correlation coefficient of the two bands is larger than 90% (Fig. 6).

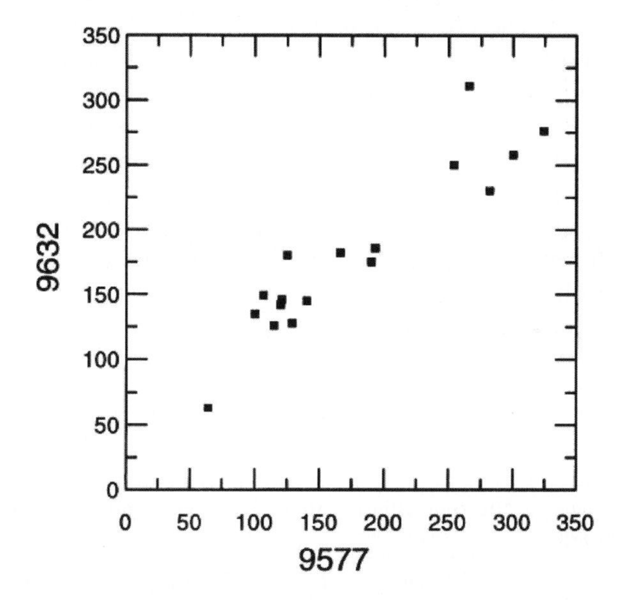

Figure 6. The correlation between the core equivalent widths of the 957.7 and 963.2 nm DIBs is larger than 90%.

Reproduced from Galatzudinov et al. (2000; *op. cit.*) with the written permission from the Monthly Notices of the Royal Astronomical Society; (bibliographic code2000NMRS.317..750G)

2.3 Ionization States for Interstellar Fullerenes

The ionization state will depend on the balance between UV photo-ionization and electronic recombination. This balance can be calculated in different interstellar environments taking into account the UV penetration in clouds of varying opacities. UV irradiation in the diffuse interstellar medium will ionize both C_{60}^+ (second ionization potential 11.3 eV) and polycyclic aromatic hydrocarbons (PAHs), but the corresponding C_{60} di-cation is very stable against fragmentation thanks to the high dissociation barrier of the fullerene cage structure. This particular cation can revert to the mono-cation C_{60}^+ through electronic recombination, which could explain a significant equilibrium abundance of C_{60}^+ in the diffuse interstellar medium. Table 1 summarizes ionization potential and electron affinity of known fullerenes. For typical "standard" diffuse interstellar clouds (ca. 100 K, n(H) = 100 cm^{-3}, standard UV field, and visual extinction, A_V the number of V magnitudes by which the star appears dimmer than ca. 1 *mag*) detailed theoretical models (Ruiterkamp et al., 2005b) predict a relative cation fraction of the total column density of 0.3 for C_{60} and 0.4 for C_{84}. The cation fraction can vary by as much as a factor two between diffuse (visual extinction, A_v, ca. 0.3 *mag*) and translucent (A_v ca. 2 *mag*) clouds. This balance is strongly affected in regions with for example a strong UV field and high electron density, such as the Orion star-forming region.

Table 1. The ionization potential and electron affinity of C_{60}, C_{70}, C_{80} and C_{84} fullerenes

Fullerenes	Symmetry group	Electron Affinity (eV)	Refs	Ionization Potential (eV)	Refs	IP^{++} (eV)
C_{60}	I_h	2.689 ± 0.008	1	7.57 ± 0.01	5	11.3
C_{70}	D_{5h}	2.676 ± 0.001	2	7.3 ± 0.2	6	
C_{80}	I_h	3.17 ± 0.06	3	6.84 ± 0.10	7	
C_{84}		3.05 ± 0.08	4	7.1	8	

References: (1) Wang et al. (1999); (2) Brink et al. (1995); (3) Boltalina et al. (1996) (4) Boltalina et al. (1993) (5) Yoo et al. (1992); (6) Hertel et al. (1992); (7) Zimmerman et al. (1991); (8) Beck et al. (1994)

3. CONCLUSIONS

The presence of soot material in C-rich stars, the spontaneous formation of fullerenes and the remarkable stability of the fullerene cage suggest that fullerene molecules should be expected to be present in interstellar space. It was discussed that the inhibition of the fullerene growth mechanisms by hydrogen would limit the formation of C_{60} to hydrogen-depleted environments in space (Goeres and Sedlmayr, 1992). However, Gerhardt et al. (1987) showed that fullerenes form also in significant quantities in the presence of H and O. Only very small quantities of C_{60} may be formed in the envelopes of R Coronae Borealis stars. But, since fullerene cages are highly resistant to fragmentation, fullerenes may still survive long enough to be cycled into the diffuse interstellar medium. Fullerene compounds may be present as neutral and ionized species in gaseous form in the diffuse interstellar medium. There is strong evidence that aromatic material in the gas phase and in macromolecular form takes up most of the carbon in the interstellar medium, comets and meteorites (Ehrenfreund et al., 2002b).

Decomposition of hydrogenated amorphous carbon in interstellar shocks may be the source of large PAHs and fullerene molecules (Scott et al., 1997). The ionization potential of C_{60} is 7.6 eV and the ionization potential of C_{60}^{+} is 11.3 eV. C_{60}^{++} can revert naturally to C_{60}^{+} while many double ionized species, such as PAHs smaller than fifty carbon atoms, have a high probability of being dissociated (Allain et al., 1996). There is strong evidence that C_{60}^{+} is present in interstellar space as observed by two near infrared absorptions around 950 nm. We have identified those bands in dusty sightlines.

Future laboratory work is awaited which will provide the gas phase spectrum of C_{60}^{+}. The study of macromolecular networks, and their relations to fullerenes and hydrogenated fullerenes should be targets of investigation. Continued astronomical observations and related laboratory data will hopefully confirm the abundance of fullerenes, and fullerene-type material in space. The detection of fullerenes in significant quantities would open a way for a fascinating interstellar catalytic surface chemistry.

Acknowledgements. The presented data are based on observations at VLT/UVES and CFHT telescopes. PE acknowledges support from NWO grant 016.023.003 and NC is supported by NOVA. We thank Dr. James Garry for graphic support.

4. REFERENCES

Allain, T., Leach, S. and Sedlmayr, E. (1996) Photodestruction of PAHs in the interstellar medium. I. Photodissociation rates for the loss of an acetylenic group. *Astron. Astrophys.*, 305, 602-615.

Beck, R.D., Weis, P., Hirsch, A. and Lamparth, I. (1994) Laser desorption mass spectrometry of fullerene derivatives: Laser-induced fragmentation and coalescence. *J. Phys. Chem.*, 98, 9683-9687.

Becker, L. and Bunch, T.E. (1997) Fullerenes, fulleranes and PAHs in the Allende meteorite. *Meteorit. Planet. Sci.*, 32, 479-487.

Bendale, R.D., Stanton, J.F. and Zerner, M.C. (1992) Investigations of the electronic structure and spectroscopy of Jahn-Teller distorted C_{60}^+. *Chem. Phys. Lett.*, 194, 467-471.

Bettens, R.P.A. and Herbst, E. (1997) The formation of large hydrocarbons and carbon clusters in dense interstellar clouds, *Astrophys. J.*, 478, 585-593.

Bondybey, V.E. and Miller, T.A. (1983) Vibronic spectroscopy and photophysics of molecular ions in low temperature matrices. *In Molecular Ions: Spectroscopy, Structure and Chemistry*, T.A. Miller and V.E. Bondybey, Eds., 125-173, North Holland Publishing Co., Amsterdam, the Netherlands.

Brink, C., Andersen, L.H., Hvelplund, P., Mathur, D. and Voldstad, J.D. (1995) Laser photodetachment of C_{60} and C_{70} ions cooled in a storage ring. *Chem. Phys. Lett.*, 233, 52-56.

Boltalina, O.V., Sidorov, L.V., Borshchevsky, A.Ya., Sukhanova, E.V. and Skokan, E.V. (1993) Electron affinities of higher fullerenes. *Rapid Comm. Mass Spectrom.*, 7, 1009-1011.

Boltalina, O.V., Dashkova, E.V. and Sidorov, L.N. (1996) Gibbs energies of gas-phase electron transfer reactions involving the larger fullerene anions. *Chem. Phys. Lett.*, 256, 253-260.

Cherchneff, I., Barker, J.R. and Tielens, A.G.G.M. (1992) Polycyclic aromatic hydrocarbon formation in carbon-rich stellar envelopes. *Astrophys. J.*, 401, 269-287.

Cernicharo, J., Heras, A., Tielens, A.G.G.M., Pardo, J.R., Herpin, F., Guélin, M. and Waters, L.B.F.M. (2001) Infrared Space Observatory's discovery of C_4H_2, C_6H_2, and benzene in CRL 618. *Astrophys. J.*, 546, L123-L126.

Cox, N.L.J., Kaper, L., Foing, B.H. and Ehrenfreund, P. (2005) Diffuse interstellar bands of unprecedented strength in the line of sight towards high-mass X-ray binary 4U 1907+09. *Astron. Astrophys.* 438, 187-199.

Dai, Z.R., Bradley, J.P., Joswiak, D., Brownlee, D.E., Hill, H.G.M. and Genge, M.M. (2000) Possible in situ formation of meteoritic nanodiamonds in the early solar system. *Nature*, 418, 157-159.

Dartois, E., Muñoz Caro, G.M., Deboffle, D. and d'Hendecourt, L. (2004) Diffuse interstellar medium organic polymers. Photoproduction of the 3.4, 6.85 and 7.25 µm features. *Astron. Astrophys.*, 423, L33-L36.

Duley, W.W. and Lazarev, S. (2004) Ultraviolet absorption in amorphous carbons: Polycyclic aromatic hydrocarbons and the 2175 Å extinction feature. *Astrophys. J.*, 612, L33-L35.

Edwards, S.E. and Leach, S. (1993) Simulated rotational band contours of C_{60} and their comparison with some of the diffuse interstellar bands. *Astron. Astrophys.*, 272, 533-540.

Ehrenfreund, P. and Charnley, S.B. (2000) Organic molecules in the interstellar medium, comets, and meteorites: A Voyage from dark clouds to the early Earth. *Ann. Rev. Astron. Astrophys.*, 38, 427-483.

Ehrenfreund, P. and Foing, B.H. (1995) Search for fullerenes and PAHs in the diffuse interstellar medium. *Planet. Space Sci.*, 43, 1183-1187.

Ehrenfreund, P. and Foing, B.H. (1996) Resolved profiles of diffuse interstellar bands: evidence for rotational contours of gas phase molecules. *Astron. Astrophys. Lett.*, 307, L25-L28.

Ehrenfreund, P. and Foing, B.H. (1997) Spectroscopic properties of polycyclic aromatic hydrocarbons (PAHS) and astrophysical implications. *Adv. Space Res.*, 7, 1023-1032.

Ehrenfreund, P., Cami, J., Jiménez-Vicente, J., Foing, B.H., Kaper, L., van der Meer, A., Cox, N., d'Hendecourt, L., Maier, J.P., Salama, F., Sarre, P.J., Snow, T.P. and Sonnentrucker, P. (2002a) Detection of diffuse interstellar bands in the Magellanic Clouds. *Astrophys. J. Lett.*, 576, L117-L120.

Ehrenfreund, P., Irvine, W., Becker, L., Blank, J., Brucato, J R, Colangeli, L., Derenne, S., Despois D, Dutrey, A., Fraaije, H., Lazcano, A., Owen, T., Robert, F., and an International Space Science Institute ISSI-Team (2002b) Astrophysical and astrochemical insights into the origin of life. *Reports Prog. Phys.*, 65, 1427-1487.

Ehrenfreund, P., Rasmussen, S., Cleaves, J. and Chen, L. (2005) Experimentally tracing the steps toward the origin of the life, *Astrobiology,* submitted.

Foing, B.H. and Ehrenfreund, P. (1994) Detection of two interstellar absorption bands coincident with spectral features of C_{60}^+. *Nature*, 369, 296-298.

Foing B.H. and Ehrenfreund, P. (1995) Diffuse interstellar bands in the near infrared - A dedicated search for polycyclic aromatic hydrocarbon and fullerene cations. *In The Diffuse Interstellar Bands*, A.G.G.M. Tielens and T.P. Snow, Eds., 65-72, Kluwer, Dordrecht, the Netherlands.

Foing, B.H. and Ehrenfreund, P. (1997) New evidences for interstellar C_{60}^+. *Astron. Astrophys. Lett.*, 317, L59-L62.

Frenklach, M. and Feigelson, E.D. (1989) Formation of polycyclic aromatic hydrocarbons in circumstellar envelopes. *Astrophys. J.*, 341, 372-384.

Fulara, J., Jakobi, M. and Maier, J.P. (1993) Electronic and infrared spectra of C_{60}^+ and C_{60}^- in neon and argon matrices. *Chem. Phys. Lett.*, 211, 227-234.

Galazutdinov, G.A., Krelowski, J., Musaev, F.A., Ehrenfreund, P., Foing, B.H. (2000) On the identification of the C_{60}^+ interstellar features. *Mon. Not. Royal Astron. Soc.*, 317, 750-758.

García-Lario, P., Manchado, A., Ulla, A. and Manteiga, M. (1999) Infrared Space Observatory observations of IRAS 16594-4656: A new proto-planetary nebula with a strong 21 micron dust feature. *Astrophys. J.*, 513, 941-946.

Gasyna, Z., Andrews, L. and Schatz, P.N. (1992) Near-infrared absorption spectra of fullerene (C_{60}) radical cations and anions prepared simultaneously in solid argon. *J. Phys. Chem.*, 96, 1525-1527.

Gerhardt, P., Loffler, S. and Homann, K.P. (1987) Polyhedral carbon ions in hydrocarbon flames. *Chem. Phys. Lett.*, 137, 306-310.

Goeres, A. and Sedlmayr, E. (1992) The envelopes of R Coronae Borealis stars. I - A physical model of the decline events due to dust formation. *Astron. Astrophys.*, 265, 216-236.

Guillois, O., Ledoux, G. and Reynaud, C. (1999) Diamond infrared emission bands in circumstellar media. *Astrophys. J.*, 521, L133-L136.

Henning, T. and Salama, F. (1998) Carbon in the Universe. *Science*, 282, 2204-2210.

Herbig, G.H. (1995) The diffuse interstellar bands. *Ann. Rev. Astron. Astrophys.*, 33, 19-74.

Herbig, G.H. (2000) The search for interstellar C_{60}. *Astrophys. J.*, 542, 334-343.

Hertel, I.V., Steger, H., de Vries, J., Weisser, B., Menzel, C., Kamke, B. and Kamke, W. (1992) Giant plasmon excitation in free C_{60} and C_{70} molecules studied by photoionization. *Phys. Rev. Lett.*, 68, 784-787.

Hill, H.G.M., Jones, A.P. and d'Hendecourt, L.B. (1998) Diamonds in carbon-rich proto-planetary nebulae. *Astron. Astrophys.*, 336, L41-L44.

Huss, G.R. and Lewis, R.S. (1995) Presolar diamond, SiC, and graphite in primitive chondrites: Abundances as a function of meteorite class and petrologic type. *Geochim. Cosmochim. Acta*, 59, 115-160.

Iglesias-Groth, S. (2004) Fullerenes and buckyonions in the interstellar medium. *Astrophys.J.*, 608, L37-L40.

Jenniskens, P. and Desert, X. (1994) A survey of diffuse interstellar bands (3800-8680 Å). *Astron. Astrophys. Suppl.*, 106, 39-78.

Krätschmer, W., Lamb, L., Fostiropoulos, K. and Huffman, D.R. (1990) Solid C_{60}: a new form of carbon. *Nature*, 347, 354-358.

Kroto, H.W. and Jura, M. (1992) Circumstellar and interstellar fullerenes and their analogues. *Astron. Astrophys.*, 263, 275-280.

Kroto, H.W., Heath, J.R., O'Brien, S.C., Curl, R.F. and Smalley, R.E. (1985) C_{60}: Buckminsterfullerene. *Nature*, 318, 162-163.

Léger, A. and d'Hendecourt, L. (1985) Are polycyclic aromatic hydrocarbons the carriers of the diffuse interstellar bands in the visible? *Astron. Astrophys.*, 146, 81-85.

Léger, A., d'Hendecourt, L., Verstraete, L. and Schmidt, W. (1988) Remarkable candidates for the carrier of the diffuse interstellar bands - C_{60}^{+} and other polyhedral carbon ions. *Astron. Astrophys.*, 203, 145-148.

Mennella, V., Colangeli, L., Bussoletti, E., Palumbo, P. and Rotundi, A. (1998) A new approach to the puzzle of the ultraviolet interstellar extinction bump. *Astrophys. J.*, 507, L177–L180.

Moutou, C., Sellgren, K., Verstraete, L. and Léger, A. (1999) Upper limit on C_{60} and $C_{60}^{(+)}$ features in the ISO-SWS spectrum of the reflection nebula NGC 7023. *Astron. Astrophys.*, 347, 949-956.

Nuth, J.A. (1985) Meteoritic evidence that graphite is rare in the interstellar medium. *Nature*, 318, 166-168.

Papike, J.J. (Ed.) (1998) *Planetary Materials,*. 36, 1052p., The Mineralogical Society of America, Washington, DC, USA.

Pendleton, Y.J. and Allamandola, L.J. (2002) The organic refractory material in the diffuse interstellar medium: Mid-Infrared spectroscopic constraints. *Astrophys. J. Suppl.*, 138, 75-98.

Petrie, S. and Bohme, D.K. (2000) Laboratory studies of ion/molecule reactions of fullerenes: Chemical derivatization of fullerenes within dense interstellar clouds and circumstellar shells. *Astrophys. J.*, 540, 869-885.

Radicati di Brozzolo, F.R., Bunch, T.E., Fleming, R.H. and Macklin, J. (1994) Fullerenes in an impact crater on the LDEF spacecraft. *Nature*, 369, 37-40.

Rietmeijer, F.J.M., Rotundi, A. and Heymann, D. (2004) C_{60} and giant fullerenes in soot condensed in vapors with variable C/H_2 ratio. *Fullerenes, Nanotubes, and Carbon Nanostructures*, 12, 659-680.

Roush, T. and Cruikshank, D.P. (2004) Observations and laboratory data of planetary organics. *In Astrobiology: Future Perspectives*, P. Ehrenfreund, W. Irvine, T. Owen, L. Becker, J. Blank, J. Brucato, L. Colangeli, S. Derenne, A. Dutrey, D. Despois, A. Lazcano and F. Robert, Eds. *Astrophys. Space Sci, Library*, 305, 149-178, Kluwer Academic Publishers, Dordrecht, Boston, London.

Ruiterkamp, R., Peeters, Z., Moore, M., Hudson, R. and Ehrenfreund, P. (2005a) A quantitative study of proton irradiation and UV photolysis of benzene in interstellar environments. *Astron. Astrophys*. 440, 391-402.

Ruiterkamp, R., Cox, N.L.J., Spaans, M., Kaper, L., Foing, B.H., Salama, F. and Ehrenfreund, P. (2005b) PAH charge state distribution and DIB carriers: Implications from the line of sight toward HD 147889. *Astron. Astrophys.*, 431, 515-525.

Salama, F., Bakes, E.L.O., Allamandola, L.J. and Tielens, A.G.G.M. (1996) Assessment of the Polycyclic aromatic hydrocarbon-Diffuse interstellar band proposal. *Astron. Astrophys.*, 458, 621-636.

Sassara, A., Zerza, G., Chergui, M. and Leach, S. (2001) Absorption wavelengths and bandwidths for interstellar searches of C_{60} in the 2400-4100 Å region. *Astrophys. J. Suppl.*, 135, 263-273.

Scott, A., Duley, W.W. and Pinho, G.P. (1997) Polycyclic aromatic hydrocarbons and fullerenes as decomposition products of hydrogenated amorphous carbon. *Astrophys. J.*, 489, 263-273.

Snow, T. (2001) The Unidentified Diffuse Interstellar Bands as evidence for large organic molecules in the interstellar medium. *Spectrochim. Acta A*, 57, 615-626.

Sollerman, J., Cox, N., Mattila, S., Ehrenfreund, P., Kaper, L., Leibundgut, B. and Lundqvist, P. (2005) Diffuse interstellar bands in NGC 1448. *Astron. Astrophys.*, 429, 559-567.

Stoldt, C.R., Maboudian, R. and Carraro, C. (2001) Vibrational spectra of hydrogenated Buckminsterfullerene: A candidate for the unidentified infrared emission. *Astrophys. J.*, 548, 225-228.

Tomita, S., Fujii, M. and Hayashi, S. (2004) Defective carbon onions in interstellar space as the origin of the optical extinction bump at 217.5 nanometers. *Astrophys. J.*, 609, 220-224.

O'Tuairisg, S., Cami, J., Foing, B.H., Sonnentrucker, P. and Ehrenfreund, P. (2000) A deep echelle survey and new analysis of diffuse interstellar bands. *Astron. Astrophys. Suppl.*, 142, 225-238.

Wang, X.B., Ding, C.F. and Wang, C.F. (1999) High resolution photoelectron spectroscopy of C_{60}. *J. Chem. Phys.*, 110, 8217-8220.

Webster, A. (1995) Fulleranes and the Diffuse Interstellar Bands. *In The Diffuse Interstellar Bands*, A.G.G.M. Tielens and T.P. Snow, Eds., 349-358, Kluwer, Dordrecht, the Netherlands.

Yoo, R., Rusic, B. and Berkowitz, J. (1992) Vacuum ultraviolet photoionization mass spectrometric study of C_{60}. *J. Chem. Phys.*, 96, 911-918.

Zimmerman, J.A., Eyler, J.R., Bach, S.B.H. and McElvany, S.W. (1991) 'Magic number' carbon clusters - Ionization potentials and selective reactivity. *J. Chem. Phys.*, 94, 3556-3562.

Chapter 5

NATURAL C_{60} AND LARGE FULLERENES: A MATTER OF DETECTION AND ASTROPHYSICAL IMPLICATIONS

ALESSANDRA ROTUNDI
Dipartimento di Scienze e Applicazioni, Università "Parthenope" di Napoli, Via A. De Gasperi 5, 80133 Napoli, Italy

FRANS J.M. RIETMEIJER
Department of Earth and Planetary Sciences, MSC03-2040, 1-University of New Mexico, Albuquerque, NM 87131-0001, USA

JANET BORG
Institut d'Astrophysique Spatiale, Bâtiment 121, Campus 91405 Orsay Cedex, France

Abstract: Fullerene was theoretically predicted and experimentally discovered, but its detection in laboratory studies is still underrepresented with respect to its theoretical abundance. Recent High Resolution Transmission Electron Microscopy (HRTEM) studies of soot samples, however, lead to single fullerene molecule detection in higher amounts than was previously established. HRTEM is able to identify fullerenes even if they are only present in small quantities that would be below the detection limit of chemical techniques. Fullerenes will probably remain largely undetected until higher signal to noise ratio measurements are used to search for them. Such studies could yield different conclusions on fullerene abundances both in terrestrial and in extraterrestrial samples. For the latter, important astrophysical implications have to be considered.

Key words: Carbon chain molecules; carbon monocyclic ring molecules; carbon vapor condensation; carbonaceous chondrites; cosmic dust; fullerenes; high-resolution transmission electron microscopy (HRTEM); interplanetary dust particles; meteorites; polycyclic aromatic hydrocarbons (PAHs); PAH molecules; Raman microspectroscopy; soot

Frans J.M. Rietmeijer (ed.), Natural Fullerenes and Related Structures of Elemental Carbon, 71–94.

1. INTRODUCTION

The prediction by Kroto et al. (1985) that fullerene molecules, including C_{60}, ought to be abundant seems not yet to be borne out by the analyses of natural soot samples but, more importantly, they are apparently also not seen in soot that was produced in the laboratory under carefully controlled conditions that should have led to fullerene formation. In fact, while Krätschmer et al. (1990) synthesized macroscopic amounts of C_{60}, this molecule remains under-represented in natural samples and it seems that experimental conditions for its production are either too peculiar to prove its theoretical pervasiveness (Taylor et al., 1991), or some other factors control the existence of natural fullerenes. Fullerene searches included circumstellar and interstellar environments, Interplanetary Dust Particles (IDPs), meteorites, lunar rocks, terrestrial hard-rocks, coal, and sedimentary rocks (Heymann et al., 2003). The only diagnostic tool available for fullerene detection in astronomical environments (e.g. Snow and Seab, 1989; Foing and Ehrenfreud, 1994, 1997; Webster, 1997; Sassara et al., 2001) relies on the analysis of electromagnetic radiation, which requires a synergy between laboratory and theoretical studies. Because of the prediction that fullerenes should exist and their subsequent detection in astronomical environments, the search for fullerenes in meteorites was a next logical step. In particular, carbonaceous chondrites rich in presolar grains, and being more carbon-rich than any other meteorites, were considered the most likely to contain fullerenes. Some chemical analyses of carbon-rich residues extracted from these meteorites gave positive results (e.g. Becker and Bunch, 1997; Becker et al., 1994, 1999) while others did not find fullerenes in other allocations of these same meteorites (e.g. Ash et al., 1993; DeVries et al., 1993; Gilmour et al., 1993; Heymann, 1997).

Chondritic aggregate IDPs and cluster IDPs from ca. 2 μm to ca. 1 mm in size are considered the least-modified debris from the solar nebula because they were not processed in small protoplanets such as the meteorite parent bodies (Rietmeijer, 2002). The carbonaceous phases in these IDPs contain D/H "hot spots" of interstellar origin that were incorporated during solar nebula dust accretion (Keller et al., 2000; Messenger, 2000) and these particles should be good candidates to contain astronomical fullerene. Carbon XANES (X-ray Absorption Near Edge Spectroscopy) spectra of carbon-rich IDP L2008F4 show a notable similarity to the C-XANES spectrum of C_{60}, but the search for fullerenes was inconclusive (Bajt et al., 1996). Flynn et al. (2003) later concluded that the C-XANES signals, which were initially interpreted as possibly C_{60}, could be due to C=O of organic carbon, result from the C-H, carboxyl, or carbonyl functional group, or could be due to an oxidized $C_{60}O$ fullerane.

Using High-Resolution Transmission Electron Microscopy (HRTEM), Wang and Buseck (1991) were the first to directly view the stacking of C_{60} molecules forming close-packed arrays in a synthetic sample of crystalline fullerite that was known to contain C_{60} and C_{70} fullerenes. Buseck et al. (1992) were also first to identify C_{60} and C_{70} fullerenes by HRTEM in naturally occurring shungite that is compositionally a coal of meta-anthracite rank albeit the origin of this unusual geological rock remains uncertain. The HRTEM results obtained on carbon films in the shungite sample were confirmed by Fourier transform mass spectroscopy and by both laser desorption and thermal desorption/electron-capture methods (Buseck et al., 1992; Buseck, 2002). Recently, using transmission electron microscopy, Rietmeijer et al. (2004) successfully identified C_{60} and higher ('giant') fullerene molecules in amorphous soot particles produced by the arc discharge technique. The presence of fullerenes in soot particles, which were predicted by Taylor et al. (1991) but that was heretofore not identified in soot, was subsequently confirmed by mass spectroscopy and HPLC chromatography analyses (Rietmeijer et al., 2004). Subsequently C_{60} and smaller fullerene molecules were identified by HRTEM analyses of flame-produced soot (Goel et al., 2004) confirming that amorphous soot is an agglomeration of fullerenes, but not of uniformly C_{60} fullerene, as both smaller (Goel et al., 2004) and larger (Rietmeijer et al., 2004) fullerenes can be present.

These studies on soot evidenced that HRTEM is able to detect and identify fullerenes when present in minute quantities that might be below the detection limits of conventional bulk chemical analyses. For example, circular objects approximately the size of C_{60} and C_{70} fullerenes can be seen in HRTEM images of soot not only from low-pressure benzene/oxygen flames well-known to contain fullerenes but also in soot produced in atmospheric-pressure ethylene/air flames (Grieco et al., 2000). The corollary being that C_{60} and other fullerenes exist in both synthetic and natural carbon samples wherein they went undetected by the particular analytical tools used in those soot studies.

Even though the Raman signature of the various elemental carbons, including C_{60} and C_{70}, is well known, apart from the pioneering work by Wopenka (1988), Raman microspectroscopy has not been commonly used to identify the presence of fullerenes in laboratory-produced and extraterrestrial carbon-containing samples. In part this situation might exist because small amounts of C_{60} and other fullerenes could not be confirmed previously by independent chemical techniques. This situation has changed after the successful HRTEM fullerene identifications by Rietmeijer et al. (2004) and Goel et al. (2004), which justifies the re-assessment of some of the published Raman carbon identifications, as was the case for published HRTEM soot images.

2. FULLERENE DETECTION

After the HRTEM identification of C_{60} and other, both smaller and larger, fullerenes in amorphous soot (Goel et al., 2004; Rietmeijer et al., 2004) we revisited the published record on HRTEM soot observations. We noticed that the same characteristic fullerene fingerprints, i.e. densely packed single-walled rings, are present in HRTEM images of soot particles produced by quenching a supersaturated vapor obtained by evaporation of a bulk elemental carbon sample using different techniques (e.g. Curl and Smalley, 1988; Ugarte, 1992; Bethune et al., 1993; De Heer and Ugarte, 1993; Wang and Kang, 1996; Jäger et al., 1999; Richter and Howard, 2000; Reynaud et al., 2001; Henning et al., 2004). With HRTEM analyses, capable of identifying fullerenes by their diameter, it is now possible to locate fullerene molecules in samples wherein they are present below the detection limits of chemical and spectroscopic techniques (e.g. Heymann et al., 2003). It seems warranted to conduct Raman micro-spectroscopy of elemental carbons on samples known to contain C_{60} and other fullerenes using a wider spectral range and a higher signal to noise ratio to resolve the 1470 cm^{-1} C_{60} peak in between the 'D' and 'G' peaks of graphitic carbon.

2.1 High Resolution Transmission Electron Microscopy

When searching for fullerene in a carbon-rich material wherein fullerene is present among many different carbon forms, it is important to consider its unique signature in HRTEM images (Wang and Buseck, 1991). This signature was not noticed previously as a unique carbon-vapor condensed texture, but it now appears that many of the amorphous soot particles in the peer-reviewed literature could actually be agglomerations of fullerene molecules that went undetected. The very nature of HRTEM analysis requires only a small amount of sample mass to identify nanometer-sized structures and as such the detection of the characteristic single-walled fullerene rings was just a matter of time.

2.1.1 Laboratory-Condensed Fullerene Identification

Soot is a form of solid carbon that is easy to make in the laboratory (Ugarte, 1992; De Heer and Ugarte, 1993; Mennella et al., 1995; Wang and Kang, 1996; Rotundi et al., 1998; Jäger et al., 1999; Reynaud et al., 2001) and found in natural environments such as in the remote marine troposphere (Pósfai et al., 1999), and it has industrial applications such as in traditional Chinese ink sticks (Osawa et al., 1997a, 1997b) and for its structural reinforcement properties (Baccaro et al., 2003). Although it is not our focus, soot and fullerenes are considered a potential health hazard from combustion processes used in transportation, manufacturing and power generation (Richter and Howard, 2000). It is thus important that the presence and lifetime of fullerene molecules in soot can be assessed properly. The first HRTEM detection of fullerenes in soot is an example of serendipity.

The study by Rotundi et al. (1998), using electron microbeam techniques to characterize the carbon phases in samples produced by arc discharge was originally intended to identify the carrier phase or phases of the astronomical 217.5-nm 'carbon' feature. These condensed samples contained many different carbon forms, viz. (1) chain-like aggregates of amorphous soot grains 7 to 70 nm in diameter, (2) multi-walled onions (10 - 40 nm in diameter) and hollow tubes (up to about 100 x 10 nm) (i.e. fullerenic carbons), (3) amorphous carbon, and (4) poorly graphitized and graphitic carbons. The individual amorphous soot grains are dense agglomerations of single-walled rings (Fig. 1) with average diameters corresponding to C_{60} fullerene and 'giant' fullerenes (Rietmeijer et al., 2004). Goel et al. (2004) also used ring diameter as the defining fullerene property and their C_{60} diameter is well matched to the one reported by Rietmeijer et al. (2004) (Table 1).

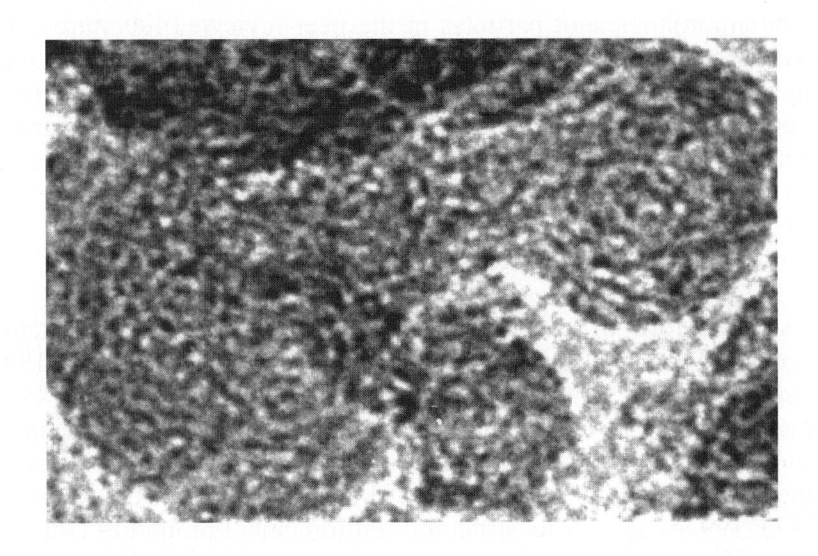

Figure 1. HRTEM image of a dense cluster of individual soot grains (20 to 38.5 nm in diameter) with a mostly random packing of the single-walled open rings of the condensed fullerenes in a typical soot grain produced by arc-discharge (applied voltage = 100 V, current = 10 Amp) between two amorphous carbon electrodes in a controlled atmosphere of argon and hydrogen at 10mbar pressure (Rietmeijer et al., 2004; unpublished image)

Table 1. Diameters of single-walled rings in soot grains from two different studies and the corresponding number of carbon atoms for the fullerene molecules

Fullerene Diameter (nm)		**Carbon Atoms**
Rietmeijer et al. (2004)	Goel et al. (2004)	
	0.52	36
0.70	0.685	60
	0.86	
1.1	1.03	
	1.2	176
3.0		540
5.5		960
8.2		1500

The soot-containing samples studied by Rotundi et al. (1998) were produced by condensation of a carbon vapor inside a chamber wherein the heat dissipation rate and efficiency may have contributed to a transient phase of autoannealing of the condensed carbons. The linear and curvi-linear structures, referred to as proto-fringes, seen inside many soot grains could be due to such post-condensation structural ordering of individually condensed fullerenes (see section 3).

2.1.2 Natural Fullerenes in Meteorites

Elemental carbon can be found in two of the three major classes of meteorites, i.e. iron meteorites and both differentiated and undifferentiated stony meteorites (see for reviews the chapters in Papike, 1998). Searches for fullerenes were directed towards the undifferentiated, carbonaceous chondrite meteorites of the stony class because they are carbon-rich meteorites with high abundances of accreted pre-solar grains that survived post-accretion thermal processing. From what we learn from laboratory work fullerene is an "easy to come and easy to go" molecule, i.e. will most likely be formed in any carbon condensation process and sensitive to solid-state evolution due to even modest thermal processing. Thus, the more pristine a sample, the higher is the probability of finding C_{60} molecules not yet evolved into giant fullerenes either in fully disordered solids or locally ordered in proto- fringes.

The presence of disordered fullerenes in extraterrestrial environments sampled by these primitive carbonaceous chondrites is supported by HRTEM images of meteoritic soot. For example, a HRTEM image of the Orgueil meteorite (Fig. 1 in Derenne et al., 2002) that is a very primitive meteorite shows a preponderance of single-walled ring structures. A HRTEM image of the Allende meteorite (Fig. 3 in Henning et al., 2004) shows circular and elongated single-walled ring structures, as well as curvilinear concentric features that we consider being evolved fullerene molecules.

To date no such single-walled ring structures or curvi-linear features (i.e. proto-fringes) were reported in amorphous carbons from carbon-rich cometary aggregate IDPs despite the fact these primitive particles would be quite likely candidates of solar system material to contain fullerenes, assuming these molecules were present in the solar nebula. However, when IDPs enter the Earth's atmosphere at velocities ranging from ca. 10 km/s to ca. 25 km/s they are decelerated by collisions with air molecules and experience a brief period (5-15s) of flash heating to temperatures between ca. 300 ^{o}C and ca. 1000 ^{o}C

(Nier and Schlutter, 1993; Rietmeijer, 1998; Flynn, 2002). It seems possible that under these conditions all traces of fullerenes could be erased. This is not the case for meteorites where the kinetic energy goes towards surface ablation that protects the interior of the meteorite against flash heating. In fact, the meteorite temperature below ca. 3mm from its ablation surface (i.e. the fusion crust) is much less than its temperature in space of ca. 500 °C (Ramdohr, 1967; Rietmeijer and Mackinnon, 1984). From these examples we conclude that condensed C_{60} and other fullerenes in carbonaceous chondrite meteorites would survive atmospheric entry with a much higher probability than those in chondritic aggregate IDPs.

2.2 Raman Microspectroscopy

Among the various analytical techniques used to characterize carbonaceous material of terrestrial and extraterrestrial origins, Raman micro-spectroscopy has been widely used to study the degree of order of polyaromatic materials (Ferrari and Robertson, 2000). However, its application to samples of astrophysical interest, e.g. IDPs (Wopenka, 1988; Raynal, 2003) and carbons in meteorites, remains quite limited. The latter include interstellar graphite grains extracted from the Murchison meteorite (Zinner et al., 1995) and carbon inclusions associated with Fe,Ni-grains in chondrites (Mostefaoui et al., 1999). A typical spectrum of such material exhibits several bands, the most intense being the first order bands, namely the G-band (peaking at ca. 1580 cm^{-1}) and the D-band (peaking at ca. 1350 cm^{-1}). These bands were intensively studied, as they are sensitive to the degree of order/disorder in the aromatic plane. The degree of structural order of the carbonaceous material modifies the parameters of these bands, i.e. their position, width or intensity in manner that is directly linked to the length of the coherent domain L_a.

Raman microspectroscopy is a relatively fast, non-destructive method that can be applied in situ on samples a few microns in size. Meteorites and IDPs are complex samples and when searching for a minor constituent, e.g. fullerene, it would be expedient to know the unique fullerene signature in Raman spectra of carbon-rich material obtained in the laboratory by carbon-vapor condensation experiments wherein fullerene is present among many different carbon forms in such analog samples.

In the late 80's, a survey of 20 chondritic IDPs showed that 75% of these particles contained "poorly crystallized carbonaceous material" with a Raman signature consistent with variable degrees of "disorder".

The mean crystallite size in the most 'ordered' of these IDP materials was estimated at ca. 3 nm but <6 nm assuming the material resembled activated charcoal or glassy carbon with a "turbostratic structure" (Wopenka, 1988). Since this pioneering work, searches for C_{60} and other fullerenes in these materials were so far neglected, even in recent studies of the organic matter in meteorites and IDPs (Quirico et al., 2003, 2005). The Raman spectra from two different areas in a sample from the same condensed carbon study with fullerenic carbons, C_{60} and larger fullerenes containing soot (Rietmeijer et al., 2004) show peaks at 396, ca. 500, 680, 1358 and 1594 cm^{-1} (Fig. 2). In the spectrum the separation of the 'D' and 'G' peaks at ca. 1350 and ca. 1580 cm^{-1}, respectively, is poorly defined with regard to this separation in a typical Raman spectrum of disordered, pre-graphic carbons, and the intensity ratio of the D/G peaks is ≥ 1, i.e. the D/G ratio expected for pre-graphitic mature carbons (Raynal, 2003).

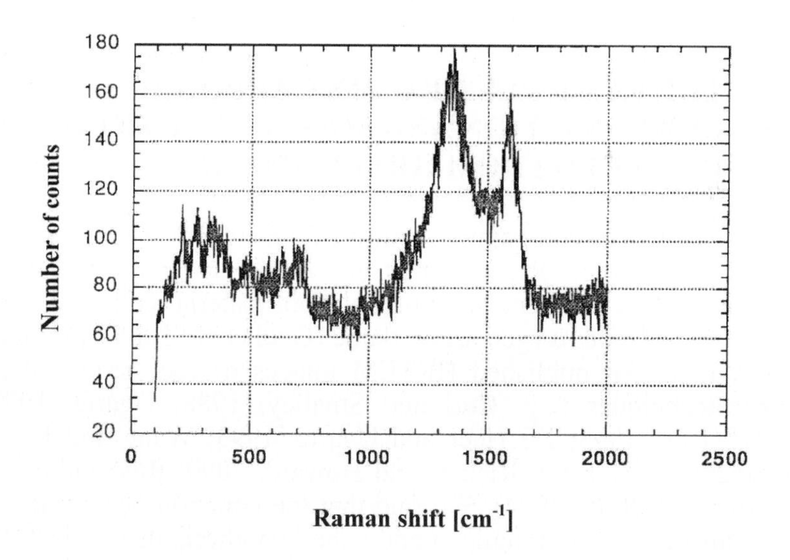

Figure 2. Raman spectrum from sample ACH_2800 (Rotundi et al., 1998) obtained in the 200 – 3000 cm^{-1} region showing signatures that could be due to the presence of fullerene and/or fullerenic carbons

This particular peak configuration could suggest an additional Raman carbon signal such as from C_{60}+C_{70} fullerene or fullerenic-carbon nanotubes present in the sample. The latter have additional peaks at <1100 cm^{-1} and >2600 cm^{-1}; the fullerenes have a major peak at 1470 cm^{-1}. Thus, we hypothesize that this particular peak configuration (Fig. 2) in this condensed carbon analog sample (Rotundi et al., 1998) is due to the presence of either one, or both, of these carbon forms.

A wider spectral range and a stronger signal to noise ratio between 300-1100 cm^{-1} and in the 'D' and 'G' regions, would allow detection of fullerene and fullerenic carbons in various extraterrestrial samples. The clear implication being that these carbon forms went undetected by previous Raman micro-spectroscopic analyses. We note that the Raman spectra of the "most 'ordered' material" in IDPs (Wopenka, 1988) show a similar 'D' and 'G' peak configuration as those shown in Fig. 2. Whereas Jelička at al. (2005) concluded that Raman microspectroscopy could not detect dispersed, low (ppb to ppm levels), fullerene concentrations in carbon-rich geological materials, the situation is much more favorable for synthetic and natural soot samples.

3. THE KEY ROLE OF C_{60} IN CARBON CONDENSATION AND SOOT EVOLUTION INFERRED FROM HRTEM IMAGES

The results of a comparative study of soot condensed in an arc discharge experiment (Rietmeijer et al., 2004), and of pure carbon black, carbon-black-toluene solution and soot material collected from a pre-mixed benzene/oxygen/argon flame (Goel et al., 2004), provided a reason to revisit published HRTEM images of soot produced by different techniques (e.g. Curl and Smalley, 1988; Ugarte, 1992; Bethune et al., 1993; De Heer and Ugarte, 1993; Wang and Kang, 1996; Jäger et al., 1999; Richter and Howard, 2000; Reynaud et al., 2001; Henning et al., 2004). We find that the common occurrence of single-wall rings in soot could support the hypothesis that metastable C_{60} is the primary condensed form of carbon. Many important details of carbon condensation still remain poorly understood.

The commonly accepted scenario of soot formation (e.g. Richter and Howard, 2000) starts from small molecules such as benzene and then proceeds to larger and larger polycyclic aromatic hydrocarbon (PAH) growth, involves both the addition of C_2, C_3 or other small units, among which acetylene has received much attention, to PAH radicals, and reactions among the growing aromatic species, such as PAH–PAH radical recombination and addition reactions. This process is followed by the nucleation or inception of small soot particles whereby mass is converted from the molecular to particulate systems, i.e. heavy PAH molecules form nascent soot particles with a molecular mass of approximately 2000 amu and an effective diameter of about 1.5 nm (Fig. 3).

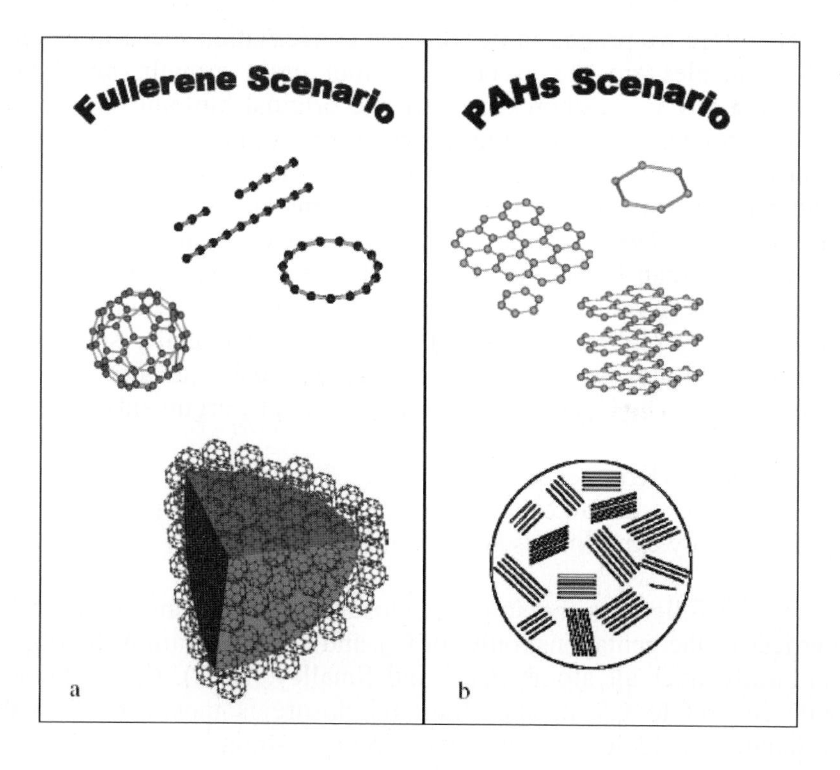

Figure 3. Carbon condensation. The initial seeds that agglomerate to form carbon grains in the "***Fullerene scenario***" (a) are molecules with different geometry depending on the number of C atoms (chains, rings, cages) and in the "***PAHs scenario***" (b) they are small molecules, e.g. benzene, that grow to larger and larger polycyclic aromatic hydrocarbons (PAHs) that can either form extended graphene layers or small aromatic nanocrystalline stacks called 'Basic Structural Units'

A slightly different condensation scenario (Henning et al., 2004) is based on the presence of small graphene sheets of few aromatic rings stacked (sub) parallel to each other, i.e. aromatic nanocrystalline stacks referred to as Basic Structural Units (*BSUs*) (Oberlin et al., 1984). *BSUs* can be differently organized depending on the experimental conditions during soot formation (Henning et al., 2004) (see Fig. 3).

It is important to mention that flat aromatic structures in a carbon vapor would have numerous dangling bonds and they would have little reason to remain flat. The physical tendency to reach the lowest energy level available would induce the sheets to eliminate the dangling bonds by curling up (Robertson et al., 1992).

Therefore we propose a new carbon condensation scenario wherein C_{60} molecules are the "seeds" for soot grain growth; not PAHs. Consequently C_{60} carbon cages are the original building blocks that upon agglomeration will yield amorphous soot grains (Fig. 3). Carbon atoms in the vapor will form short chains that assume a monocyclic ring geometry for a number of carbon atoms larger than 10 (Weltner and Van Zee, 1989). In the C_{30}-C_{40} region closed structures become more stable than linear chains and planar ring structures (von Helden et al., 1993). Even when small fullerenes appeared around C_{30} they will grow by sequential addition of C_2 units, eventually stopping at C_{60} when further C_2 additions become improbable in the gas phase (Heat, 1991). The C_{60} scenario is substantiated by arguments of

1. Symmetry (Kroto et al., 1985),
2. Electronic structure calculations (Newton and Stanton, 1986), and
3. Laboratory experiments (von Helden et al., 1993).

For all fullerenes the strain of closure tends to concentrate at the vertices of the pentagons; only for C_{20} and C_{60} this strain is distributed uniformly over all atoms (Curl and Smalley, 1988). C_{60} is favored with respect to C_{20} as the strain of closure is independent of the fullerene molecule size but the average strain per carbon atom increases for smaller clusters (Schmaltz et al., 1988).

When separating *proper* carbon condensation features from secondary, evolved features due to variable quench rates and post-condensation heat dissipation, Rotundi et al. (1988) and Rietmeijer et al. (2004) found that C_{60} fullerene was the original carbon condensate that had agglomerated in soot grains. The C_{60} molecule has a key role in solid-state modification of soot, that is, concentric structures in soot grains represent re-adjustments of condensed metastable C_{60} molecules that had coalesced into higher fullerenes (Yeretzian et al., 1992; Zhao et al., 2002a, 2002b; Kim et al., 2003; Rietmeijer et al., 2004) (Fig. 4).

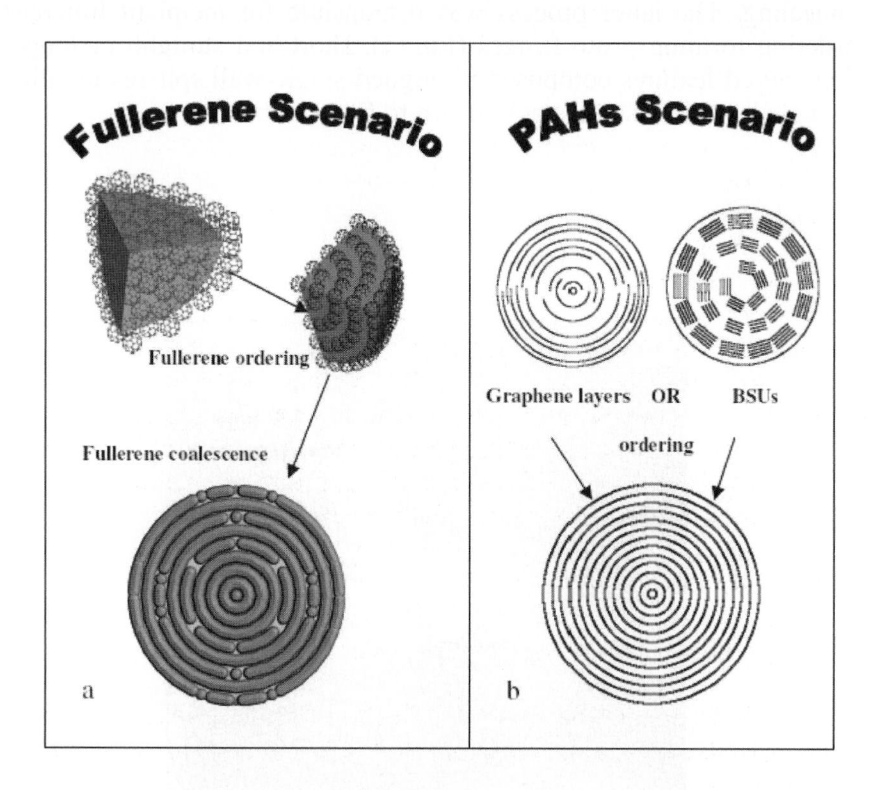

Figure 4. Ordering processes of the original molecules drives graphitization and in the "***Fullerene condensation scenario***" (a) it will be driven by the coalescence of closed carbon cages in ever-larger fullerenes finally forming the oft-seen concentric graphitized structure. Graphitization in the "***PAHs condensation scenario***" (b) will be caused by the arrangement of PAH molecules on ever-larger curved planes (i.e. graphene layers) or by enlargement of *BSUs*

Fullerene coalescence is kinetically controlled and depends on the combination of the reigning thermal regime and vapor density. Larger fullerenes are formed by a reaction of the type,

$$m\mathrm{C}60 \rightarrow \mathrm{C}60m$$

which would yield an orderly sequence of fullerenes. Giant fullerene growth can occur within the vapor or during post-condensation auto-annealing by uncontrolled endogenic thermal annealing. The latter process was responsible for incipient fullerene ordering forming proto-fringes (Fig. 5): short and straight, or longer and curved features composed of aligned single-wall spheres in either a single layer or a small stack of two to five layers.

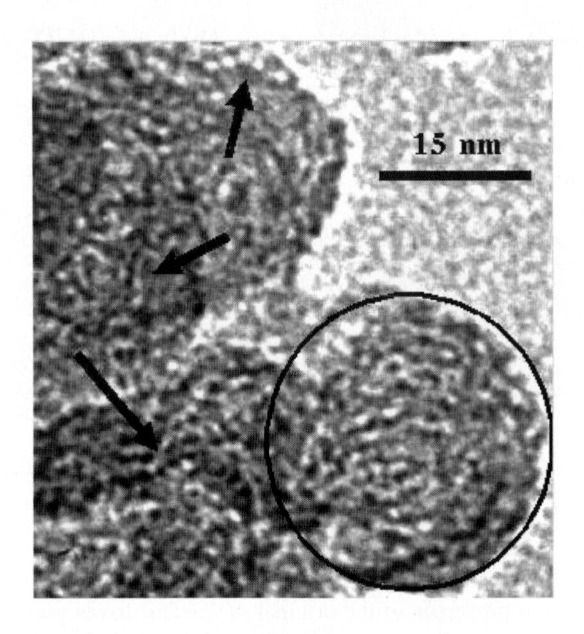

Figure 5. High magnification TEM image showing aligned (arrows) and densely packed single-walled rings of individual fullerene molecules in a soot sample. The encircled grain shows concentric proto-fringes that could evolve towards the typically concentric circular onion structure. (Rietmeijer et al., 2004; unpublished image)

We depicted a scenario of continuous ordering from amorphous soot formed by agglomerated fullerenes that coalesced into ever-larger fullerenes that finally formed the often-seen concentric graphitized structure (Fig. 4). Giant fullerenes may have properties of a single graphitic sheet (Yeretzian et al., 1992). For comparison with the new proposed soot ordering process based on fullerenes growth (Fig. 4) we show the classical carbon graphitization processes led by the arrangement of PAH molecules on curved and wide planes (i.e. graphene layers) or by the enlargement of *BSUs* (Reynaud et al., 2001; Henning et al., 2004).

In support of the fullerene coalescence scenario we mention the studies of electron-beam irradiation (Luzzi and Smith, 2000) and heat treatment (Bandow et al., 2001) on carbon composites that induced coalescence of fullerenes into a single-wall nanotube, 0.71 nm in diameter. Bandow et al. (2001) found that coalescence started at 800 °C and the transformation to a single-wall nanotube was completed at 1200 °C.

Confirmation for the post-condensation secondary nature of curvilinear features, that represent very small-scale structural ordering of individual fullerene molecules, comes from HRTEM images of soot samples subjected to post-condensation heat-treatment, ion bombardment or prolonged electron-beam irradiation. These images show

1. Dense agglomerations of single-wall rings in the pre-treated soot, and
2. Linear, curvilinear and concentric circular features that were invariably present in thermally processed soot (see, Fig. 4 in Ugarte, 1992; Figs. 1 and 2 in De Heer and Ugarte, 1993; Fig. 3 in Reynaud et al., 2001).

In addition, Grieco et al. (2000) found that the formation of highly ordered carbon nanostructures, such as nanotubes and onion structures, requires long residence times, seconds or minutes depending on the temperature, in the flame environment, while the formation of amorphous and fullerenic carbon occurs in milliseconds. Internal rearrangement processes are required for more ordered structures to form rather than gas-phase molecular-weight growth pathways as occur at flame temperatures (Grieco et al., 2000). These observations support the hypothesis that autoannealing could induce better-ordered structural forms of carbon in condensed soot.

Such evolved carbon features may well form in natural environments as a result of interactions between fullerene-laden amorphous soot with some type of natural processes that could induce heat treatment of the material, e.g. energetic particles, UV irradiation, or atmospheric entry. Proto-fringe textures are seen in extraterrestrial samples, e.g. the Allende meteorite (e.g. Fig. 3 in Henning et al., 2004). Different stages in the evolution of secondary features are visible in this published figure ranging from single-wall rings, to elongated fullerenes and long curvilinear structures.

4. ASTROPHYSICAL IMPLICATION

Astronomical observations from the far-UV to the millimeter spectral region indicate the presence of carbon-based materials in space. Amorphous carbon, hydrogenated amorphous carbon, coal-like matter, soot, quenched-carbonaceous condensates, nanodiamonds, and fullerenes were all proposed as possible candidates for interstellar carbon dust (Thaddeus, 1994; Mennella et al., 1995; Papoular et al., 1996; Cherchneff et al., 2000; Ehrenfreud and Charnley, 2000; Henning et al., 2004). The origin of the interstellar UV absorption spectrum with the most prominent feature at 217.5 nm, although it is commonly assigned to carbonaceous material, does not yet have a definitive explanation. Similarities between experimental and observed absorption spectra indicated that carbon onions are very strong candidates for the origin of this interstellar UV absorption peak (Chhowalla et al., 2003).

Iglesis-Groth (2004) showed that photoabsorption by fullerenes and bucky-onions (i.e. multi-shell fullerenes) could explain the shape, width, and peak energy of the UV absorption feature. Comparing theoretical photoabsorption cross-sections of individual and multi-shell fullerenes and astronomical data, Iglesis-Groth (2004) estimated a density of fullerenes and bucky-onions in the diffuse interstellar medium of 0.1-0.2 ppm. In particular, for a mixture based on single fullerenes she estimated that about 80 carbon atoms per million hydrogen atoms would be locked in these molecules. If as expected the cosmic carbon abundance is close to the solar atmosphere value, individual fullerenes may lock up 20% to 25% of the total carbon in diffuse interstellar space. When bucky-onions are also considered in the mixture, carbon in fullerene-based molecules could reach 50% of the cosmic carbon abundance. That is, fullerene-based molecules appear to be a major carbon reservoir in the interstellar medium.

Carbon vapor condensation experiments are conducted for the purpose of constraining the composition, morphology and structure of dust formed around C-type stars residing in the diffuse and dense regions of the interstellar medium (De Heer and Ugarte, 1993; Mennella et al., 1995; Papoular et al., 1996; Rotundi et al., 1998; Jäger et al., 1999; Ehrenfreud and Charnley, 2000; Reynaud et al., 2001; Henning et al., 2004).

Circumstellar regions range from the stellar outflow winds of very young stars to dust envelopes around highly evolved stars. Within the dense circumstellar envelopes around carbon–rich stars interesting chemistry involving a plethora of carbon molecules and dust formation occur. Carbon grains form in the vicinity of the star, but one of the major open problems in circumstellar dust studies remains to define the chemical pathways from small radical clusters towards carbon grain formation. It was proposed that the route to carbon grains passes through PAH molecules (e.g. Léger and Puget, 1984; Herbst, 1991). Other circumstellar condensation scenarios took the path to grain formation from linear chains, to rings and to fullerenes depending on the number of carbon atoms (Cherchneff et al., 2000). These carbon molecules could coexist in various proportions and pass from one geometry to another in the ambient radiation field because of small energetic separations (less than 1–2 eV) (Thaddeus, 1994; Pascoli and Polleux, 2000).

If C_{60} fullerene molecules are shown to be the primary carbon condensates available for agglomeration and structural evolution, there will be a number of implications for astrophysics:

1. The scenario of carbon chains, rings and cages will be more plausible than the one starting with PAHs.
2. While it is almost impossible to avoid metastable C_{60} formation, once formed it may be short-lived as an isolated molecule whose existence in astrophysical environments depends on competition between the efficiency of the condensation process and the destruction rate due to cosmic radiation, among others. The experimental data show that C_{60} has a very high formation rate and the degree of C_{60} decomposition due to prolonged γ-irradiation was shown to be less than 15%, which actually demonstrates extreme stability (Basiuk et al., 2004).
3. Once agglomerated in amorphous soot grains, C_{60} will be shielded from interacting with the environment. The C_{60} molecules near the soot grain surface would evolve prior to the inner part and evolved

larger fullerenes and other secondary features from solid-state processing will be preferentially found in the soot near-surface zone.

4. Graphitization of carbon grains in astrophysical environments (Papoular et al., 1996; Mennella et al., 1997; Henning et al., 2004) will be driven by structural evolution of condensed C_{60} molecules once closely-packed inside a soot grain. Progressive fullerene ordering would lead to proto-fringes concentrically arranged within the soot grain. Once the original fullerenes so arranged had unfolded the new arrangement could eventually lead to the formation of nested fullerenes that could ultimately evolve into well-ordered carbon onions.

5. CONCLUSIONS

Although C_{60} appears to be the dominant condensate in laboratory experiments wherein carbon had agglomerated into soot grains, the published literature consistently reported an overall dearth of this and other fullerene molecules among laboratory-condensed carbon samples. Fullerenes appear also to be missing from carbon-rich extraterrestrial samples, as the collected primitive meteorite and IDP samples that should contain the products of carbon condensation in astrophysical environments. These samples were analyzed using mostly chemical techniques that may not have been sensitive enough to detect very small quantities of C_{60} and other fullerenes. A positive result means that fullerenes are present in the sample but a negative result cannot be interpreted as proof of absence of fullerenes. In such 'negative outcome' samples HRTEM analyses would be able to detect individual fullerene molecules and fullerene clusters in an amorphous carbon matrix. Recent Raman microspectroscopy analyses of condensed soot showed that fullerenes are detectable when the "traditional" experimental conditions are modified to enhance the signal in the region between the 'D' and 'G' carbon region. Recent HRTEM studies of condensed carbons found that soot grains were agglomerations of C_{60} and other fullerenes ranging from C_{36} to C_{1500} that were recognizable as distinctive single-wall rings with variable diameters between 0.52 nm and 8.2 nm. The very nature of such TEM analyses allows for the detection of individual nanostructures without constraints of abundance.

The apparent dearth of fullerenes in natural and synthetic carbon samples may just be a matter of detection. Finding a technique able to detect small quantities of fullerenes was only a matter of time. Some studies searching for fullerenes among meteoritic carbons were positive while others were negative. Such studies did not analyze the same sample allocations and the possibility remains that sample heterogeneity might be a contributing factor, in particular when fullerene concentrations were low.

The recognition that C_{60} was the major primary carbon condensate has implications for the interpretation of secondary carbon textures of fullerene and soot evolution in synthetic samples used to identify the carbon phase most likely to be responsible for the 217.5-nm astronomical absorption feature. Fullerene molecules as primary condensates prior to carbon ageing, instead of the scenarios involving PAHs or *BSUs*, might offer an alternative to assess the physical conditions for carbon-condensation within the dense circumstellar envelopes around carbon–rich stars and an opportunity to re-assess the presence of low fullerene abundances in primitive amorphous meteoritic carbons. Such low abundances would not rule out the possibility that C_{60} and other fullerene molecules might be the dominant condensed carbon in circumstellar environments and in the interstellar medium. It will require additional laboratory analyses of carbons from carbonaceous chondrite meteorites and interplanetary dust particles and of in situ collected asteroid and comet nucleus samples before the anticipated pervasiveness of natural extraterrestrial fullerenes will be a firmly established fact.

Acknowledgements. We thank Dr. Nuth for constructive suggestions. We are grateful to Dr. V. Della Corte for kind help in drawing Figs. 3 and 4. This work was supported by ASI and MURST research contracts (AR). One author (J.B.) is grateful to G. Montagnac of the *Laboratoire des Sciences de la Terre* (ENS-Lyon, France) who helped in performing the Raman measurements, using a LabRaman Jobin-Yvon confocal micro-spectrometer. FJMR was supported by NAG5-11762 from the National Aeronautics and Space Administration.

6. REFERENCES

Ash, R.D., Russell, S.S., Wright, I.P. and Pillinger, C.T. (1993) Minor high temperature components confirmed in carbonaceous chondrites by stepped combustion using a new sensitive static mass spectrometer (abstract). *Lunar Planet. Sci.*, 22, 35-36, Lunar and Planetary Institute, Houston, Texas, USA.

Baccaro, S., Cataldo, F., Cecilia, A., Cemmi, A., Padella, F and Santini, A. (2003) Interaction between reinforce carbon black and polymeric matrix for industrial applications. *Nuclear Instr. Methods Phys. Res. B*, 208, 191-194.

Bandow, S., Takizawa, M., Hirahara, K., Yudasaka, M. and Iijima, S. (2001) Raman scattering study of double-wall carbon nanotubes derived from the chains of fullerenes in single-wall carbon nanotubes. *Chem. Phys. Lett.*, 337, 48-54.

Bajt, S., Chapman, H.N., Flynn, G.J. and Keller, L.P. (1996) Carbon characterization in interplanetary dust particles with a scanning transmission X-ray microscope (abstract). *Lunar Planet. Sci.*, 27, 57-58, Lunar and Planetary Institute, Houston, Texas, USA.

Basiuk, V.A., Albarrána, G., Basiuk, E.V. and Saniger, J.-M. (2004) Stability of interstellar fullerenes under high-dose γ-irradiation: New data. *Adv. Space Res.*, 33, 72-75.

Becker, L. and Bunch, T.E. (1997) Fullerenes, fulleranes and polycyclic aromatic hydrocarbons in the Allende meteorite. *Meteorit. Planet. Sci.*, 32, 479-487.

Becker, L., Bada, J.L., Winas, R.E. and Bunch, T.E. (1994) Fullerenes in the Allende meteorite. *Nature*, 372, 507-507.

Becker, L., Bunch, T.E. and Allamandola, L.J. (1999) Higher fullerenes in the Allende meteorite. *Nature*, 400, 227-228.

Bethune, D.S., Kiang, C.H., de Vries, M.S., Gorman, G., Savoy, R., Vazquez, J. and Beyers, R. (1993) Cobalt-catalyzed growth of carbon nanotubes with single-atomic-layer walls. *Nature*, 363, 605-607.

Buseck, P.R. (2002) Geological fullerenes: review and analysis. *Earth Planet. Sci. Lett.*, 203, 781-792.

Buseck, P.R., Tsipurski, S.J. and Hettich, R. (1992) Fullerenes from the geological environment. *Nature*, 247, 215-217.

Cherchneff, I., Le Teuff, Y.H., Williams, P.M. and Tielens, A.G.G.M. (2000) Dust formation in carbon-rich Wolf-Rayet stars? I. Chemistry of small carbon clusters and silicon species. *Astron. Astrophys.*, 357, 572-580.

Chhowalla, M., Wang, H., Sano, N., Teo, K.B.K. and Amaratunga, G.A.J. (2003) Carbon onions: Carriers of the 217.5-nm interstellar absorption feature. *Phys. Rev. Lett.*, 90, 155504-1-155504-4.

Curl, R.C. and Smalley, R.E. (1988) Probing C_{60}. *Science*, 242, 1017-1022.

De Heer, W.A. and Ugarte, D. (1993) Carbon onions produced by heat treatment of carbon soot and their relation to the 217.5 nm interstellar absorption feature. *Chem. Phys. Lett.*, 207, 480-486.

DeVries, M.S., Reihs, K., Wendt, H.R., Golden, W.G., Hunziker, H.E., Fleming, R., Peterson, E. and Chang, S. (1993) A search for C_{60} in carbonaceous chondrites. *Geochim. Cosmochim. Acta*, 57, 933-935.

Derenne, S., Robert, F., Binet, L., Gourier, D., Rouzaud, J.-N. and Largeau, C. (2002) Use of combined spectroscopic and microscopic tools for deciphering the chemical structure and origin of the insoluble organic matter in the Orgueil and Murchison meteorites. *Lunar Planet. Sci.*, 33, abstract #1182, Lunar and Planetary Institute, Houston, Texas, USA, CD-ROM.

Ehrenfreud, P. and Charnley, S.B. (2000) Organic molecules in the interstellar medium, comets, and meteorites: a voyage from dark clouds to the early Earth. *Ann. Rev. Astron. Astrophys.*, 38, 427-483.

Ferrari, A.C. and Robertson, J. (2000) Interpretation of Raman spectra of disordered and amorphous carbon. *Phys. Rev. B*, 61, 14095-14107.

Flynn, G.J. (2002) Extraterrestrial dust in the near-Earth environment. In Meteors in the Earth's Atmosphere, E. Murad and I.P. Williams, Eds., 77-94, Cambridge University Press, Cambridge, United Kingdom.

Flynn, G.J., Keller, L.P., Feser, M., Wirick, S. and Jacobsen, C. (2003) The origin of organic matter in the solar system: Evidence from the interplanetary dust particles. *Geochim. Cosmochim. Acta*, 67, 4791-4806.

Foing, B.H. and Ehrenfreud, P. (1994) Detection of two interstellar absorption bands coincident with spectral features of C_{60}^+. *Nature*, 369, 296-298.

Foing, B.H. and Ehrenfreud, P. (1997) New evidence for interstellar C_{60}^+. *Astron. Astrophys.*, 317, L59-L62.

Gilmour, I., Russell, S.S, Newton, J., Pillinger, C.T., Arden, J.W., Dennis, T.J., Hare, J.P., Kroto, H.W., Taylor, R. and Walton, D.R.M. (1993) A search for the presence of C_{60} as an interstellar grain in meteorites (abstract). *Lunar Planet. Sci.*, 22, 445-446, Lunar and Planetary Institute, Houston, Texas, USA.

Goel, A., Howard, J.B. and Vander Sande, J.B. (2004) Size analysis of single fullerene molecules by electron microscopy. *Carbon*, 42, 1907-1915.

Grieco, W.J., Howard, J.B., Rainey, L.C. and Vander Sande, J.B. (2000) Fullerenic Carbon in combustion-generated soot. *Carbon*, 38, 597-614.

Heat, J.R. (1991) Synthesis of C_{60} from small carbon clusters: a model based on experiment and theory. *In Fullerenes, Synthesis, Properties and Chemistry of Large Carbon Clusters*, G.S. Hammond and V.J. Kuck, Eds., *Am. Chem. Soc. Symp. Series*, 481, 1-27.

Henning, T., Jäger, C. and Mutschke, H. (2004) Laboratory studies of carbonaceous dust analogs. *In Astrophysics of Dust*, A.N. Witt, G.C. Clayton and B.T. Draine, Eds., *ASP Conf. Series*, 309, 603-628, Astronomical Society of the Pacific, San Francisco, California, USA.

Herbst, E. (1991) In situ formation of large molecules in dense interstellar clouds. *Astrophys. J.*, 366, 133-140.

Heymann, D. (1997) Fullerenes and fulleranes in meteorites revisited. *Astrophys. J.*, 489, L111-L114.

Heymann, D., Jenneskens, L.W., Jehlička, J., Koper, C. and Vlietstra, E. (2003) Terrestrial and extraterrestrial fullerenes. *Fullerenes, Nanotubes, and Carbon Nanostructures*, 11, 333-370.

Iglesias-Groth, S. (2004) Fullerenes and buckyonions in the interstellar medium., *Astrophys. J.*, 608, L37-L40.

Jäger, C., Henning, Th., Schlögl, R. and Spillecke, O. (1999) Spectral properties of carbon black. *J. Crystal Growth*, 258, 161-179.

Jelička, J., Frank, O., Pokorný, J. and Rouzaud, J.-N. (2005) Evaluation of Raman spectroscopy to detect fullerenes in geological materials. *Spectrochim. Acta A* 61, 2364-2367.

Keller, L.P., Messenger, S. and Bradley, J.P. (2000) Analysis of a deuterium-rich interplanetary dust particle and implications for presolar materials in IDPs. *J. Geophys. Res. Space Phys.*, 105, 10397-10402.

Kim, Y-H., Lee, I-H., Chang, K.J. and Lee, S. (2003) Dynamics of fullerene coalescence. *Phys. Rev. Lett.*, 90, 065501-065501.

Krätschmer, W., Lamb, L.D., Fostiropoulos, K. and Huffman, D. (1990) Solid C_{60}: a new form of carbon. *Nature*, 347, 354-358.

Kroto, H.W., Heat, J.R., O'Brien, S.C., Curl, R.F. and Smalley, R.E. (1985) C_{60}: Buckminsterfullerene. *Nature*, 318, 162-163.

Léger, A. and Puget, J.L (1984) Identification of the 'unidentified' IR emission features of interstellar dust? *Astron. Astrophys.*, 137, L5-L8.

Luzzi D.E. and Smith B.W. (2000) Carbon cage structures in single wall carbon nanotubes: a new class of materials. *Carbon*, 38, 1751-1756.

Mennella, V., Colangeli, L., Bussoletti, E., Monaco, G., Palumbo, P. and Rotundi, A. (1995) On the electronic structure of small carbon grains of astrophysical interest. *Astrophys. J. Suppl.*, 100, 149-157.

Mennella, V., Baratta, G., Colangeli, Palumbo, P., Rotundi, A. and Bussoletti, E. (1997) Ultraviolet spectral changes in amorphous carbon grains induced by ion irradiation. *Astrophys. J.*, 481, 545-549.

Messenger, S. (2000) Identification of molecular-cloud material in interplanetary dust particles. *Nature*, 404, 968-971.

Mostefaoui, S., Perron, C., Zinner, E. and Sagon, G. (1999) Metal-associated carbon in primitive chondrites: structure, isotopic composition and origin. *Geochim. Cosmochim. Acta*, 64, 1945-1964.

Newton, M.D. and Stanton, R.E. (1986) Stability of buckminsterfullerene and related carbon clusters. *J. Am. Chem. Soc.*, 108, 2469-2470.

Nier, A.O. and Schlutter, D.J. (1993) The thermal history of interplanetary dust particles collected in the Earth's stratosphere. *Meteoritics*, 28, 675-681.

Oberlin, A., Goma, J. and Rouzaud, J.N. (1984) Techniques d'étude des structures et textures (microtextures) des materiaux carbones (in French). *J. Chemie Physique*, 81, 701-710.

Osawa, E., Hirose, Y., Kimura, A., Shibuya, M., Gu, Z. and Li, F.M. (1997a) Fullerenes in Chinese ink. A correction. *Fullerene Sci. Techn.*, 5, 177-194.

Osawa, E., Hirose, Y., Kimura, A., Shibuya, M., Kato, M. and Takezawa, H. (1997b) Seminatural occurence of fullerenes. *Fullerene Sci. Techn.*, 5, 1045-1055.

Papike, J.J., (Ed.) (1998) *Planetary Materials*, *Revs. Mineral.*, 3**6**, 1052p., The Mineralogical Society of America, Washington, D.C., USA.

Papoular, R., Conard, J., Guillois, O., Nenner, I., Reynaud, C. and Rouzard, J.N. (1996) A comparison of solid-state carbonaceous models of cosmic dust. *Astron. Astrophys.*, 315, 222-236.

Pascoli, G. and Polleux, A. (2000) Condensation and growth of hydrogenated carbon clusters in carbon-rich stars. *Astron. Astrophys.*, 359, 799-810.

Pósfai, M., Anderson, J.R., Buseck, P.R. and Sievering, H. (1999) Soot and sulfate aerosol particles in the remote marine troposphere. *J. Geophys. Res.*, 104(D17), 21685-21693.

Quirico, E., Raynal, P.I. and Bourot-Denise, M. (2003) Metamorphic grade of organic matter in six unequilibrated ordinary chondrites. *Meteorit. Planet. Sci.*, 38, 795-812.

Quirico, E., Borg, J., Raynal, P.I., Montagnac, G. and d'Hendecourt, L. (2005) A micro-Raman survey of 10 IDPs and 6 carbonaceous chondrites. *Planet. Space Sci.*, in press; *on line* doi:10.1016/j.p.s.s.2005.07.09

Ramdohr, P. (1967) Die Schmelzkrüste der Meteoriten (in German). *Earth. Planet. Sci. Lett.*, 2, 197-209.

Raynal, P.I. (2003) Étude en laboratoire de matière extraterrestre: implications pour la physico-chimie du Système Solaire primitif (in French). PhD thésis, Université Paris 6, France.

Reynaud, C., Guillois, O., Herlin-Boime, N., Rouzaud, J-N., Galvez, A., Clinard, C., Balanzat, E. and Ramillon, J-M (2001) Optical properties of synthetic carbon nanoparticles as model of cosmic dust. *Spectrochim. Acta, A*, 57, 797-814.

Richter, H. and Howard, J.B. (2000) Formation of polycyclic aromatic hydrocarbons and their growth to soot - a review of chemical reaction pathways. *Progr. Energy Comb. Sci.*, 26, 565-608.

Rietmeijer, F.J.M. (1998) Interplanetary Dust Particles. *In Planetary Materials*, J.J. Papike, Ed., Revs. Mineral., 36, 2-1–2-95, The Mineralogical Society of America, Washington, D.C., USA.

Rietmeijer, F.J.M. (2002) The earliest chemical dust evolution in the solar nebula. *Chemie der Erde*, 62, 1-45.

Rietmeijer, F.J.M. and Mackinnon, I.D.R. (1984) Melting, ablation and vapor phase condensation during atmospheric passage of the Bjurböle meteorite. *J. Geophys. Res.*, 87, *Suppl.*, B597-B604.

Rietmeijer, F.J.M. and Nuth, J.A. (1991) Tridymite and maghémite formation in a Fe-SiO smoke. *Proc. Lunar Planet. Sci.*, 21, 591-599.

Rietmeijer, F.J.M., Nuth III, J.A., Karner, J.M. and Hallenbeck S.L. (2002) Gas to solid condensation in a Mg-SiO-H_2-O_2 vapor: Metastable eutectics in the MgO–SiO_2 phase diagram. *Phys. Chem. Chem. Phys.*, 4, 546-551.

Rietmeijer, F.J.M., Rotundi, A. and Heymann, D. (2004) C_{60} and giant fullerenes in soot condensed in vapors with variable C/H_2 ratio. *Fullerenes, Nanotubes, and Carbon Nanostructures*, 12, 659-680.

Robertson, D.H., Brenner, D.W. and White, C.T. (1992) On the way to fullerenes: molecular dynamics study of the curling and closure of graphitic ribbons. *J. Phys. Chem.*, 96, 6133-6135.

Rotundi, A., Rietmeijer, F.J.M., Colangeli, L., Mennella, V., Palumbo, P. and Bussoletti, E. (1998) Identification of carbon forms in soot materials of astrophysical interest. *Astron. Astrophys.*, 329, 1087-1096.

Sassara, A., Zerza, G., Chergui, M. and Leach, S. (2001) Absorption wavelengths and bandwidths for interstellar searches of C_{60} in the 2400-4100 Å, region. *Astrophys. J. Suppl.*, 135, 263-273.

Schmaltz, T.G., Seitz, W.A. and Hite, G.E. (1988) Elemental carbon cages. *J. Am. Chem. Soc.*, 110, 1113-1127.

Snow, T.P. and Seab, C.G. (1989) A search for interstellar and circumstellar C_{60}. *Astron. Astrophys.*, 213, 291-294.

Stephan, O., Bando, Y., Dussarrat, C., Kurashima, K., Sasaki, T., Tamiya, T. and Akaishi, M. (1997) Onionlike structures and small nested fullerenes formation under electron irradiation of turbostratic BC2N. *Appl. Phys. Lett.*, 70, 2383-2385.

Taylor, R., Parsons, J.P., Avent, A.G., Rannard, S.P., Dennis, T.J., Hare, J.P., Kroto, H.W. and Walton, D.R.M. (1991) Degeneration of C_{60} by light. *Nature*, 351, 277.

Thaddeus, P. (1994) On the large organic molecules in the interstellar gas. *In Molecules and Grains in Space*, I. Nenner, Ed., *AIP Conf. Proc.*, 312, 711-728, The American Institute of Physics Press, Woodbury, New York, USA..

Ugarte, D. (1992) Curling and closure of graphitic networks under electron-beam irradiation. *Nature*, 359, 707-709.

von Helden, G., Gotts, N.G. and Bowers, M.T. (1993) Experimental evidence for the formation of fullerenes by collisional heating of carbon rings in the gas phase. *Nature*, 363, 60-63.

Wang, S. and, Buseck, P.R. (1991) Packing of C_{60} molecules and related fullerenes in crystals: a direct view. *Chem. Phys. Lett.*, 182, 1-3.

Wang, Z.L. and Kang, Z.L. (1996) Pairing of pentagonal and heptagonal carbon rings in the growth of nanosize carbon spheres synthesized by a mixed-valent oxide-catalytic carbonization process. *J. Phys. Chem.*, 100, 17725-17731.

Webster, A.S. (1997) The interstellar extinction curve and the absorption spectra of two fulleranes. *Mon. Not. R. Astron. Soc.*, 288, 221-224.

Weltner, W. and Van Zee, R.J. (1989) Carbon molecules, ions, and clusters. *Chem. Rev.*, 89, 1713-1747.

Wopenka, B. (1988) Raman observations on individual interplanetary dust particles. *Earth Planet. Sci. Lett.*, 88, 221-231.

Yeretzian, C., Hansen, K., Diederich, F. and Whetten, R.L. (1992) Coalescence reactions of fullerenes. *Nature*, 359, 44-47.

Zhao, Y., Smalley, R.E. and Yakobson, B.I. (2002a) Coalescence of fullerene cages: Topology, energetics, and molecular dynamics simulation. *Phys. Rev. B*, 66, 195409-1-195409-9.

Zhao, Y., Yakobson, B.I. and Smalley, R.E. (2002b) Dynamic topology of fullerene coalescence. *Phys. Rev. Lett.*, 88, 185501-1-185501-4.

Zinner, E., Amari, S., Wopenka, B. and Lewis, R.S. (1995) Interstellar graphite in meteorites: isotopic compositions and structural properties of single grains from Murchison. *Meteoritics*, 30, 209-226.

Chapter 6

FULLERENES IN METEORITES AND THE NATURE OF PLANETARY ATMOSPHERES

LUANN BECKER
Department of Geological Sciences, Institute of Crustal Studies, University of California, Santa Barbara, California 93106, USA.

ROBERT J. POREDA
Department of Earth and Environmental Sciences, University of Rochester, Rochester, New York 14627, USA.

JOSEPH A. NUTH
Goddard Space Flight Center, Greenbelt, Maryland 20771, USA.

FRANK T. FERGUSON
Chemistry Department, The Catholic University of America, Washington, D.C. 20064, USA.

FENG LIANG
Rice Chemistry and Biochemistry Department, Rice University, Houston, Texas, USA.

W. EDWARD BILLUPS
Rice Chemistry and Biochemistry Department, Rice University, Houston, Texas, USA.

Abstract: We address the hypothesis that fullerenes are an important carrier phase for noble gases in carbonaceous chondrite meteorites. Unlike other proposed carbon carriers, nanodiamond, SiC, graphite and phase *Q*, fullerenes are extractable in an organic solvent. It is this unique property, in fact, this may be why fullerene molecules or fullerene-related compounds were overlooked as a carrier phase of noble gases in meteorites. To further evaluate how fullerenes trap noble gases within their closed-cage structure, we compared the natural

Frans J.M. Rietmeijer (ed.), Natural Fullerenes and Related Structures of Elemental Carbon, 95–121.

meteorite fullerenes to synthetic “Graphitic Smokes” soot. High Resolution Transmission Electron Microscopy used to directly image the fullerene extracted residues clearly showed that C_{60} and higher fullerenes, predominantly C > 100, are indeed the carrier phase of the noble gases measured in the Tagish Lake, Murchison and Allende carbonaceous chondrite meteorites, and synthetic “Graphitic Smokes” material. The implication for the role of fullerenes, which trap noble gases condensed in the atmosphere of carbon-rich stars, is that the true nature of terrestrial planetary atmospheres is presolar in origin. Fullerene, like other carbon carriers, were then transported to the solar nebula, accreted into carbonaceous chondrites and delivered to the terrestrial planets.

Key words: C_{60}; C_{70}; carbonaceous meteorites; carbon-rich stars; graphitic smokes; higher fullerenes; high-resolution transmission electron microscopy (HRTEM); laser desorption-mass spectrometry (LDMS); noble gas-mass spectrometry; phase *Q*; planetary atmospheres; planetary noble gas component; soot

1. INTRODUCTION

1.1 Fullerenes in the Cosmos

The suggestion that the fullerene molecule, C_{60} might be widely distributed in the Universe, particularly in the outflows of carbon stars, was first proposed after the discovery of their exceptional thermal stability and photochemical properties (Kroto et al., 1985). This hypothesis soon led to the search for C_{60}, or its ions (C_{60}^{+}), in carbonaceous chondrites (de Vries et al., 1993) and interstellar spectra (Kroto, 1988; Webster, 1991, 1993) for diffuse interstellar bands (*DIBs*) and infrared (IR) emission bands. Initial studies of fullerenes in carbonaceous chondrites led to negative results and the suggestion that the synthesis of fullerenes in the interstellar medium (ISM) might be inhibited in environments that contain a very high abundance of molecular and atomic hydrogen. It prompted others to suggest that fulleranes ($C_{60}H_X$) might be responsible for certain *DIBs* in the ISM (Kroto, 1989; Webster, 1991; Kroto and Jura, 1992). Foing and Ehrenfreund (1994, 1997) and Ehrenfreund et al. (1997) were the first to show evidence that two diffuse interstellar bands may be due to C_{60}^{+} but, to date, their is no evidence for fulleranes.

Trace amounts of C_{60} and C_{70} (5 to 100 ppb) and possibly fulleranes $C_{60}H_X$ were first reported in the Allende carbonaceous chondrite meteorite (Becker et al., 1994a; Becker and Bunch, 1997).

The fullerenes and fulleranes were detected in separate samples by laser desorption, (linear) time-of-flight, mass spectrometry (LDMS). Fullerenes were also detected in impact residues on the Long Duration Exposure Facility (LDEF) Spacecraft (Radicati di Brozolo et al., 1994). The LDEF fullerenes were either already present in space or may have formed on impact with the satellite when part of the carbonaceous component in the impacting meteoroid was transformed into fullerenes. Using a scanning transmission X-ray microscope for carbon XANES (X-ray Absorption Near-Edge Spectroscopy) analyses, Bajt et al. (1996) reported the possible identification of fullerenes in interplanetary dust particles collected in the stratosphere. However, the strong C-XANES peaks at 285 eV and ~288.6 eV were later reallocated to carbon rings (amorphous carbon, PAHs, graphite) or C=O bonds (Flynn et al., 2003).

The Allende and Murchison CM carbonaceous chondrite meteorites contain several well-ordered graphitic particles that are remarkably similar in size and appearance to carbon onions and nanotubes (Smith and Buseck, 1981, Becker et al., 1993; Zinner et al., 1995) and to the concentric structures in amorphous soot made by electron beam heating (Ugarte, 1992) that could be nested fullerenes. The well-ordered carbon onions in the Allende meteorite (Smith and Buseck, 1981) were 10-50 nm in diameter (cf. Becker et al., 1993), which compares to the diameters of the nested fullerene grains ranging from ca. 5 nm to ca. 50 nm (Ugarte, 1992).

1.2 Trapped Noble Gases in Meteorites

Lewis et al. (1975) were the first to demonstrate that most of the noble gases in the Allende carbonaceous chondrite were retained in a carbon (C)-rich, acid resistant residue generated by acid demineralization of a sample from this meteorite. Lewis et al. (1975) and a companion paper by Anders et al. (1975) suggested that roughly half of the noble gases were trapped in amorphous carbon and that the rest was retained in an unknown Cr,Fe-mineral provisionally called phase *Q*. In fact, Heymann (1986) suggested that fullerenes C_{60} and C_{70} might be a 'carbon-rich phase *Q*' based on their unique three dimensional closed-caged structure. For the next decade several debates about the nature of the carrier phases for noble gases in meteorites continued with some researchers promoting both carbon and the 'chromite' (phase *Q*) carrier phases while others claimed that only C-rich phases were involved in the trapping process.

A consensus that C-rich phases are the predominant carriers for noble gases has now emerged with the recognition of several discrete phases with exotic or presolar noble gases such as nanodiamond, SiC, graphite (Lewis et al., 1987; Huss, 1990; Zinner et al., 1995). Unlike the presolar noble gas carriers, the nature of the planetary atmosphere, noble gas carrier phase has remained somewhat elusive. It was postulated that the terrestrial planet atmosphere noble gases predominantly formed in the solar nebula and were trapped at a set of adsorption sites in the amorphous carbon-rich acid residue, which we here will refer to as the C-rich residue, generated from the bulk material of carbonaceous chondrite meteorites.

Wieler et al. (1991, 1992) proposed that an oxidizable 'C-rich *Q*-phase' is likely the major carrier for the planetary gases. By further oxidizing the C-rich residue of the Murchison and Allende meteorites with HNO_3 using a closed-system stepped etching (CSSE) approach, they suggested that previous measurements of the planetary gases (Anders et al., 1975; Lewis et al., 1975) correlated with a 'C-rich *Q*-phase'. Wieler et al. (1991, 1992) also noted that the isotopic compositions for noble gases in ureilites (a group of differentiated meteorites; see, Kerridge and Matthews, 1988) and several other carbon-rich meteorite types had similar or identical compositions to phase *Q*-gases, suggesting that this component is widespread in the solar nebula.

However, '*Q*' was only identified by the stepped-release pattern of the CSSE residue and so it remains unclear whether '*Q*' is a discrete phase, the result of elemental fractionation, or a mixture of planetary and solar gases released during analysis. For example, the typical $^{20}Ne/^{22}Ne$ value (ca. 10.5) obtained for '*Q*' clearly suggests that it is not a discrete phase but a mixture of solar (ca. 12-13) and planetary (ca. 8.2) components.

We explore herein the role that natural fullerenes played as a unique carrier phase for the encapsulation of the noble gases (He, Ne, Ar, Kr, Xe) formed in the atmosphere of carbon-rich stars and their delivery to the terrestrial planets as accreted components in asteroid parent bodies and carbonaceous chondrite meteorites.

2. THEORIES ON THE ORIGIN OF PLANETARY ATMOSPHERES

Over the past few decades, several issues have dominated the discussion of planetary noble gas patterns:

1. The general resemblance of the noble gas abundances in carbonaceous chondrites to those measured in the Earth's atmosphere,
2. Atmospheric inventories of argon and neon that fall-off significantly with increasing distance from the Sun,
3. Neon isotopic ratios for the Earth's atmosphere that resemble neither the solar nor meteoritic ratio, and
4. Earth's atmosphere has a significantly lower $^{130}Xe/^{36}Ar$ ratio in comparison to carbonaceous chondrite meteorites.

The very low abundances of xenon and argon on Earth have led to the paradox that the planetary noble gas component, which was once thought to be delivered by meteorites to the terrestrial planets, is not found on these planets. The failure to explain the planetary noble gases, especially Ar, Kr and Xe, by measuring the bulk release of these noble gases in carbonaceous chondrite meteorites has led some researchers to conclude that the planetary signature is not linked to a real or discrete phase (e.g. Hunten et al., 1988; Zahnle, 1990; pepin, 1991; Porcelli and Pepin, 1997).

In particular, the inability to explain the missing xenon reservoir, once thought sequestered in rocks of the terrestrial planetary crust (Fanale and Cannon, 1972), has been extremely troublesome. Some models have focused on various fractionation mechanisms of a solar wind component rather than on the accumulation of gas-rich meteorites at planetary surfaces to explain the noble gas inventories in the terrestrial planets. However, these models cannot explain the observed heliocentric gradient of the gases (Hunten et al., 1988; Swindle, 1988) nor do they account for the similar Ne/Ar ratios and the dissimilar planetary Ar/Kr ratios among the terrestrial planets.

More recent studies have focused on hydrodynamic escape to explain the fractionation of gases like neon in the planetary atmosphere and the planet's mantle (Hunten et al., 1988, Pepin, 1991; Porcelli and Pepin, 1997). The escape theory also seems to explain, at least in part, the isotopically heavy argon on Mars but it does not explain the discrepancies observed for the abundances of argon and neon on Venus and the Earth (Pepin, 1991). This situation led to the assumption that some combination of solar wind implantation, absorption and hydrodynamic escape is needed to explain the nature of the observed planetary noble gases.

On the other hand, the true nature of the terrestrial planetary atmospheres may reside in a discrete carrier phase that has not been previously recognized in the bulk release pattern of noble gases from the C-rich residue of carbonaceous chondrite meteorites. Thus, we propose that fullerenes (C_{60} to C_{300}) may be the unique carbon carrier phase for the terrestrial planetary noble gases. We further test our hypothesis by isolating and concentrating fullerenes from the C-rich residue of carbonaceous chondrites and identifying these closed-caged molecules using high-resolution transmission electron microscopy.

3. SYNTHETIC FULLERENES AND THE ENCAPSULATION OF NOBLE GASES

Insight on the possible role of natural fullerenes as a carrier of planetary noble gases can be ascertained from laboratory investigations. Research on synthetic fullerenes, such as produced by carbon-arc evaporation and soot deposition, suggests that the encapsulation of the noble gases in C_{60} and C_{70} is proportional to the partial pressure of the gas at the time of fullerene formation (Saunders et al., 1993; Giblin et al., 1997). The noble gas abundances of endohedral C_{60} adducts relative to C_{60} (i.e. empty fullerenes) were 0.4% He, 0.2% Ne, 0.3% Ar and 0.3% for Kr, respectively, and 0.2% for the three C_{70} adducts. The yield for C_{70} relative to C_{60} is only 15% using the arc evaporation process resulting in fewer abundance measurements for C_{70} (Giblin et al., 1997).

This unique encapsulation of the noble gases into the fullerene-cage structure also has important implications for the origin of the planetary gas signature. If fullerene had formed in a solar gas reservoir then the abundances of ^{3}He to ^{36}Ar would be closer to 1 compared to the abundances in a He-depleted carbon-rich star that would have much more ^{36}Ar to ^{3}He. Furthermore, C_{60} and C_{70} exclude Xenon by a factor of 10-30 times relative to He or Ar because Xe is too large to fit efficiently into the C_{60} cage (Giblin et al., 1997). The experimentally observed abundances of Xe@C_{60} relative to C_{60} was ca. 0.008%, and the abundance of Xe@C_{70} relative to C_{70} was ca. 0.04% (Giblin et al., 1997).

Only the smaller fullerenes, C_{60}, C_{70}, C_{76}, C_{84}, exclude Xe but carbon molecules >C_{100} have sufficient interior diameter to accommodate Xe. This preferential exclusion of Xe could explain the deficiency of Xe in the Earth's atmosphere relative to Ne, Ar and Kr (Fanale and Cannon, 1972) when fullerene is a major carrier of the planetary noble gases. Our previous investigations of fullerenes in carbonaceous chondrites showed that carbon cages >C_{100} are the dominant component in the solvent-extracted fullerene residue and the acid residual meteorite fractions. That is, LDMS analyses of the fullerene residues (not the acid residue of the meteorite) indicate that these molecules are present in abundance in carbonaceous chondrite meteorites (Becker et al., 2000). One further question that may be difficult to answer remains. Namely did the CI carbonaceous chondrites, that are widely considered to represent the solar nebula composition, originally condense with a large fraction of C_{60} and C_{70} that was ultimately lost during 4.56 Ga of processing on the meteorite parent body? Or, are most of the fullerenes formed in stellar processes at carbon-rich stars dominated by large cages $>C_{100}$?

4. FULLERENE AS A NOBLE GAS CARRIER IN METEORITES

The major focus of the fullerene research by Becker and Poreda has been to measure the noble gases encapsulated within the caged structure of this new carbon carrier phase (Becker et al., 2000, 2001; Poreda and Becker, 2003) and compare it to the myriad of components found in the bulk meteorite acid residues of the carbonaceous chondrite meteorites Allende and Murchison and Tagish Lake that is a new intermediate CI2 (Brown et al., 2000) meteorite. These meteorites have abundant noble gases, typically with a planetary signature that dominates the low temperature (<800 °C) step-release of the bulk acid residue (Table 1).

Table 1. Trapped noble gases in meteorites: Isotropic compositional data for Helium, Neon and Argon

	$^{3}He/^{4}He$ (x10^{-4})	$^{20}Ne/^{22}Ne$	$^{21}Ne/^{22}Ne$	$^{36}Ar/^{38}Ar$	$^{40}Ar/^{36}Ar$
Solar	3.9-4.3	12-13	0.033-0.042	5.20-5.37	~1
Planetary A	1.43	8.2-8.9	0.024-0.035	5.31	0.001
Earth Atmosphere	0.0140	9.80	0.0290+	5.32	295
Phase Q	1.46-1.50	10.3	0.029	5.29	~1
Fullerene Murchison	1.42-1.66	8.5-9.2*	0.027-0.028	5.31	6.5*
Fullerene Allende	1.28-1.55	7.5-8.8	0.028	5.32	3.2*
Fullerene Tagish Lake	1.42-1.50	8.5-9.9*	0.027-0.029	5.31	0.6

Notes.

1. Helium, Ne and Ar ratios for Solar, 'Planetary A', Earth Atmosphere and Q were taken from Swindle (1988). The fullerene noble gases are from the low temperature stepped release (400 °C to 800 °C) of the extracted fullerene separate.
2. The star (*) indicates $^{20}Ne/^{22}Ne$ ratios measured for Murchison and Tagish Lake that vary between 'Planetary A' and 'Earth Atmosphere' ratios, which we attribute to absorption of the atmospheric component to our fullerene separate (i.e. most of the atmosphere is released in the lower temperature steps indicative of an absorbed component). Better handling of the Allende fullerene separate resulted in $^{20}Ne/^{22}Ne$ ratios that are closer to the planetary signature. Neon might also reside in the smaller fullerene cages (C_{60} to C_{100}) and is less efficiently trapped in the higher fullerenes ($>C_{100}$ up to C_{300}) in comparison to the heavier noble gases.
3. The star (*) indicated for $^{40}Ar/^{36}Ar$ in Murchison and Allende also reflect an absorbed atmospheric component, hence values >1, however, Tagish Lake had a value of <1. Clearly, Ar is more abundant and, thus, less affected by an absorbed component in comparison to Ne. Future studies of individual fullerene molecules are needed to further evaluate the absorption and/or encapsulation of noble gases in fullerenes.

These meteorites also contain an extractable fullerene component that can be isolated and purified from the same bulk material. Several papers reported on the finding of non-atmospheric He and Ar in the Murchison and Allende meteorites (Becker et al., 2000, 2001) and the Tagish Lake meteorite (Pizzarello et al., 2001) with $^{3}He/^{36}Ar^*$ ratios that approach the planetary ratio of 0.01 compared to a solar ratio of ca. unity, calculated according to the formula

$$^{36}Ar^* = {}^{36}Ar_m\,[1-(40/36)_m/295.5]$$

In addition, the measured neon isotopic composition ($^{20}Ne/^{22}Ne$) of the fullerene fraction was at, or below, the present Earth atmospheric ratio of 9.8 (range: 8.5-8.9) most consistent with a planetary gas carrier ($^{20}Ne/^{22}Ne$ = 8.2). The higher Ne isotope ratios are due to mixing with air. New, improved techniques for handling, measuring and separating fullerenes should result in lower experimental blanks and $^{20}Ne/^{22}Ne$ values closer to the planetary ratio of ca. 8.2, thus lending support to the suggestion that fullerene is also a carrier of the planetary neon.

5. THE FULLERENE CARRIER HYPOTHESIS

Some researchers (e.g. Heymann, 1995, 1997) have cast doubt on the possible role of fullerene as a carrier of planetary noble gases, in part due to the inability to obtain a measurement of the fullerene mass spectrum and the encapsulated noble gases on a single sample. The analytical techniques we have adapted for our studies require two separate mass spectrometers. Thus, an isolated fullerene extract is dissolved in toluene and a small aliquot (ca. 10%) of the solution goes to a laser desorption mass spectrometer while the remaining (90%) is used for the noble gas analyses. The LDMS spectrum (Becker et al., 2000, 2001) of the larger fullerene cages does not allow for isotopic resolution of endohedral fullerenes; not even for Ar that is the most abundant noble gas. It was suggested by Buseck (2002) that minor amounts of other carbon carrier phases (e.g. nanodiamond) might have survived extraction and thus could account for the noble gases in residues of the Murchison and Allende meteorites studied by Becker et al. (2000; 2001). Until it will be demonstrated that nanodiamonds are soluble, or can be suspended as a colloid, in a toluene solution, we submit that nanodiamonds are not a noble gas carrier in these meteorite fractions. As we will show below, high-resolution transmission electron microscopy (HRTEM) can be used to resolve this issue. The identification of fullerenes using LDMS was questioned, although Buseck et al. (1992) and Buseck (2002) have used a similar technique to identify C_{60}^+ and C_{70}^+ in shungite and for fullerene identification in fulgurite (Daly et al., 1993). Mossman et al. (2002) used LDMS to replicate the report of C_{60}^+ in the Sudbury Onaping breccia made by Becker et al. (1994b). In each case, the fullerene identifications were supported by additional techniques (e.g. Electron Impact-Mass Spectroscopy, HRTEM). Thus, the notion that fullerenes were an artifact of the desorption process itself does not appear to be a tenable argument. LDMS is the method of choice for the identification of large fullerene cages in synthetic soot yielding similar mass spectra when extracted in organic solvents or when functionalized (Ying et al., 2003) to release fullerenes from the bulk soot material (Sadana et al., 2005). Thus, we formulated the hypothesis that fullerenes formed in the atmosphere of carbon-rich stars are the major carrier of the observed planetary atmosphere noble gases in carbonaceous chondrite meteorites and therefore fullerenes played a key role in the evolution of planetary atmospheres.

6. FULLERENES IN SYNTHETIC GRAPHITIC SMOKES

To further demonstrate the validity of LDMS fullerene detection and isolation (solvent extraction), we carried out new studies of a synthetic, potential noble gas carbon carrier referred to as Graphitic Smokes (*GS*). Slightly modified from the original experimental setup (Olsen et al., 2000), these smokes are produced by electrically heating, at high currents, a ca. 5 cm-long graphite rod thinned to increase its electrical resistance. For each experiment this graphite rod was heated fairly quickly to produce 'hot cylinder' of carbon vapor in a pre-mixed noble gas atmosphere. The condensed smoke formed a deposit on an aluminum collection plate that was positioned in the condensation chamber 15.2 cm above the continuously heated graphite rod. This thin graphite rod allows carbon to condense more uniformly in comparison to the arc-discharge evaporation process typically used in fullerene synthesis (among many others, Krätschmer et al., 1990; Rotundi et al., 1998). In the arc-discharge process bits of the solid graphite rod can be sputtered off to the walls of the apparatus resulting in lower soot yields. We note that Kimura et al. (1995) used the arc-discharge technique for the preferential production of higher fullerenes that were was also produced by Rotundi et al. (1998) wherein soot contained higher fullerenes (Rietmeijer et al., 2004). The arc-discharge process is highly non-linear in comparison to the simple evaporative heating method (Olsen et al., 2000) that we have adapted for our studies of synthetic noble gas carbon carriers. Mixtures of known compositions of noble gases can be introduced into the *GS* apparatus at the time soot is condensing from the carbon/noble gas vapor. This material can then be step-heated in a noble gas analyzer to evaluate the compositions of the gases in the bulk material.

Previous investigations of the bulk *GS* sample that was analyzed for Xe yielded values as high as 13.7 x 10^{-6} cm^3 STP/gm of ^{132}Xe (Olsen et al., 2000). This value is ca. 2 times magnitude greater than typical ^{132}Xe amounts for other synthesized carbonaceous residues (Ott et al., 1981). While this result is certainly exciting, the origin of the carrier for *GS* bulk material remained unknown.

We obtained *GS* soot material synthesized in a noble gas (*NG*) mixture of 49% Ne, 49% Ar, 1% Xe and 1% Kr with the balance provided from He for a 300-torr atmosphere maintained during *GS* soot condensation (Olson et al., 2000). For our fullerene investigations (Becker et al., 2000, 2001; Poreda and Becker, 2003) we developed a two-step extraction method for isolating fullerenes from natural samples,

1. Toluene extraction that separated predominately C_{60}, C_{70} up to C_{100}, and
2. Second extraction step with a high boiling solvent (e.g. *1,2,4 trichlorobenzene* or *1,2,3,5 tetramethylbenzene*) to separate higher fullerenes in the C_{100} to C_{300} range.

This method was successful at isolating fullerenes in the clay sediments from the Cretaceous-Tertiary boundary and the Permian-Triassic boundary (Becker et al., 2000, 2001) and in carbonaceous chondrites (Becker et al., 2000; Pizzarello et al., 2001). Both LDMS spectra and noble gas measurements of He in small fractions of toluene extracted *GS* residue (100 micrograms) indicated that fullerene, C_{60}^+ up to C_{100}^+, are the carrier for the *GS* noble gases. The yield for He in the toluene fullerene residue (Fig. 1a) corresponds to one ^4He per 880,000 fullerene molecules with a similar stepped-release pattern and the same abundances measured in synthetic fullerenes produced in the arc-discharge experiments (Krätschmer et al., 1990; Saunders et al., 1993).

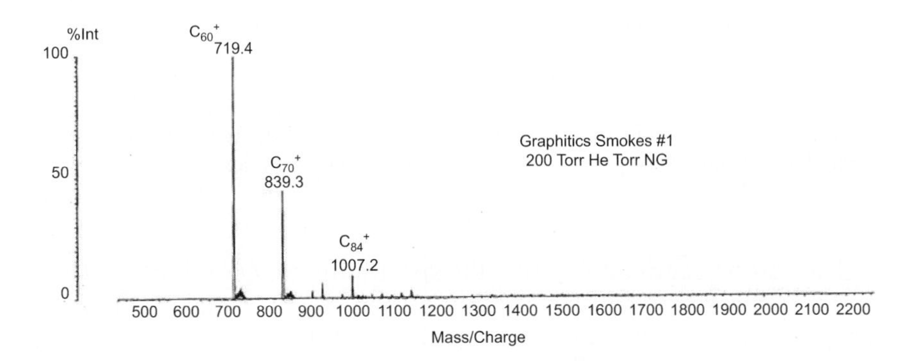

Figure 1a. Laser desorption-mass spectrum (% intensity vs. mass/charge ratio) for the toluene extraction of "Graphitic Smokes" material obtained at 0 Torr He and 200 Torr noble gas mixture (*NG*) that yielded mostly C_{60}, C_{70} and some larger fullerene cages up to C_{96}

The LDMS spectrum of *GS* bulk material extracted a second time using *1,2,4 trichlorbenzene* (*TCB*) revealed higher fullerenes up to C_{300}^+ in this material (Fig. 1b).

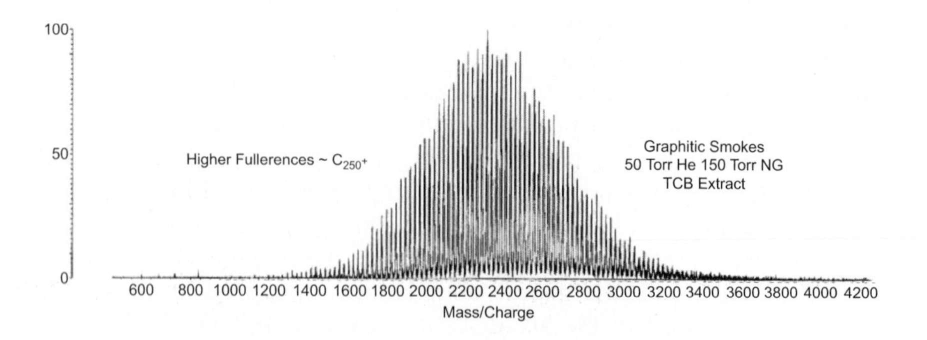

Figure 1b. Laser desorption-mass spectrum (% intensity vs. mass/charge ratio) for the *TCB* extraction of "Graphitic Smokes" material obtained at 50 Torr He and 150 Torr *NG* mixture showing fullerene cages ranging from C_{100} to C_{300}

Varying the partial pressure of the noble gas (*NG*) mixture in the He atmosphere resulted in different abundances of the smaller fullerene cages in these toluene-extracted residues. Preliminary measurements of He and Ne in a small amount of toluene-extracted *GS* residue (100 micrograms) indicate that fullerenes C_{60} and C_{70} are the predominant carrier for these gases. The higher fullerenes (see, Fig. 1b) likely retain most of the heavier noble gases (i.e. Kr, Xe) and experiments are planned to examine these gases in the separated fullerene residues.

The carbon cages dominating the spectra for the *GS* fullerenes (Fig. 1) are remarkably similar to the carbon cages isolated in some carbonaceous chondrites (Becker et al., 2000, Pizzarello et al., 2001). These results for *GS* fullerenes using LDMS indicate that fullerenes are present in the *GS* extracted residues and are not an artifact of the desorption process. We were able to sublime fullerenes directly from the solid *GS* soot material using a simple quartz tube apparatus under a partial vacuum that is also lending support to the suggestion that fullerenes are generated in the *GS* apparatus. Finally, this extraction methodology for the isolation of fullerenes in this size range is quite efficient. Therefore we propose here that the data in Fig. 1 when taken in combination the observed very efficient trapping of Xe in condensed *GS* soot points to a significant role for the larger fullerene cages to encapsulate the heavy noble gases.

7. TESTING FOR OTHER CARRIERS IN EXTRACTED FULLERENE METEORITE RESIDUES

In order to test the hypothesis that other carriers may be responsible for the noble gas signatures measured in fullerene extracted residues we began a systematic investigation of the low temperature (100-800 °C) release of gases from Murchison, Allende and Tagish Lake fullerenes and the remaining bulk acid resistant residues. As stated earlier, previous research (Wieler et al., 1991, 1992) demonstrated that digestion of the bulk meteorite acid residues with an oxidizing acid releases copious quantities of planetary gas with a unique ^{20}Ne/^{22}Ne ratio of ca. 10.5 (phase *Q*). Becker et al. (2000, 2001) and Poreda and Becker (2003) suggested that because oxidizing acids destroy the fullerene cages, the release of the encapsulated gases during Wieler's acid digestion experiments could be taken as indirect evidence for the existence of a fullerene component that has been mixed with solar gases during analysis. Other studies have either

1. Ignored the low temperature fraction by beginning the step heating of meteorite residue at 800 °C (Wieler et al., 1991, 1992; Huss, 1990),
2. Destroyed the fullerene carrier with an oxidizing acid in the search for presolar diamonds (i.e. Lewis et al., 1987), or
3. Sublimated away fullerenes during extended 'bake-outs' (Lewis et al., 1975) since fullerenes sublime from the bulk meteorite 'C-rich' acid residue at low temperatures under a partial vacuum.

Initial results for the fullerene fractions show the release of a planetary atmosphere component at temperatures from 500-800 °C in the Murchison, Allende and Tagish Lake meteorites. The ^{3}He, ^{22}Ne, ^{36}Ar and ^{84}Kr noble gas ratios (Table 1) follow the same general trend as these planetary noble gas ratios (Swindle, 1988) with the ^{20}Ne/^{22}Ne ratios being a mixture of air and planetary Ne-A to a range of 8.2-8.9 (Table 1). These trends are shown for ^{3}He release from the Murchison and Tagish Lake fullerenes and the Tagish Lake meteorite acid resistant residue that represents the 'planetary' pattern (Fig. 2).

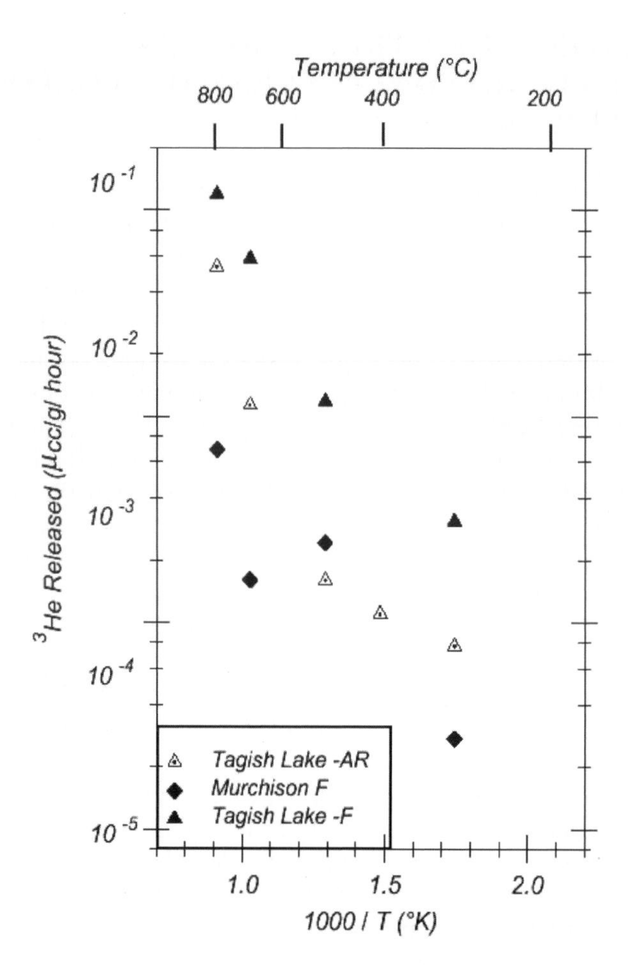

Figure 2. Temperature (T)-dependent ^{3}He release for the extracted Murchison (solid diamonds) Tagish Lake fullerenes (solid triangles) samples and the Tagish Lake meteorite acid resistant residue (open triangle with dot). Similar patterns for ^{22}Ne, ^{36}Ar and ^{84}Kr noble gas releases for the Murchison and Tagish Lake fullerenes also trend in the general direction as the 'Planetary' Tagish Lake meteorite acid resistant residue suggesting fullerenes are the carrier of planetary atmosphere gases in this meteorite

The much lower value for Ne-A supports the notion that *Q* gases are not a discrete phase, rather the *Q* gases are a mixture of planetary and solar gases released upon analysis. The measured ^{36}Ar content is >75% non-atmospheric ($^{40}Ar/^{36}Ar \approx$ 1-10; measured Earth atmosphere is 295; Table 1) and the $^{38}Ar/^{36}Ar$ ratio is atmospheric (ca. 5.31), which is also consistent with a planetary atmosphere signature. The stepped-release curve for the bulk meteorite residue (Fig. 3) is similar to the pure fullerene release (Fig. 2) suggesting that fullerene is likely present in the gas release of the C-rich meteorite acid residues (Fig. 3).

Throughout the low-temperature step-release, the abundances of the noble gases and the planetary signature in the isolated fullerene fraction remained the same. Thus, an absorbed component or gases trapped in other carriers (nanodiamond, silicates, chromite) cannot explain this release pattern characteristic of the fullerene fraction. Only at 800 °C and higher do the other gas carriers become dominant in these bulk meteorite acid residues, as the amount of gas released at each temperature step drops dramatically. The overall signature of the high temperature release is dominated by "Ne-A" typified by the diamond carrier (Lewis et al., 1987).

The low temperature analyses for $\delta^{136}Xe(\%)_{Air} \approx$ -10-12 ‰ in the Tagish Lake and Murchison bulk acid resistant residues is clearly consistent with the composition of the Earth's early atmosphere (Ozima and Podesek, 1983). The amount of Xe relative to Ar in these low temperature steps is some 20-times lower than the planetary ratio of $^{132}Xe/^{36}Ar$ and it is at or below the ratio in air. If this finding could be confirmed, there would be no need to invoke a "missing Xe" reservoir on Earth that, for example, was suggested being adsorbed on shales (Fanale and Cannon, 1972), it was simply never present in the first place.

Instead the Earth's atmosphere was supplied with noble gases from a fullerene component that was Xe-deficient. Subsequent mixing with solar-type gases from the interior of the Earth ultimately produced the atmosphere that exists today. This new theory would support a late veneer from meteoritic matter to the early Earth (Hunten et al., 1988) that would provide the fullerene planetary component needed instead of invoking two distinct noble gas reservoirs, i.e. one reservoir for the fractionation of Xe and another for the remaining noble gases (Pepin, 1991).

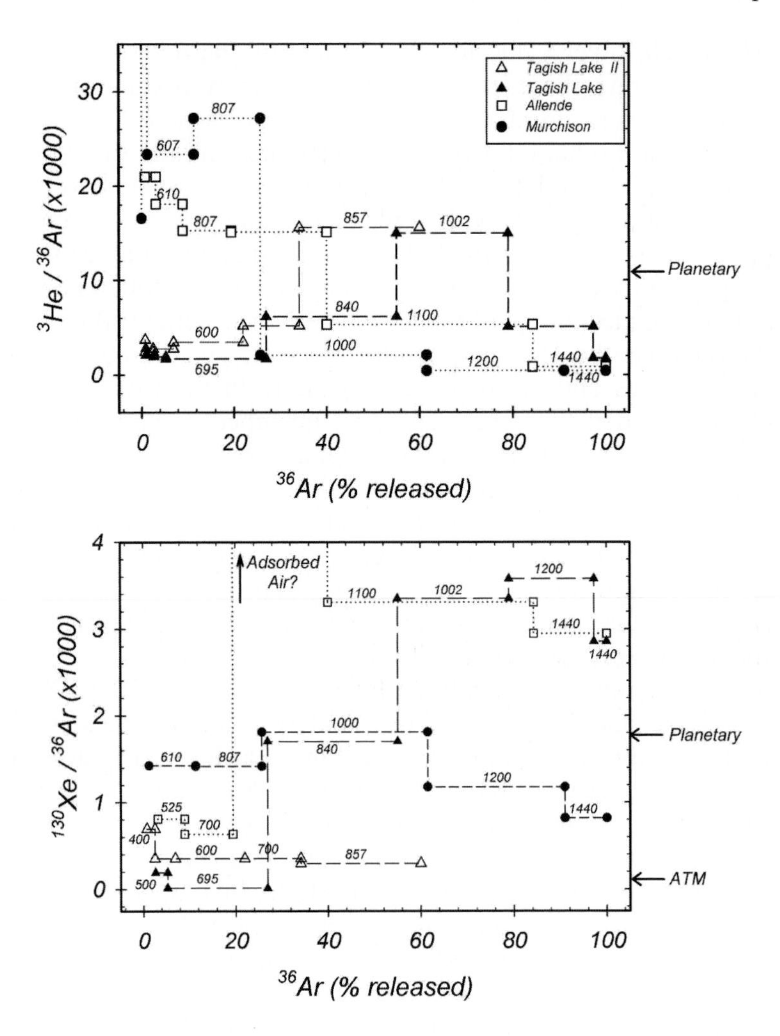

Figure 3. The stepped-heating release patterns for Xe and Ar presented as the $^{130}Xe/^{36}Ar$ versus ^{36}Ar released in C-rich acid resistant residues of the Tagish Lake (#1 and #2), Murchison and Allende carbonaceous chondrite meteorites. Between 600 °C and 800 °C, all samples had sub-planetary $^{130}Xe/^{36}Ar$ ratios. For example, Tagish Lake heating steps are characterized by $\delta^{136}Xe$ anomalies of -10%, $^{40}Ar/^{36}Ar$ ratios of 1-10 and sub-planetary $^{3}He/^{36}Ar$. The "sub-planetary" $^{130}Xe/^{36}Ar$ released at low temperature could be an indicator of a fullerene component since smaller fullerenes exclude Xe from their cages (*ATM*: atmosphere, Table 1)

8. HRTEM OF FULLERENES RESIDUES REVISITED

Recent HRTEM studies have interpreted single-walled and multiple-walled structures in acid resistant residues of the Allende meteorite as closed fullerenic carbon nanostructures (Harris et al., 2000, Harris and Vis, 2003). These structures are ca. 2-5 nm in diameter (Harris and Vis, 2003) which would be consistent with $>C_{800}$ and thus larger then the fullerenes C_{60-300} in some carbonaceous chondrites (Becker et al., 2000; Pizzarello et al., 2001), noting that a single-walled ring of 1.3 nm in diameter is roughly equivalent to 200 carbon atoms, or C_{200}. Fullerenic carbons such as multi-walled nanotubes and onions were observed using HRTEM in arc-discharge produced soot and the presence of C_{60} fullerene, extracted with toluene, was verified using high performance liquid chromatography (Rietmeijer et al., 2004). The larger cages that were not observed using HPLC were deemed to be too large, up to C_{1500}, fullerenes (Rietmeijer et al., 2004). Goel et al. (2004) using HRTEM identified single walled ring structures in a high temperature (benzene/argon) flame-generated soot having a range of diameters consistent with fullerenes C_{36} to C_{176}, including C_{60}. The range of fullerenes in these studies, C_{36} to C_{1500}, includes the size range for the higher fullerenes reported in the Tagish Lake meteorite, C_{60}^{+} to C_{160}^{+} (Becker et al., 2000; Pizzarello et al., 2001).

Encouraged by the results obtained by Harris et al. (2000), Harris and Vis (2003), Goel et al. (2004) and Rietmeijer et al. (2004, 2005), on demineralized meteorite solid matter and condensed soot, we used HRTEM to search for similar, single-walled, fullerene nanocarbon structures in our *isolated fullerenes* from the acid resistant residue of carbonaceous chondrites (Becker et al., 2000; Poreda and Becker, 2003), as well as in fullerene extracts of the synthetic *GS* soot samples. First, it is worth mentioning that the demineralization process that we adapted from Robl and Davis (1993) for an enhanced solvent extraction of the higher fullerenes has proven successful in generating a much cleaner residue that led to the observation of these single-walled fullerenes and double-walled fullerenic carbons in the Allende meteorite (Harris et al., 2000; Harris and Vis, 2003). Previous investigations of the bulk Allende acid residue used a traditional HF-HCL demineralization method. It is known that this procedure will form insoluble salts that could trap carbonaceous material (Robl and Davis, 1993) possibly causing a bias in the complete fullerene content of acid resistant residues when studied by HRTEM. Harris et al.

(2000) and Harris and Vis (2003) added the HF-H_3BO_3 step we also use to remove the insoluble residues resulting in a much more-abundant yield of fullerene carbons.

9. HRTEM OF FULLERENES IN *GS* SMOKE AND THE TAGISH LAKE METEORITE

9.1 Synthetic *GS* Smoke Fullerene

The fullerene residue was first reconstituted into chloroform and then deposited as a drop of solution onto a lacey carbon thin-film that supports the sample during electron microscope analyses. Upon evaporation of the chloroform, samples were analyzed using a JEOL 2010 transmission electron microscope operating at an accelerating voltage of 100 keV. The HRTEM images of the *GS* higher fullerene residue show several single-walled carbon structures clumped together such as in the *TCB* and Toluene fullerene extracts of the *GS* smoke produced at 150 Torr He and 50 Torr of noble gas mixture (Fig. 4).

The black arrows in Fig. 4a indicate discrete structures indicative for the higher fullerenes that are easily discernable from the lacey carbon film. Because the fullerene structures are fairly tightly packed, it was difficult to get any measurements of the actual sizes of these ring structures. Still, all ring structures imaged (Fig. 4a) are at least larger than C_{100} since the LDMS spectrum indicates no fullerenes $<C_{100}$ present in the extract.

The procedure was repeated for the *GS* toluene-extracted residue that contains mostly C_{60} and C_{70}. The HRTEM image (Fig. 4b) shows several single-wall structures comparable to the size of C_{60} and some larger closed structures. One of the structures in Fig. 4b is very similar to the calculated image for C_{60} in the [100] direction in the flame-generated soot studied by Goel et al. (2004). It is indicated by the arrow and labeled in Fig. 4b.

The identification of single-walled C_{60} ring structures in the *GS* toluene extract is not surprising considering the very high concentration of C_{60} and other, smaller fullerene cages in the LDMS mass spectrum (Fig. 1a). The HRTEM results confirm that (1) our extraction methodology is preferentially extracting C_{60} and the higher fullerenes and (2) these fullerenes are likely the carrier of the noble gases measured in the *GS* material.

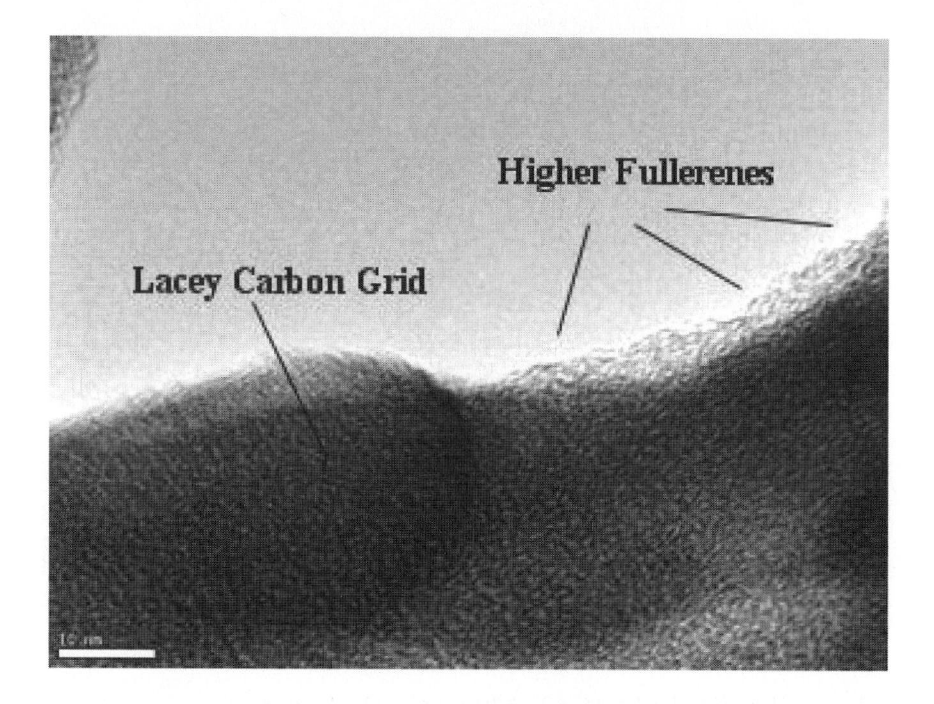

Figure 4a. High Resolution Transmission Electron Microcopy image of the *TCB* fullerene extract of a *GS* smoke. The fullerene cage sizes in Fig. 4a are larger than C_{100} as indicated by the LDMS spectrum (Fig. 1b). The fullerene-containing sample is shown draped over the lacey carbon grid that cases the darker band across this image. Scale bar is 10 nm

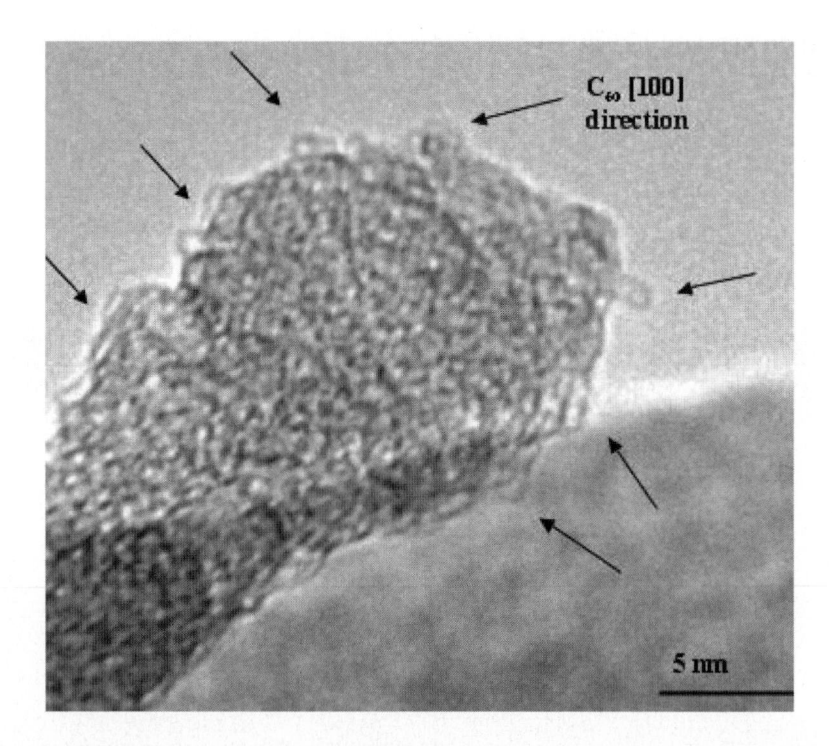

Figure 4b. High Resolution Transmission Electron Microcopy images of the Toluene fullerene extract of the *GS* smoke. Toluene fullerene-extract shows several smaller single-walled and some larger structures as indicated in the LDMS spectrum (Fig. 1a). C_{60} in the [100] direction can be clearly seen in the image

9.2 Tagish Lake (CI2) Meteorite Fullerene

The HRTEM image for the Tagish Lake *TCB*-extracted residue (Fig. 5) shows small single-walled rings both individually and in clusters that are consistent with the previous LDMS identification by Pizzarello et al. (2001) of fullerenes between C_{60}^+ and C_{180}^+ with a predominance of C_{60} in the Tagish Lake extracted residue. The higher fullerenes tend to form tight clusters making it difficult to get precise size measurements of the ring structures. To our knowledge, this is the first time natural fullerenes were identified in an extracted meteorite fullerene separate using the combination of HRTEM and LDMS.

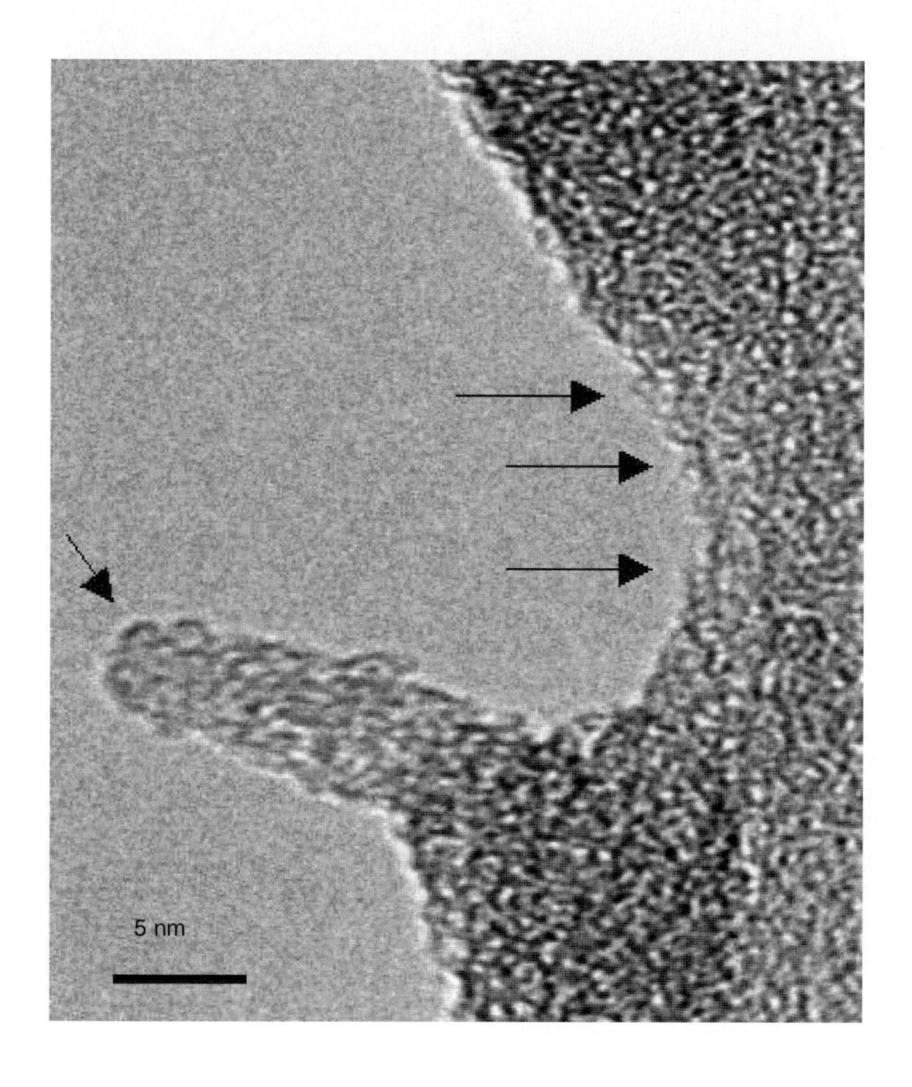

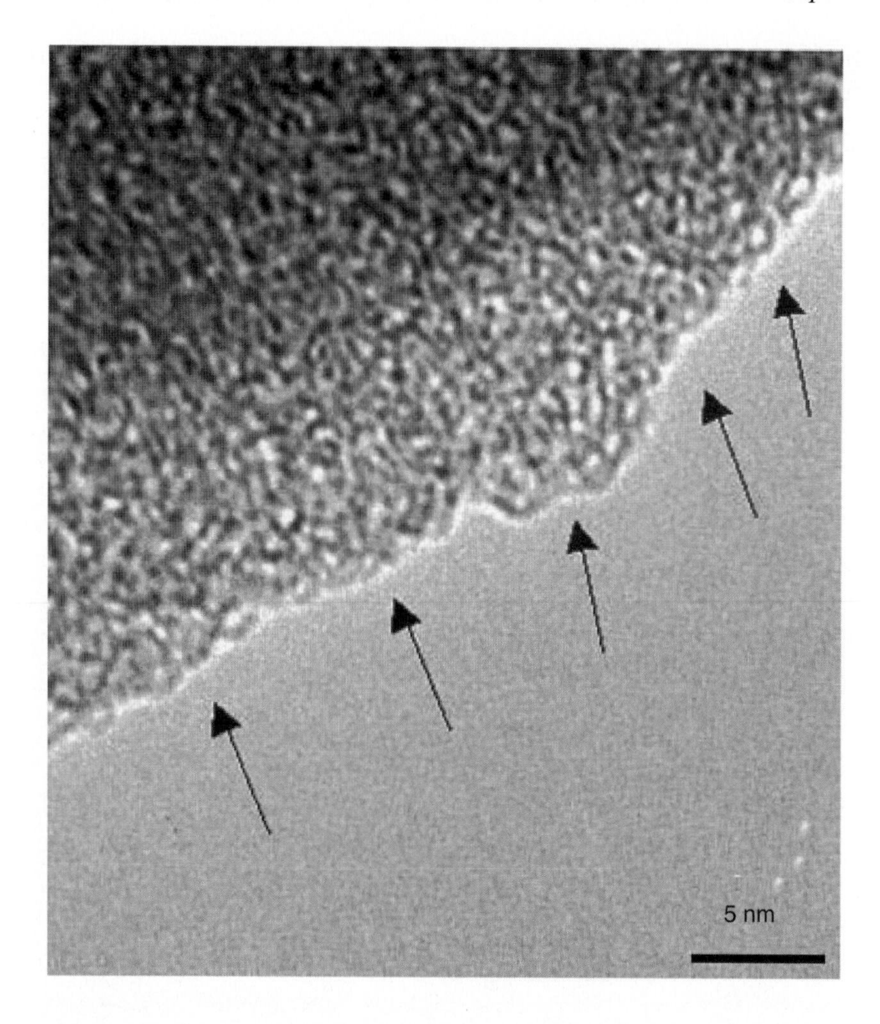

Figure 5. High-resolution Transmission Electron Microscopy images of Tagish Lake (meteorite) fullerenes in the *TCB*-extracted residue taken with a JEOL 2010 microscope operating at 100kV

As can be seen in both images, several single walled structures (arrows) are observed along the fringe of the lacey carbon film. The sizes of these fullerenes are dominated by small, 0.8 to 1.2 nm, ring structures as indicated in the LDMS spectrum of the same fullerene extract (Pizzarello et al., 2001).

The obvious advantage to imaging a discrete phase (i.e. extracted/isolated component) rather than a bulk residue is that the identification of single-walled or double-walled closed carbon structures can only be fullerene. We have begun a HRTEM search for some single-walled ring carbon structures in the Allende and Murchison meteorites with encouraging preliminary results (Becker et al., this volume).

10. SIGNIFICANCE OF FULLERENES AS THE PLANETARY NOBLE GAS COMPONENT

The HRTEM results support our hypothesis that fullerenes played a key role in the evolution of planetary atmospheres. We then postulate that fullerenes in primitive carbonaceous chondrite meteorites potentially resolve the mystery related to a planetary gas reservoir because it links this putative reservoir, and fullerene formation, to carbon-rich stars that predated our own solar system. The first studies conducted to search for fullerenes in Allende meteorite led to the discovery of nanodiamonds predicted to have formed in a presolar (or interstellar) environment (Lewis et al., 1987). The procedures used in these searches would have destroyed any fullerenes that might have been present in the nanodiamond acid resistant residue. The identification of fullerenes as a noble gas carrier suggests that previous estimates of the abundances of other noble gas carriers such as graphite and amorphous carbon were overestimated.

The stepped-release patterns of the C-rich meteorite acid residues (Fig. 3) and the fullerene separates (Fig. 2) is comparable to the amount of gas released from the Allende meteorite that was used by Lewis et al. (1987) to estimate its amount of nanodiamonds at 100 to 400 ppm. As pointed out by Nuth (1985), the most abundant meteoritic carbon component is in the form of kerogen that is estimated to be (2-3) x 10^4 ppm of the total organic carbon content, rather then graphite. It is unclear how kerogen could retain noble gases within its complex macromolecular structure during post-formation processing in the ISM and in the solar nebula. Because fullerene would be the only soluble carbon carrier, it is likely that fullerenes were embedded in kerogen material that is not being efficiently extracted from the meteorite acid resistant residues. If this is indeed the case then fullerene rather than any other solid form of carbon is the predominant carrier of the planetary atmosphere noble gas component.

There remain many unanswered questions about the role of fullerene as a carrier of the planetary atmosphere component and about the possible delivery of these carbon molecules to the terrestrial planets such as to the early Earth. For example, it is particularly important to fully assess the presence and abundances of fullerenes C $>C_{100}^{+}$ since it would now appear that these large carbon cages, that initially formed in carbon-rich stellar atmospheres of stars other than the sun, do play a significant role in the encapsulation of the noble gases and their delivery to the atmospheres of the terrestrial planets.

11. CONCLUSIONS

The combination of LDMS and HRTEM analyses reveal the presence of C_{60}, C_{70} and higher fullerenes up to C_{300} in synthetic "Graphitic Smokers" soot and in the Tagish Lake CI2 carbonaceous chondrite. The extraction methodology and chemical analytical techniques we have summarized allow the identification of indigenous fullerenes in natural, geologic and meteorite, samples. The common presence and high abundances of fullerenes C_{60} to C_{400} in carbonaceous chondrite meteorites lend support to our hypothesis that fullerenes, more than any other carbon phase, are the carriers of the noble gases of planetary atmospheres.

Acknowledgements: We thank the Rick Smalley at Rice University, Houston, Texas, for the use of the High Resolution Transmission Electron Microscope for imaging fullerenes in our samples. We thank the NASA Exobiology program and National Science Foundation (CHE-0011486) and the Robert A. Welch Foundation (C-0490) for supporting the research.

12. REFERENCES

Anders, E., Higuchi, H., Gros, J., Takahashi, H. and Morgan, J.W. (1975) Extinct superheavy elements in the Allende meteorite. *Science*, 190, 1262-1268.

Bajt, S., Chapman, H.N., Flynn, G.J. and Keller, L.P. (1996) A possibility of the presence of C_{60} in interplanetary dust particles (abstract). *Meteorit. Planet. Sci.*, 31 (suppl.), A11.

Becker, L. and Bunch, T.E. (1997) Fullerenes, fulleranes and polycyclic aromatic hydrocarbons in the Allende Meteorite. *Meteorit. Planet. Sci.*, 32, 479-487.

Becker, L., McDonald, G.D. and Bada, J.L. (1993) Carbon onions in the Allende meteorite. *Nature*, 361, 595.

Becker, L., Bada, J.L., Winans, R.E. and Bunch T.E. (1994a) Fullerenes in the Allende meteorite. *Nature*, 372, 507-508.

Becker, L., Bada, J.L., Winans, R.E., Hunt, J.E., Bunch, T.E. and French, B.E. (1994b) Fullerenes in the 1.85 Billion-Year-Old Sudbury Impact Structure. *Science,* 265, 642-645.

Becker, L., Bunch, T.E. and Allamandola, L. (1999) Higher fullerenes in the Allende meteorite. *Nature,* 400, 227-228.

Becker, L., Poreda, R.J. and Bunch, T.E. (2000) Fullerene: A new extraterrestrial carbon carrier phase for noble gases. *Proc. Natl. Acad. Sci.*, 97, 2979-2983.

Becker, L., Poreda, R.J., Hunt, A.G., Bunch, T.E. and Rampino, M. (2001) Impact event at the Permian-Triassic boundary: Evidence from extraterrestrial noble gases in fullerenes. *Science*, 291, 1530-1533.

Brown, P.G., Hildebrand, A.R., Zolensky, M.E., Grady, M., Clayton, R.N., Mayeda, T.K., Tagliaferri, E., Spalding, R., MacRea, N.D., Hoffman, E.L., Mittlefehldt, D.W., Wacker, J.F., Bird, J.A., Campbell, M.D., Carpenter, R., Gingerich, H., Glatiotis, M., Greiner, E., Mazur, M.J., McCausland, P. JA., Plotkin, H., and Mazur, T.R. (2000) The fall, recovery, orbit, and composition of the Tagish Lake meteorite: a new type of carbonaceous chondrite. *Science*, 90, 320-325.

Buseck, P.R. (2002) Geological fullerenes: review and analysis. *Earth Planet. Sci. Lett.*, 203, 781-792.

Buseck, P.R., Tsipursky, S.J. and Hettich, R. (1992) Fullerenes from the geological environment. *Science*, 257, 215-217.

Daly, T.K., Buseck, P.R., Williams, P. and Lewis, C.F. (1993) Fullerenes from a fulgurite. *Science*, 259, 1599-1601.

de Vries, M.S., Wendt, H.R., Hunziker, H., Peterson, E. and Chang, S. (1993) A search for C_{60} in carbonaceous chondrites. *Geochim. Cosmochim. Acta*, 57, 933-940.

Ehrenfreund P., Cami, J., Dartois, E. and Foing, B.H. (1997) Diffuse interstellar bands towards BD63+ 1964 A new reference target. *Astron. Astrophys.*, 318, L28-L31.

Fanale, F.P. and Cannon, W.A. (1972) Origin of planetary primordial rare gas: the possible role of adsorption. *Geochim. Comochim. Acta*, 36, 319-328.

Flynn, G.J., Keller, L.P., Feser, M., Wirick, S. and Jacobsen, C. (2003) The origin of organic matter in the solar system: Evidence from the interplanetary dust particles. *Geochim. Cosmochim. Acta*, 67, 4791-4806.

Foing, B.H. and Ehrenfreund, P. (1994) Detection of two interstellar absorption bands coincident with spectral features of C_{60}^+. *Nature*, 369, 296-299.

Foing, B.H. and Ehrenfreund, P. (1997) New evidence for interstellar C_{60}^+. *Astron. Astrophys.*, 317, L59-63.

Giblin, D.E., Gross, M.L., Saunders, M., Jimenez-Vazquez, H.A. and Cross, R.J. (1997) Incorporation of helium and endohedral complexes of C_{60} and C_{70} containing noble-gas atoms: A tandem mass spectrometry study. *J. Am. Chem. Soc.*, 119, 9883-9890.

Goel, A., Howard, J.B., Vander Sande, J.B. (2004) Size analysis of single fullerene molecules by electron microscopy. *Carbon*, 42, 1907-1915.

Harris, P.J.F. and Vis, R.D. (2003) High-Resolution transmission electron microscopy of carbon and nanocrystals in the Allende meteorite. *Proc. R. Soc. London A*, 459, 2069-2076.

Harris, P.J.F., Vis, R.D. and Heymann, D. (2000) Fullerene-like carbon nanostructures in Allende meteorite. *Earth Planet. Sci. Lett.*, 183, 355-359.

Heymann, D. (1986) Buckminsterfullerenes its siblings and soot: Carriers of trapped inert gases in meteorites? *J. Geophys. Res.*, 91, E135-E138.

Heymann, D. (1995) Search for extractable fullerenes in the Allende meteorite. *Meteoritics*, 30, 436-438.

Heymann, D. (1997) Fullerenes and fulleranes in meteorites revisited. *Astrophys. J.*, 489, L111-L114.

Hunten, D.M., Pepin, R.O. and Owen, T.C. (1988) Planetary Atmospheres. *In Meteorites and the Early Solar System*, J.F. Kerridge and M.S. Matthews, Eds., 565-591, University of Arizona Press, Tucson, Arizona, USA.

Huss, G.R. (1990) Ubiquitous interstellar diamond and SiC in primitive chondrites: Abundances reflect metamorphism. *Nature*, 347, 159-162.

Kerridge, J.F. and Matthews, M.S., Eds. (1988) *Meteorites and the Early Solar System*, 1269p., University of Arizona Press, Tucson, Arizona, USA.

Kimura, T., Sugai, T., Shinohara, H., Goto, T., Tohji, K. and Matsuka, I. (1995) Preferential arc-discharge production of higher fullerenes. *Chem. Phys. Lett.*, 246, 571-576.

Krätschmer, W., Lamb, L.D., Fostiropoulos, K. and Huffman, D.R. (1990) Solid C_{60}: A new form of carbon. *Nature*, 347, 354-357.

Kroto, H.W., Heath, J.R., O'Brien, S.C., Curl, R.F. and Smalley, R.E. (1985) C_{60}: Buckminsterfullerene. *Nature*, 318, 162-165.

Kroto, H.W. (1988) Space, stars, C_{60} and soot. *Science*, 243, 1139-1142.

Kroto, H.W. (1989) The role of linear and spherodal carbon molecules in interstellar grain formation. *Ann. Phys.*, 14, 169-173.

Kroto, H.W. and Jura, M. (1992) Circumstellar and interstellar and their analogues. *Astron. Astrophys.*, 263, 275-280.

Lewis, R.S., Srinivasin, B. and Anders, E. (1975) Host phase of a strange xenon component in Allende. *Science*, 190, 1251-1262.

Lewis, R.S., Ming, T., Wacker, J.F., Anders, A. and Steel, E. (1987) Interstellar diamonds in meteorites. Nature, 326,160-162.

Mossman, D., Eigendorf, G., Tokaryk, D., Gauthier-Lafaye, F., Guckert, K.D., Melezhik, V. and Farrow, C.E.G. (2003) Testing for fullerenes in geologic materials: Oklo carbonaceous substances, Karelian shungites, Sudbury Black Tuff. *Geology*, 31, 255-258.

Nuth, J.A. (1985) Meteoritic evidence that graphitic carbon is rare in the ISM. *Nature*, 318, 166-168.

Olsen, E.K., Swindle, T.D., Nuth, J.A. and Ferguson, F. (2000) Noble gases in graphitic smokes. *Lunar Planet. Sci.*, 31, abstract #1479, Lunar and Planetary Institute, Houston, Texas, USA.

Ott, U., Mack, R. and Chang, S. (1981) Trapping noble gas-rich separates from the Allende meteorite. *Geochim. Cosmochim. Acta*, 45, 1211-1237.

Ozima, M. and Podosek, F.A., Eds. (2001) *Noble Gas Geochemistry*, *2nd Ed.*, 300p., Cambridge University Press, Cambridge, United Kingdom.

Pepin, R.O. (1991) On the origin and early evolution of terrestrial planet atmospheres and meteoritic volatiles. *Icarus*, 92, 2-79.

Pizzarello, S., Hang Y., Becker, L., Podera R.J., Nieman R.A., Cooper, G. and Williams, M. (2001) The organic content of the Tagish Lake meteorite. *Science*, 293, 2236-2239.

Porcelli, D. and Pepin, R.O. (1997) Rare gas constraints on early earth history. *In Origin of the Earth and Moon*, R.M. Canup and K. Righter, Eds., 435-458, University of Arizona Press, Tucson, Arizona, USA.

Poreda, R.J. and Becker, L. (2003) Fullerenes and interplanetary dust at the Permian-Triassic boundary. *Astrobiology*, 3, 120-136.

Radicati di Brozolo, F., Bunch, T.E., Fleming, R.H. and Macklin, J. (1994) Fullerenes in an impact crater on the LDEF spacecraft. Nature, 369, 37-40.

Rietmeijer, F.J.M., Rotundi, A. and Heymann, D. (2004) C_{60} and giant fullerenes in foot condensed in vapors with variable C/H_2 ratios. *Fullerene, Nanotubes and Carbon Nanostructures*, 12, 659-680.

Rietmeijer, F.J.M., Borg, J. and Rotundi, A. (2005) Revisiting C_{60} and fullerenes in carbonaceous chondrites and interplanetary dust particles: HRTEM and Raman spectroscopy. *Lunar Planet. Sci.*, 36, abstract #1225, Lunar and Planetary Institute, Houston, Texas, USA.

Robl, T.L. and Davis, B.H. (1993) Comparison of the HF-HCL and $HFBF_3$ maceration techniques and the chemistry of the resultant organic concentrates. *Org. Geochem.*, 20, 249-255.

Rotundi, A., Rietmeijer, F.J.M., Colangeli, L., Mennella, V., Palumbo, P. and Bussoletti, E. (1998) Identification of carbon forms in soot materials of astrophysical interest. *Astron. Astrophys.*, 329, 1087-1096.

Sadana, A.K., Liang, F., Brinson, B., Arepalli, S., Farhat, S., Hauge, R.H., Smalley, R.E. and Billups, W.E. (2005) Functionalization and extraction of large fullerenes and carbon-coated metal formed during the synthesis of single wall carbon nanotubes by laser oven, direct current arc, and high-pressure carbon monoxide production methods. *J. Phys. Chem. B*, 109, 4416-4418.

Saunders, M., Jimenez-Vasquez, H.A., Cross, R.J. and Poreda, R.J. (1993) Stable compounds of helium and neon $He@C_{60}$ and $He@C_{70}$. *Science*, 259, 1428-1431.

Smith, P.P.K. and Buseck, P.R. (1981) Graphitic carbon in the Allende meteorite a microstructural study. *Science*, 212, 322-324.

Swindle, T. (1988) Trapped noble gases in meteorites. *In Meteorites and the Early Solar System*, J. F. Kerridge and M.S. Matthews, Eds., 535-564, University of Arizona Press, Tucson, Arizona, USA.

Ugarte, D. (1992) Curling and closure of graphitic networks under electron beam irradiation. *Nature*, 359, 707-709.

Webster, A. (1991) Comparison of a calculated spectrum of $C_{60}H_{60}$ with the unidentified astronomical infrared emission features. *Nature*, 352, 412-416.

Webster, A. (1993) Fullerenes, fulleranes and diffuse interstellar bands. *Monthly Not. Roy. Astron. Soc.*, 255, 41p.

Wieler, R., Anders, E., Baur, H., Lewis, R.S. and Signer, P. (1991) Noble gases in "phase Q": Closed-system etching of an Allende residue. *Geochim. Cosmochim. Acta*, 55, 1709-1722.

Wieler, R., Anders, E., Baur, H., Lewis, R.S. and Signer, P. (1992) Characterization of Q-gases and other noble gas components in the Murchison meteorite. *Geochimica. Cosmochimica. Acta*, 56, 2907-2921.

Ying, Y., Saini, R.K., Liang, F., Sadana, A.K and Billups, W.E. (2003) Funtionalization of the carbon nanotubes by free radicals. *Org. Lett.*, 5, 1471-1473.

Zahnle, K. (1990) Xenon fractionation in porous planetesimals. *Geochim. Cosmochim.* Acta, 54, 2577-2586.

Zinner, E., Amari, S., Wopenka, B. and Lewis, R.S. (1995) Interstellar graphite in meteorites: Isotopic and structural properties of single graphite grains from Murchison. *Meteoritics*, 30, 209-226.

Chapter 7

FULLERENES AND NANODIAMONDS IN AGGREGATE INTERPLANETARY DUST AND CARBONACEOUS METEORITES

FRANS J.M. RIETMEIJER
Department of Earth and Planetary Sciences, MSC03-2040 1-University of New Mexico, Albuquerque, NM 87131-0001, USA.

Abstract: If fullerenes are a common carbon phase in circumstellar dust and the presolar dust of the dense molecular wherein our solar system had formed, they should be present in the most primitive samples that still contain the vestiges of the accreting dust in the solar nebula 4.56 Ga ago. Such dust would be expected to have survived in comet nuclei and in the most primitive asteroids. They would be represented by collected chondritic, aggregate, interplanetary dust particles. As yet, there is no evidence of fullerenes in these particles but C_{60} and higher fullerenes are present in several carbonaceous chondrite meteorites. Metastable fullerenes may not survive the complex natural processing of comet and asteroid debris in the parent body, during solar system sojourn and atmospheric entry and laboratory storage. The possibility of fullerene modification to nanodiamonds in primitive asteroids is discussed.

Key words: C_{60}; C_{60}^{+}; $C_{60}O_n$; C=O bonds; asteroids; carbonaceous chondrite meteorites; carbon condensation experiments; comets; cosmic dust; cubic diamond; flash heating; fullerenes; icy planetesimals; interplanetary dust particles (IDPs); Long Duration Exposure Facility (LDEF); meteoroids; meteors; nanodiamonds; non-hydrostatic compression; ozone; presolar dust; shock metamorphism; solar nebula; soot

Frans J.M. Rietmeijer (ed.), Natural Fullerenes and Related Structures of Elemental Carbon, 123–144.

1. INTRODUCTION

1.1 Sources of Collected Extraterrestrial Material

Is there any reason why C_{60} or other, larger or smaller, fullerenes should be present in comets and asteroids that were the most primitive small bodies to form in the solar system by dust accretion 4.56 Ga ago? Would these fullerenes, and perhaps other coexisting carbons, have survived post-accretion processes of these small solar system planetesimals and protoplanets? Is it all just a matter of fullerene survival in the primitive materials that had a complex history before they reached our laboratories? The collected materials from these bodies are

1. Chondritic aggregate interplanetary dust particles (IDPs), and
2. CI1, CI2 and CM carbonaceous chondrite meteorites with an ultrafine-grained matrix (Rietmeijer, 2005).

The parent bodies of these materials initially contained water ice when they accreted beyond the "snowline" located between Mars and Jupiter. In reality the "snowline" was a transition zone across which condensed ice, including small (<one meter) accreted dirty-ice/icy-dirt objects, could move towards the sun and water vapor could move away from the sun. This was probably a cyclic process. The "snowline" separates the outer asteroid belt with its very primitive objects from the middle and inner (closest to the sun) zones of the asteroid belt. The big picture of the asteroid belt (Bell et al., 1989) has three zones of organic-bearing to organic-rich asteroids as a function of increasing heliocentric distance from ca. 3 AU:

1. Hydrated CM and CI1 carbonaceous chondrite asteroids that in terms of their infrared classification (IR) are referred to as *C* (chondritic) type bodies,
2. Hydrated very-carbonaceous chondrite asteroids that are *P* (primitive) (IR) type bodies, and
3. Icy, ultra-carbonaceous chondrite asteroids that are *D* (dark) (IR) class objects where dark refers to low albedo due to processed organics at the surface (Bell et al., 1989).

Active, dormant and extinct comet nuclei are *P*- and *D*-class objects. Chondritic aggregate IDPs are *P*- and *D*-class debris (Bradley et al., 1992, 1996). The icy-planetesimals known as comets (Fig. 1) either formed in the Jupiter-Neptune zone from which they were subsequently ejected into the Oort cloud [a shell with a radius of 10^4 Astronomical Units (AU) surrounding the solar system] or they originally accreted in the Kuiper belt beyond Pluto's orbit at ca. 45 AU.

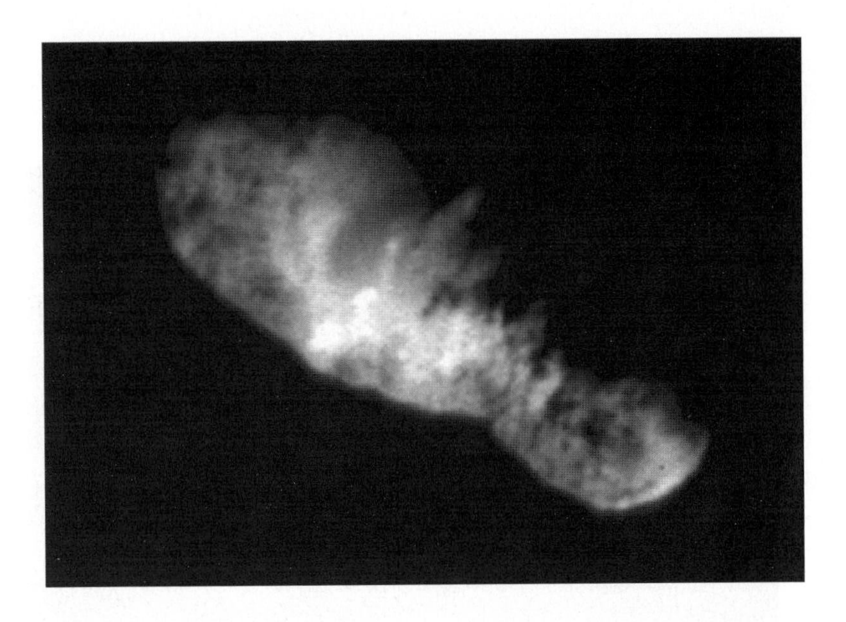

Figure 1. Comet Borrelly observed on Sept. 22, 2001 during fly-by of the NASA Deep Space 1 spacecraft. The pockmarked nucleus is about 8 x 4 km with variable albedo between 0.01 and 0.03 that makes Borrelly one of the darkest objects in the solar system. Although not visible in this image, dust jets were seen to escape the nucleus and a cloudlike "coma" of dust and gases surrounding the nucleus. In these jets particles 10-100 microns in size were ejected into interplanetary space. Comet Borrelly is classified as a C-type body. Its surface is in this regard indistinguishable from that of the main belt asteroid Mathilda shown in Fig. 2 (Soderblom et al., 2002) Image credit: NASA/JPL

In terms of early solar system evolution there are no fundamental differences between icy-planetesimals in heliocentric orbit in the outer asteroid belt (Fig. 2) and inactive comet nuclei. The latter develop a coma and dust tail when they approach the sun that is turned off after perihelion when at a distance where the sun's radiation can no longer cause ice sublimation. An icy asteroid at constant distance from the sun will not display this comet-like activity but it might once it was ejected from the asteroid belt and became a near-Earth or Earth-crossing asteroid (Rietmeijer, 2000a). In fact some of these asteroids do develop transient comet activity when they evolve a coma.

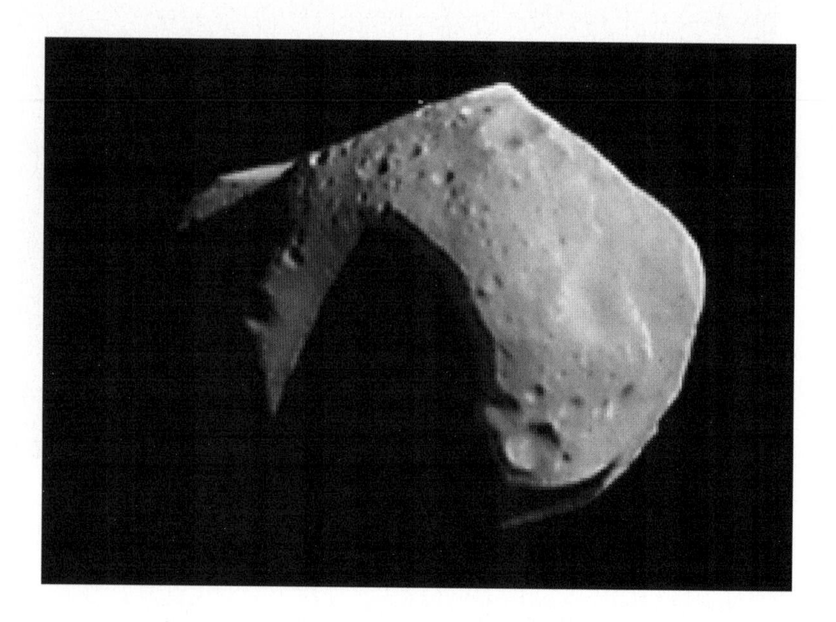

Figure 2. A mosaic of the main belt asteroid 253 Mathilda as viewed by the NEAR spacecraft during its flyby on 27 June 1997 from a distance of 2400 km. The visible portion of the asteroid is about 59 x 47 km. The large craters, the abundance of smaller craters, and the angular shape point to a history of heavy bombardment. Lighting by the sun is from the upper right. Mathilda is an infrared C-type asteroid, i.e. a primitive, undifferentiated asteroid with a [bulk] density of 1.4 gm cm^{-3}, which means it is a highly porous body such as expected for a rubble-pile planetesimal

The sun formed in a part of a dense molecular cloud that had collapsed, perhaps triggered by a nearby supernova event but the cause(s) of this early event remains shrouded. Once the dust from this cloud was gravitationally bound to the young sun it became solar nebula dust at ca. 10 K that in the outer regions of the nebula accreted into icy planetesimals (Messenger and Walker, 1997; Keller et al., 2000; Messenger et al., 2003). At such temperatures interstellar and presolar dust will be preserved virtually unmodified in comet nuclei. Truly solar nebula dust condensed from a hot gas in the innermost part of the nebula when for several hundreds of years the temperatures were high enough to (almost) completely evaporate the presolar dust. Another later high-temperature event during the T-Tauri thermal phase of the young sun might have had a similar effect (for a review, see, Cassen, 2001). The primitive, *C*-, *P*- and *D*-class objects in the asteroid belt broadly delineate a transition from mostly solar nebula dust condensates to gravitationally-bound molecular cloud dust and interstellar dust that, at larger heliocentric distance, co-accreted into primitive planetesimals and protoplanets, some of which were then subjected to parent body processes that modified the original minerals.

1.2 Natural Delivery of Extraterrestrial Samples to Earth

Copious amounts of solar system dust are produced by catastrophic and non-catastrophic collisions among airless asteroids and by ice sublimation at the surface of active comets and some near-earth asteroids. The low-mass dust is sensitive to solar radiation pressure but not to gravitational perturbations by planets. Hence, dust ejected from all sources in the solar system will spiral inwards to the sun by a process called Poynting-Robertson drag. It will take IDPs ca. 10^5 years to cross the Earth's orbit and replenish the Zodical dust cloud (Brownlee, 1985). In contrast massive debris ejected from the outer asteroid belt close to Jupiter will be ejected from the solar system, although their dust could reach the Earth. Massive fragments generated in the main and inner asteroid belts will feel the sun's gravitational pull and could eventually reach the Earth's surface as meteorites. The first ca. 30-Ma history of asteroids included dust accretion followed by post-accretion compaction, impact and shock processes, thermal and aqueous alteration leading to dust lithification, i.e. the formation of rocks. The intensity of these processes varied among and even within individual parent bodies resulting in the different meteorite classes, groups and petrographic types (Brearley and Jones, 1998).

Dust entering the Earth's atmosphere includes the debris from comets. Most comet nuclei are probably rubble piles of pebbles, (sub-) meter boulders and small (ca. 1 km?) planetesimals held together by self-gravitation, "Whipple glue" (dirty ice/icy dirt), or both. On a larger scale comet nuclei probably range from massive rubble piles to dirty snowballs or icy dirtballs (Rietmeijer, 2002). Active comets are prolific dust producers but also of sub-meter meteoroids (Rietmeijer, 2000a) that could deliver cometary meteorites to the Earth's surface (Campins and Swindle, 1998). Dust blown off a nucleus at velocities of 10-100 m/s during ice sublimation ends up in the curved comet tail during each perihelion passage. Old and newly produced debris will track the comet in its orbit. When an orbiting "debris package" intersects the Earth's orbit, the result is a meteor shower or a rare 'storm' (cf. Jenniskens et al., 2000) (Fig. 3).

The comet ejection ages of individual "debris packages" in such storms are known for many comets, such as comet Temple-Tuttle that is the parent body of Leonid meteors (Jenniskens et al., 2000). The release ages for such packages range up to several hundreds of years. Therefore, dust in meteor streams and storms will have spent much less time in interplanetary space than dust in the Zodiacal cloud. This shower and storm debris will have experienced much less on-orbit (thermal) aging. Larger debris in such packages will be comminuted by mutual collisions and eventually merge with this permanent cloud that continuously loses dust by in-fall to the sun.

The interactions of large meteoroids with the Earth's atmosphere cause surface ablation (melting; evaporation). This thermal process forms a black fusion crust on stony meteorites that can be up to 3 millimeters thick. Underneath this crust the interior meteorite temperature remains at its temperature in space, which preserves all pre-entry properties. The ablation of millimeter-size asteroid debris produces ca. 20 μm to ca. 500 μm (mostly) micrometeorite spheres that were severely heated or melted and are collected at the earth's surface (Kurat et al., 1994; Maurette et al., 1987; Taylor and Brownlee, 1991; Taylor et al., 2000). Most cometary dust will instantly evaporate in the atmosphere ('shooting stars') but this behavior critically depends on the combination of dust density, cross-section and entry velocity.

Figure 3. Artist rendition of the November 12-13, 1833, Leonid storm over the eastern United States that was the greatest meteor storm in recorded history. The engraving made many years after the event captures both its extreme intensity with up to ca. 40 meteors per second at the storm's peak, and the characteristic radiant of a meteor shower or storm. The radiant is the point in the sky from which the meteors appeared to originate, that for this storm is "located" in the constellation Leo. Source: www.arm.ac.uk/leonid/

Surviving dust experiences flash heating with complete thermal equilibrium between the surface and interior (Brownlee, 1985). Dust that is collected before it reaches the Earth's surface is called interplanetary, or cosmic, dust particles; not micrometeorites.

If C_{60} fullerene, larger or smaller fullerenes, fullerenic carbons and other related carbon dusts that had condensed in carbon-rich astronomical environments existed in the solar nebula, it would be reasonable that they accreted in comet nuclei and asteroids. Thus, carbon-rich aggregate IDPs and primitive meteorites will be good sources to look for them. I will discuss an apparent lack of C_{60} and C_{70} in chondritic aggregate IDPs followed by a discussion of a possible fullerene-nanodiamond connection in the matrix of CI, CI2 and CM meteorites.

2. IDPs

2.1 General

The mostly ca. 2 μm to ca. 55 μm IDPs are collected in the lower stratosphere (17-19 km) on inertial-impact, flat-plate, collectors coated with a viscous silicone oil carried aloft underneath the wings of high-flying aircraft. For reviews of collection procedures, the mineralogical and chemical properties, and extraterrestrial signatures, I refer to the papers by Mackinnon and Rietmeijer (1987), Sandford (1987), Rietmeijer (1998a) and Zolensky et al. (1994). There are IDPs with a unique chondritic bulk composition and non-chondritic IDPs. The former include two major classes:

1. Non-aggregate IDPs have no vestiges of a prior aggregate texture due to modification and lithification in a parent body. Their petrological properties closely resemble the lithified matrix of CI1, CI2 and CM meteorites. They probably share the same parent bodies, and
2. Aggregates of a small number of different dust types with highly variable porosity ranging from 95% in highly fluffy aggregates to close to zero porosity in collapsed aggregates. It is generally true that the anhydrous aggregate IDPs have a high porosity. Hydrated aggregates tend to be of low porosity due to aggregate compaction.

The combination of aggregate morphology (Fig. 4) and chondritic bulk composition is a good, first-order, indicator of a primordial extraterrestrial origin. Their chondritic relative element abundances that resemble those in the hydrated CI1 carbonaceous chondrites and in the solar photosphere indicate the absence of post-(dust) accretion modification. However, the H, C, N and O in meteorites and IDPs are typically lower than in the solar photosphere (Holweger, 1977) as most of these elements were originally present in ices and highly volatile hydrocarbons. The aggregate morphology (Fig. 4) shows that these IDPs were not substantially modified during 4.56 Ga residence in icy planetesimals such as very primitive asteroids or comets at temperatures below ca. 50 K. Aggregate IDPs contain the least-modified crystalline and amorphous dusts available for laboratory studies (Rietmeijer, 2002). In fact, their texture, mineralogy, stable isotope chemistry (D/H, $^{15}N/^{14}N$), noble gases (He, Ne, Ar), and high carbon contents show that aggregate IDPs are uniquely different from any of the collected meteorites (Mackinnon and Rietmeijer, 1987).

Petrographic analyses of grain size, and chemical and mineral properties of aggregate IDP constituents gave rise to the hypothesis of hierarchical dust accretion which began with ca. 100 nm principal components (PCs) that are the smallest constituents of these IDPs:

1. Carbonaceous [C,H,O,N ± S] PCs that are often fused into clumps and patches,
2. Carbon bearing ferromagnesiosilica PCs, and
3. Ferromagnesiosilica PCs; they include a Mg-rich sub-group (Rietmeijer et al., 1999) and a Fe-rich sub-group (Rietmeijer, 2002) that includes units that are already fused aggregates of condensed dust (Rietmeijer and Nuth, 2005) and presolar GEMS (glass with embedded metals and sulfides) (Bradley, 1994; Westphal and Bradley, 2004).

Hierarchical dust accretion led to aggregates of PCs, to aggregate IDPs, to cluster IDPs (ca. 500 μm in size) and probably larger aggregates that could be among comet meteoroids (Rietmeijer, 2000a; 2002). The hypothesis predicts that elemental carbon and carbonaceous phases were incorporated very early during solar nebula dust accretion but at what scale is uncertain, viz. the entire solar nebula or "domains" within the nebula.

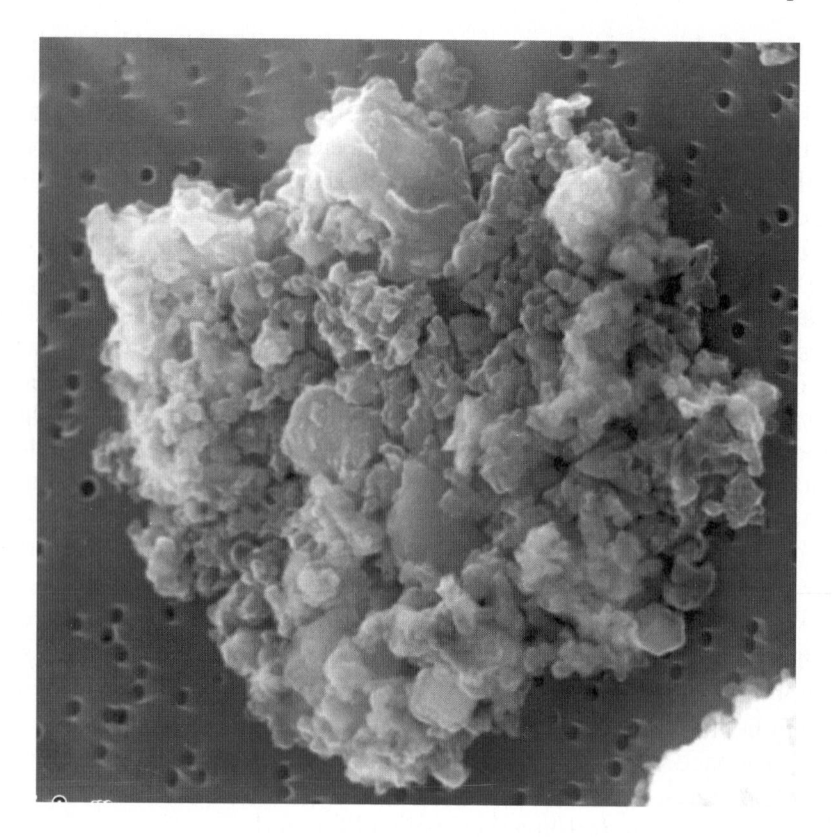

Figure 4. Scanning electron microscope image of porous (fluffy) chondritic aggregate IDP W7029B13 (13.5 x 12 microns) of mostly ultra-fine grains. It is shown on a nuclepore-filter (background). The matrix shows, partially fused, principal components with embedded thin, platy silicate grains. Hierarchical dust accretion began with PCs forming 'matrix aggregates' that co-accreted with micron-sized silicate and Fe,Ni-sulfide grains into an aggregate IDP wherein the major rock-forming elements Si, Mg, S, Ca and Fe occur in approximately chondritic proportions. Most aggregate IDPs are carbon-rich. What now are the holes in the aggregate structure were originally filled with ice (Rietmeijer, 2000b). Source: NASA/JSC Cosmic Dust Catalog 2(1), #S-82-27575; reproduced by courtesy of the ational Aeronautics and Space Administration

2.2 Carbon Content and Phases in Aggregate IDPs

Porous, initially icy, aggregate IDPs are the best link yet to interstellar dust that could include organic substances, amorphous carbon, soot, 'quenched carbon condensates', diamond and C_{60} fullerene (cf. Ehrenfreund et al., 2002). The exact nature of interstellar carbons still needs to be confirmed by spectroscopic observation. Fullerene has yet to be identified in active comets but this may be related to the ambiguity of interpreting the 217.5-nm ultraviolet (UV) carbon signature that also hampers the identifications in the interstellar medium. There is no evidence for C_{60} or other fullerenes in dust detected at comet Halley (Jessberger et al., 2001). Does it mean there is no C_{60} fullerene? We don't know, but the collected IDPs don't provide an unambiguous answer yet.

The bulk carbon content of aggregate IDPs is ca. 2 to 3 times the CI abundance. Among different IDPs and within individual IDPs the carbon content can have a wide range with values as high as ca. 13 times CI (Thomas et al., 1996). One of the more interesting features of these IDPs is that carbon, iron and sulfur occur with their full range of possible oxidation states but without any apparent correlation. The variable redox states may not be too surprising considering the varied histories in different physiochemical conditions of collected IDPs, including flash heating during deceleration in the Earth's atmosphere (Rietmeijer, 1998b) (Table 1). Flash heating tends to cause oxidation but in C-rich IDPs it might induce reduction. Oxidized carbon, i.e. carbonates, could be asteroid parent body alteration or interstellar dust (Sandford, 1986; Kemper et al., 2002).

3. C_{60} AND GIANT FULLERENES: EXPERIMENTS AND EARLY OBSERVATIONS

3.1 Astronomical Condensed Carbon Analogs

Rotundi et al. (1998) undertook a series of arc-discharge, carbon-vapor condensation experiments under controlled conditions to constrain the carrier carbon phase of the 217.5 nm astronomical UV feature. The experiments produced multi-walled fullerenic (cf. Chowdhury et al., 1996) carbon nanotubes and onions, poorly graphitized carbon, graphitic carbon and amorphous soot grains,

Table 1. Hydrocarbons, elemental and organic carbons and C-bearing minerals found in aggregate IDPs (Rietmeijer, 1992a, 1998a, and references therein)

Reduced	Neutral	Oxidized
Epsilon Fe,Ni carbide Cohenite Cementite		
	Hydrocarbons including aliphatic C-H bonds [1] Carbonyl C=O bonds [1] PAHs (C=C bonds) [1,2] S-bearing amorphous carbon Carbon filaments Amorphous carbon and graphitic rims on non-carbonaceous grains Amorphous and highly-disordered carbons [3] Poorly graphitized [4] and turbostratic pre-graphitic [5] carbons Nanodiamonds [6] Lonsdaleite (carbon-2H) [7]	
		Magnesite [$MgCO_3$] Breunnerite [8] [$(Mg,Fe,Ca)CO_3$]

Original sources: (1) Flynn et al. (2003), (2) Clemett et al. (1993), (3) Raynal et al. (2001), (4) Rietmeijer and Mackinnon (1985), (5) Rietmeijer (1992b), (6) Dai et al. (2002), (7) Rietmeijer and Mackinnon (1987), (8) Joswiak and Brownlee (2001)

which was identified as the most likely carbon solid of this spectral feature (Rotundi et al., 1998). Soot was a dense agglomerate of single-wall rings 0.7 nm to 8 nm in diameter that were identified as C_{60} fullerene and giant, C_{540}, C_{960} and C_{1500}, fullerenes (Rietmeijer et al., 2004).

The diameters of similar single-wall ring structures in condensed flame-soot were identified as fullerenes C_{36} to C_{176} including C_{60} fullerene (Goel et al., 2004). Both (high-resolution) transmission electron microscope studies demonstrated that small quantities of C_{60} and other fullerene molecules could be detected and identified in amorphous carbon. The new capability offers the possibility to search in situ for very small, perhaps residual, fullerene C_{60} and both smaller and larger (higher) fullerenes in amorphous carbons from aggregate IDPs and from CI1, CI2 and CM meteorites, wherein they may have gone unnoticed for reasons of instrument detection limits, sample heterogeneity, or both. I note that C_{60} and a range of higher fullerenes were identified in carbonaceous chondrite meteorites (Becker et al., 2000; Pizzarello et al., 2001).

3.2 IDPs: Not Yet C_{60}

Bajt et al. (1996) studied carbon-rich regions in a chondritic porous aggregate IDP using a scanning transmission X-ray microscope for carbon X-ray Absorption Near-Edge Spectroscopy (XANES) analyses. Although there were slight variations in the relative intensities of C-XANES peaks at energies 284.8, 286.3 and 288.5 eV, these peak position energies resembled those in the C_{60} XANES spectrum at 284.3, 285.6 and 286.0 eV. While the C_{60} spectrum has two peaks at 285.6 and 286.0 eV, there was only a single peak centered on 286.3 eV. This difference was within the error of measurement but the possibility that the material was an unspecified C-N-O compound remained (Bajt et al., 1996).

The strong C-XANES peaks at 285 eV and ca. 288.6 eV were later reallocated to carbon rings (amorphous carbon, PAHs, graphite) or C=O bonds (Flynn et al., 2003). The XANES data were inconclusive with regard to the presence of C_{60} in this IDP.

3.3 C_{60}^+ and C_{70}^+ in a Micrometeoroid

The NASA Long-Duration Exposure Facility (LDEF) was a passive impact-collector for anthropogenic debris and meteoroids in near-Earth space between 500 and 570 km altitudes from February 14, 1980 to April 12, 1984. One impact crater contained a ropey melt residue with a (Si, Mg, Fe) signature suggesting the hypervelocity (ca. 19 km/s) impact of a (sub)millimeter chondritic meteoroid. Using laser ionization, time-of-flight, mass spectrometer (LIMS) analyses Radicati di Brozolo et al. (1994) found C_{60} fullerene associated with this impact crater. They could rule out sample contamination. The interpretation of the LIMS C_{60} signature was ambiguous due to the fact that hypervelocity impact experiments simulating C_{60} behavior under such conditions were limited to 6.1 km/s and less (Radicati di Brozolo et al., 1994). They found that even at this high impact velocity fullerene could survive but it would not form by transformation of a precursor such as amorphous carbon, diamond, graphite, phthalic acid crystals and Murchison meteorite matrix material. The LDEF study neither confirms nor denies C_{60} as an indigenous micrometeoroid constituent.

4. SURVIVAL OF C_{60} FULLERENE

4.1 Natural Processes

The seeming ease with which C_{60} can be degraded (e.g. Taylor et al., 1991) may be a reason for its elusiveness in natural terrestrial environments. So far searches for C_{60} and other fullerenes have not produced unambiguous evidence for these molecules in aggregate IDPs and meteorites but accepting that small amounts of fullerenes could have gone undetected by the analytical techniques used prior to searches that used Transmission Electron Microscopy (cf. Goel et al., 2004; Rietmeijer al., 2004). Let's assume that C_{60} was present in collected extraterrestrial samples from comets and carbonaceous chondrite asteroids. It is unlikely that C_{60} could have condensed in the oxidizing environments of the solar nebula. Could it be possible that presolar C_{60} was efficiently destroyed by:

1. Post-accretion processes in asteroids such as shock metamorphism, thermal metamorphism, aqueous alteration, or all three,
2. Thermal alteration when decelerating in the Earth's atmosphere, or
3. Interactions with high energy solar and cosmic ray particles at the surface of comet nuclei or debris in transit in the solar system.

The absence or dearth of C_{60} and other fullerenes would then be a secondary feature due to natural processing.

4.2 A Fullerene - Nanodiamond Connection?

4.2.1 Stability of C_{60}

There is a matter of C_{60} stability. Zwanger et al. (1996) suggested that fullerenes are metastable carbons that when in contact with graphite or poorly crystalline (disordered) carbon may want to equilibrate with the ambient, local free-energy field. The rate of C_{60} "merging" may be very slow allowing metastable C_{60} and other fullerenes to persist until sudden changes in the physiochemical environment will force non-equilibrium reactions. Núñez Regueiro et al. (1992) reported the highly efficient, rapid transformation of C_{60} to (f.c.c.) diamond under 20 ± 5 GPa non-hydrostatic compression in a mélange of well-crystallized diamond, smaller diamond crystallites (2-100 nm) and amorphous, possibly sp^3 carbon. A contributing factor to this fast reaction is the predominant sp^3 hybridization of C_{60} pentagons, which means that C_{60} is intermediate between graphite (sp^2) and diamond (Núñez Regueiro et al., 1992). Similarly, a mixture of cubic diamond and disordered graphitic carbon was obtained by shock compression of fullerene between 27.2 GPa and 51.2 GPa (Donnet et al., 2000).

4.2.2 Shock on Asteroids

Nanodiamonds were found in the matrix of the Orgueil (CI1), Tagish Lake (CI2) and Murchison (CM) meteorites, in CM-like micrometeorites and in asteroidal cluster IDPs, but not in cometary aggregate IDPs (Lewis et al., 1989; Dai et al., 2002; Gradie et al., 2002). Nanodiamonds, 1 nm to ca. 10 nm in size (mode = 3 nm), are embedded in amorphous carbon (Dai et al., 2002). This size range is much smaller than for experimentally produced nanodiamonds (Núñez Regueiro et al., 1992) but the associations of diamond and amorphous carbon are similar. Non-hydrostatic compression could occur during non-catastrophic impacts on asteroids. Such impacts were necessary to eject CI1, CI2, and CM fragments and smaller debris into orbit. Shock experiments of such events on hydrated CM asteroids showed that pressures between ca. 20 GPa and 35 GPa would cause 'explosive' dispersal of meteoroid debris (Tomeoka et al., 2003). These shock levels resemble those of the experimental C_{60} to nanodiamond transformation. Assuming C_{60} and larger fullerene molecules existed in asteroids that are parent bodies of these meteorites, micrometeorites and the IDPs, it seems possible that the presence of nanodiamonds but a dearth or even absence of fullerenes could be a shock related property. Comet dust is released by ice sublimation and this C_{60} to nanodiamond would not occur. C_{60} and larger fullerenes were identified in the Tagish Lake and Murchison meteorites (Pizzarello et al., 2001) but not yet in cometary aggregate IDPs. The corollary is that C_{60} could be common in asteroidal debris that escaped non-hydrostatic pressures above ca. 20 GPa and wherein nanodiamonds would then be absent. Nanodiamonds are inferred to be presolar dust because they carry implanted noble gases, such as Xe (Lewis et al., 1987, 1989).

When this proposed diffusion-less, solid-state, fullerene-nanodiamond transformation took place it presumes that the noble gases were originally present in fullerene molecules of sufficient cage size to accommodate these gases. Such molecules would have to be larger than C_{60}, i.e. they are higher fullerenes, such as those that were found in condensed soot (Rietmeijer et al., 2004) and in some meteorites (Becker et al., 2000; Pizzarello et al., 2001).

4.2.3 Unshocked Asteroids and Comet Nuclei

Meteoritic nanodiamonds may well be interstellar dust (Lewis et al., 1989; Gradie et al., 2002). Anders and Zinner (1993) noted that the nanodiamonds might come only from supernovae. Diamond is the only thermodynamically stable form of crystalline carbon grains <5 nm (Nuth, 1987a; Nuth, 1987b). Fullerenes <5 nm would then be metastable phases and sensitive to a rapid diffusion-less C_{60} to nanodiamond transformation. It predicts a correlation between size range and size distributions of precursor fullerene molecules and nanodiamonds. Interstellar nanodiamonds range from ca. 0.5 nm to ca. 6.5 nm and have an unusual lognormal size distribution (Lewis et al., 1989; Dai et al., 2002). The lower value resembles the 0.52 nm diameter of C_{36} (Goel et al., 2004). The nanodiamond size range is similar to fullerenes C_{36} to C_{1500}, including C_{60} (Rietmeijer et al., 2004; Goel et al., 2004) and the median 5 nm size of nanodiamonds is similar to the giant fullerene modal value 5.5 ± 1.0 nm (Rietmeijer et al., 2004). While no proof, these similar size characteristics lend support to a possible genetic link between fullerenes and nanodiamonds. Ironically the initial search for a carrier of isotopically anomalous Xe, Kr and nitrogen in meteorites was initiated to confirm that C_{60}, and not nanodiamonds, carried the anomalies (Lewis et al., 1987).

Lewis et al. (1987, 1989) suggested that metastable nanodiamond was a carbon-rich vapor condensate in the atmospheres of red giants. Tielens et al. (1987) hypothesized shock waves >30 km/s generated by supernova explosions produced interstellar nano-diamonds by a diffusion-less transformation of interstellar graphite or amorphous carbon dust. They offered no argument why precursor carbon grains existed with this specific size range. Interstellar nanodiamonds often form compact clusters of tiny grains, suggesting that graphite and amorphous carbon grain-grain collisions in the interstellar medium were involved (Blake et al., 1988).

Taken at face value and not considering possible sampling bias when comparing "large" meteoritic samples with picogram IDPs, cometary dust (i.e. aggregate IDPs) does not seem to contain either fullerene molecules or nanodiamonds. These IDPs would experience higher flash-heating temperatures than asteroidal IDPs during atmospheric entry, while the meteorite interior temperature will not be raised during this event. It is possible that nanodiamond or fullerene graphitization is an efficient process to erase all traces of these particular carbons during atmospheric entry. To solve this conundrum it will require direct sampling of icy protoplanets before we can make definitive statements on the state of fullerene molecules in comets.

4.3 Fullerene Oxidation in IDPs

Chibante and Heymann (1993) found that oxidation of C_{60} to $C_{60}O_n$ (n = 1, 2, 3) by ozone is orders of magnitude faster than by O_2 in an oxidizing atmosphere or from moisture in the ambient atmosphere. They pointed out that O_3 is required to initiate the oxidation process until cage rupture has occurred at which time both O_2 and H_2O could complete the process. Ozone-triggered fullerene degradation in IDPs could be initiated in the mesopause region around 90 km where IDP deceleration begins, when IDPs settled through the stratospheric ozone layer, or both. The former seems unlikely, as $C_{60}O_n$ did not form when C_{60} was heated in ambient air at 230 °C and 250 °C (Chibante and Heymann, 1993). Fullerene oxidation might also occur during IDP storage and handling when no protective measures are in place to avoid this process. A "degraded C_{60}" with C=O bonds in aggregate IDPs (Flynn et al., 2003) could be oxidized fullerene but does not constrain where and when oxidation occurred.

5. EXTRATERRESTRIAL FULLERENES: STILL UP IN THE AIR

So where does it leave us? Chondritic aggregate IDPs contain polycyclic aromatic hydrocarbons (PAHs) (Table 1). Given the unstable nature of C_{60} and C_{70} co-existing with PAHs (Murata et al., 1999, among many others) it would be surprising if fullerenes in these IDPs resisted equilibration with the ambient, local free-energy field. Chondritic IDPs are not good candidates for fullerene searches, also for other reasons discussed here. The STARDUST mission captured dust from comet Wild-2 and laboratory studies of this dust will hopefully shed light on the presence of C_{60} in active comets. Fullerenes fared better in CI1, C12 and CM parent bodies judging by the meteorites we have from these parent bodies that are mostly of the infrared C-type asteroids but also the *D*-type Tagish Lake CI2 meteorite (Brown et al., 2000). Still, high concentrations of C_{60} and higher fullerenes in carbonaceous chondrite meteorites are unlikely. On circumstellar and interstellar fullerenes in meteorites, Heymann (1997) commented that if such fullerenes existed they were destroyed in the molecular cloud from which the sun formed, or that they did not survive the formation of the solar system, or did not survive the formation of meteorites. Perhaps, no search for fullerenes in chondritic IDPs, micrometeorites and primitive meteorites has been sufficiently systematic to assess heterogeneous distributions of these metastable carbon molecules.

6. CONCLUSIONS

To unravel the complex histories of meteorites, earth and planetary scientists have to work backwards in time by removing all signatures of the youngest event first, then the one before, and so on, to excavate the oldest information from the time of formation. Natural processes do not occur with predictable directionality or sequentially. On Earth a hand specimen will always have a geological context. The same is not true for meteorites, let alone for aggregate IDPs. Even if we could identify the exact asteroid or comet for each collected meteorite or IDP, we still would not have this context. It is a serious matter when chasing the history of natural C_{60} and other fullerenes and related nanocarbons. The original metastable fullerenes experienced a long and chaotic history before we could analyze samples from their planetesimals in the laboratory. Fullerene molecules are present in the primitive carbonaceous chondrites but not in the abundances one might expect if they were a common carbon dust in the earliest solar nebula. Perhaps there is an intimate genetic relationship between natural fulleneres and nanodiamonds and is it entirely coincidental that nanondiamonds occur in some Cretaceous-Tertiary boundary clays (Carlisle and Braman, 1991) but not in others wherein fullerenes were found? Is it pure coincidence that carbonaceous meteorites contain C_{60} and higher fullerenes as well as nanodiamonds, or should the presence of fullerenes or nanodiamonds be mutually exclusive as a function of impact energy at the time of release from the asteroid parent body? With regard to natural extraterrestrial fullerenes, sampling missions to comets and carbonaceous meteorite asteroids might solve the question: What was the nature of carbon solids among the interstellar and solar nebula dusts at the time of solar nebula formation?

Acknowledgements: I greatly appreciate the insightful review by Rhian Jones. I thank Dieter Heymann and Joe Nuth for stimulating discussions. The material is based upon work supported by the National Aeronautics and Space Administration under Grant NAG5-11762 issued through the Office of Space Science and by RTOPS from the Cosmochemistry and Origins of Solar Systems Research Programs.

7. REFERENCES

Anders, E. and Zinner, E. (1993) Interstellar grains in primitive meteorites: Diamond, silicon carbide, and graphite. *Meteoritics*, 28, 490-514.

Bajt, S., Chapman, H.N., Flynn, G.J., Keller, L.P. (1996) Carbon characterization in interplanetary dust particles with a scanning transmission X-ray microscope (abstract). *Lunar Planet. Sci.*, 27, 57-58, Lunar and Planetary Institute, Houston, Texas, USA.

Becker, L., Poreda, R.J. and Bunch, T.E. (2000) Fullerene: A new extraterrestrial carbon carrier phase for noble gases. *Proc. Natl. Acad. Sci.*, 97, 2979-2983.

Bell, J.F., Davis, D.R., Hartmann, W.K. and Gaffey, M.J. (1989) Asteroids: The big Picture. *In Asteroids II*, R.P Binzel, T. Gehrels and M.S. Matthews, Eds., 921-945, University of Arizona Press, Tucson, Arizona, USA.

Blake, D.F., Freund, F., Krishnan, K.F.M., Echer, C.J., Shipp, R., Bunch, T.E., Tielens, A.G., Lipari, R.J., Hetherington, C.J.D. and Chang, S. (1988) The nature and origin of interstellar diamond. *Nature*, 332, 611-613.

Bradley, J.P. (1994) Chemically anomalous, pre-accretionally irradiated grains in interplanetary dust from comets. *Science*, 265, 925-929.

Bradley, J.P., Humecki, H.J. and Germani, M.S. (1992) Combined infrared and analytical electron microscope studies of interplanetary dust particles. *Astrophys. J.*, 394, 643-651.

Bradley, J.P., Keller, L.P., Brownlee, D.E. and Thomas, K.L. (1996) Reflectance spectroscopy of interplanetary dust particles. *Meteorit. Planet. Sci*, 31, 394-402.

Brearley, A.J. and Jones, R.H. (1998) Chondritic meteorites. *In Planetary Materials*, J.J. Papike, Ed., *Revs. Mineral.* 36, 3-1 – 3-398, Mineralogical Society of America, Washington, D.C., USA.

Brown, P.G., Hildebrand, A.R., Zolensky, M.E., Grady, M., Clayton, R.N., Mayeda, T.K., Tagliaferri, E., Spalding, R., MacRea, N.D., Hoffman, E.L., Mittlefehldt, D.W., Wacker, J.F., Bird, J.A., Campbell, M.D., Carpenter, R., Gingerich, H., Glatiotis, M., Greiner, E., Mazur, M.J., McCausland, P. JA., Plotkin, H., and Mazur, T.R. (2000) The fall, recovery, orbit, and composition of the Tagish Lake meteorite: a new type of carbonaceous chondrite. *Science*, 90, 320-325.

Brownlee, D.E. (1985) Cosmic dust: Collection and research. *Ann. Rev. Earth Planet. Sci.*, 13, 147-173.

Campins, H. and Swindle, T.D. (1998) Expected characteristics of cometary meteorites. *Meteorit. Planet. Sci.*, 33, 1201-1211.

Carlisle, D.B. and Braman, D.R. (1991) Nanometre-size diamonds in the Cretaceous/Tertiary boundary clay of Alberta. *Nature*, 352, 708-709.

Cassen, P. (2001) Nebular thermal evolution and the properties of primitive planetary materials. *Meteorit. Planet. Sci.*, 36, 671-700.

Chibante, L.P.F. and Heymann, D. (1993) On the geochemistry of fullerenes: Stability of C_{60} in ambient air and the role of ozone. *Geochim. Cosmochim. Acta*, 57, 1879-1881.

Chowdhury, K.D., Howard, J.B. and VanderSande, J.B. (1996) Fullerenic nanostructures in flames. *J. Mat. Res.*, 11, 341-347.

Clemett, S.J., Maechling, C.R., Zare, R.N., Swan, P.D. and Walker, R.M. (1993) Identification of complex aromatic molecules in individual interplanetary dust particles. *Science*, 262, 721-772.

Dai, Z.R., Bradley, J.P., Joswiak, D., Brownlee, D.E., Hill, H.G.M. and Genge, M.M. (2002) Possible *in situ* formation of meteoritic nanodiamonds in the early solar system. *Nature*, 418, 157-159.

Donnet, J.B., Fousson, E., Wang, T.K., Samirant, M., Baras, C. and Pontier-Johnson, M. (2000) Dynamic synthesis of diamonds. *Diamond and Related Materials*, 9, 887-892.

Ehrenfreund, P., Irvine, W., Becker, L., Blank, J., Brucato, J.R., Colangeli, L., Derenne, S., Despois, D., Dutrey, A., Fraaije, H., Lazcano, A., Owen ,T. and Robert, F., an International Space Science Institute ISSI-team (2002) Astrophysical and astrochemical insights into the origin of life. *Reports Progress Phys,* 65: 1427-1487.

Flynn, G.J., Keller, L.P., Feser, M., Wirick, S. and Jacobsen, C. (2003) The origin of organic matter in the solar system: Evidence from the interplanetary dust particles. *Geochim. Cosmochim. Acta*, 67, 4791-4806.

Gradie, M.M., Verchovsky, A.B., Franchi, I.A., Wright, I.P. and Pillinger, C.T. (2002) Light element geochemistry of the Tagish Lake CI2 chondrite: Comparison with CI1 and CM2 meteorites. *Meteorit. Planet. Sci.*, 37, 713-735.

Goel, A., Howard, J.B. and Vander Sande, J.B. (2004) Size analysis of single fullerene molecules by electron microscopy. *Carbon*, 42, 1907-1915.

Heymann, D. (1997) Fullerenes and fulleranes in meteorites revisited. *Astrophys. J.*, 489, L111-L114.

Holweger, H. (1977) The solar Na/Ca and S/Ca ratios: A close comparison with carbonaceous chondrites. *Earth Planet. Sci. Lett.*, 34, 152-154.

Jenniskens, P., Rietmeijer, F.J.M., Brosch, N. and Fonda. M. (Eds.) (2000) *Leonid Storm Research*, 606p., Kluwer Academic Publishers, Dordrecht, the Netherlands.

Jessberger, E.K., Stephan, T., Rost, D., Arndt, P., Maetz, M., Stadermann, F.J., Brownlee, D.E., Bradley, J. and Kurat, G. (2001) Properties of interplanetary dust information from collected samples. *In Interplanetary Dust*, E. Grün, B.Å.S. Gustafson, S.F. Dermott and H. Fechtig, Eds., 253-294, Springer Verlag, Berlin, Germany.

Joswiak, D.J. and Brownlee, D.E. (2001) Carbonate mineralogy in stratospheric IDPs: Compositions, co-existing smectite and comparison to CI carbonaceous chondrites. *Lunar Planet. Sci.*, 32, abstract #1998, Lunar and Planetary Institute, Houston, Texas, USA (CD-ROM).

Keller, L.P., Messenger, S. and Bradley, J.P. (2000) Analysis of a deuterium-rich interplanetary dust particle and implications for presolar materials in IDPs. *J. Geophys. Res. Space Phys.*, 105(A5), 10397-10402.

Kemper, F., Jäger, C., Waters, L.B.F.M., Henning, Th., Molster, F.J., Barlow, M.J., Lim, T. and de Koter, A. (2002) Detection of carbonates in dust shells around evolved stars. *Nature*, 415, 295-297.

Kurat, G., Koeberl, C., Preper, T., Brandstätter, F. and Maurette, M. (1994) Petrology and geochemistry of Antarctic micrometeorites. *Geochim. Cosmochim. Acta*, 58, 3879-3904.

Lewis, R.S., Ming, T., Wacker, J.F., Anders, A. and Steel, E. (1987) Interstellar diamonds in meteorites. *Nature*, 326, 160-162.

Lewis, R.S., Anders, A. and Draine, B.T. (1989) Properties, detectability and origin of interstellar diamonds in meteorites. *Nature*, 339, 117-121.

Mackinnon, I.D.R. and Rietmeijer, F.J.M. (1987) Mineralogy of chondritic interplanetary dust particles. *Revs. Geophys.*, 25, 1527-1553.

Maurette, M., Jéhanno, C., Robin, E. and Hammer, C. (1987) Characteristics and mass distribution of extraterrestrial dust from the Greenland ice cap. *Nature*, 328, 699-702.

Messenger, S. and Walker, R.M. (1997) Evidence for molecular cloud material in meteorites and interplanetary dust. *In Astrophysical implications of the laboratory study of presolar materials*, T.J. Bernatowicz and E.K. Zinner, Eds., *Amer. Inst. Phys. Conf. Proc.*, 402, 545-564, American Institute of Physics Press,. Woodbury, New York, USA.

Messenger, S. Keller, L.P., Stadermann, F, Walker, R.M. and Zinner, E. (2003) Samples of stars beyond the solar system: Silicate grains in interplanetary dust. *Science*, 200, 105-108.

Murata, Y., Kato, N., Fujiwara, K., and Komatsu, K. (1999) Solid-state {4+2} cycloaddition of fullerene C_{60} with condensed aromatics using a high-speed milling technique. *J. Org. Chem.*, 64, 3483-3488.

Núñez Regueiro, M., Monceau, P. and Hodeau, J-L. (1992) Crushing C_{60} to diamond at room temperature. *Nature*, 355, 237-239.

Nuth, J.A. (1987a) Small-particle physics and interstellar diamonds. *Nature*, 329, 589.

Nuth, J.A. (1987b) Are small diamonds thermodynamically stable in the interstellar medium? *Astrophys. Space Sci.*, 139, 103-109.

Pizzarello, S., Hang, Y., Becker, L., Podera, R.J., Nieman, R.A., Cooper, G. and Williams, M. (2001) The organic content of the Tagish Lake meteorite. *Science*, 293, 2236-2239.

Radicati di Brozolo, F., Bunch, T.E., Fleming, R.H. and Macklin, J. (1994) Fullerenes in an impact crater on the LDEF spacecraft. *Nature*, 369, 37-40.

Raynal, P.I., Quirico, E., Borg, J., d'Hendecourt, L. (2001) Micro-Raman survey of the carbonaceous matter structure in stratospheric IDPS and carbonaceous chondrites. *Lunar Planet. Sci.*, 32, abstract #1341, Lunar and Planetary Institute, Houston, Texas, USA (CD-ROM).

Rietmeijer, F.J.M. (1992a) Carbon petrology in cometary dust. *In Asteroids, Comets, Meteors 1991*, A. Harris, E. Bowell, Eds., 513-516, Lunar and Planetary Institute, Houston, Texas, USA.

Rietmeijer, F.J.M. (1992b) Pregraphitic and poorly graphitised carbons in porous chondritic micrometeorites. *Geochim. Cosmochim. Acta*, 56, 1665-1671.

Rietmeijer, F.J.M. (1998a) Interplanetary Dust Particles. *In Planetary Materials*, J.J. Papike, Ed., *Revs. Mineral.*, 36, 2-1 – 2-95, Mineralogical Society of America, Washington, D.C., USA.

Rietmeijer, F.J.M. (1998b) Interplanetary Dust. *In Adv Mineral.*, Vol. 3, A.S. Marfunin, Ed., 22-28, Springer Verlag, Berlin-Heidelberg, Germany.

Rietmeijer, F.J.M. (2000a) Interrelationships among meteoric metals, meteors, interplanetary dust, micrometeorites, and meteorites. *Meteorit. Planet. Sci.*, 35, 1025-1041.

Rietmeijer, F.J.M. (2000b) Interplanetary dust particles. *In McGraw-Hill Yearbook of Science & Technology 2001*, 208-211, The McGraw-Hill Companies Inc.

Rietmeijer, F.J.M. (2002) The earliest chemical dust evolution in the solar nebula. *Chemie der Erde*, 62, 1-45.

Rietmeijer, F.J.M. (2005) Iron-sulfides and layer silicates: A new approach to aqueous processing of organics in interplanetary dust particles, CI and CM meteorites. *Adv. Space Res.,* in press; *available on-line*. doi:10.1016/j.asr.2004.11.024

Rietmeijer, F.J.M. and Mackinnon, I.D.R. (1985) Poorly graphitized carbon as a new cosmothermometer for primitive extraterrestrial materials. *Nature*, 316, 733-736.

Rietmeijer, F.J.M. and Mackinnon, I.D.R. (1987) Metastable carbon in two chondritic porous interplanetary dust particles. *Nature*, 326, 162-165.

Rietmeijer, F.J.M. and Nuth III, J.A. (2005) Laboratory simulation of Mg-rich ferro-magnesiosilica dust: The first building blocks of comet dust. *Adv. Space Res*., in press; *available on-line.* doi:10.1016/j.asr.2005.03.113

Rietmeijer, F.J.M., Nuth III, J.A. and Karner, J.M. (1999) Metastable eutectic condensation in a Mg-Fe-SiO-H_2-O_2 vapor: Analogs to circumstellar dust. *Astrophys. J.*, 527, 395-404.

Rietmeijer, F.J.M., Rotundi, A. and Heymann, D. (2004) C_{60} and giant fullerenes in soot condensed in vapors with variable C/H_2 ratio. *Fullerenes, Nanotubes, and Carbon Nanostructures*, 12, 659-680.

Rotundi, A., Rietmeijer, F.J.M., Colangeli, L., Mennella, V., Palumbo, P. and Bussoletti, E. (1998) Identification of carbon forms in soot materials of astrophysical interest. *Astron. Astrophys*., 329, 1087-1096.

Sandford, S.A. (1986) Acid dissolution experiments: Carbonates and the 6.8-micrometer bands in interplanetary dust particles. *Science*, 231, 1540-1541.

Sandford, S.A. (1987) The collection and analysis of extraterrestrial dust particles. *Fundamentals Cosmic Phys*., 12, 1-73.

Soderblom, L.A., Becker, T.L., Bennett, G., Boice, D.C., Britt, D.T., Brown, R.H., Buratti, B.J., Isbell, C., Giese, B., Hare, T., Hicks, M.D., Howington-Kraus, E., Kirk, R.L., Lee, M., Nelson, R.M., Oberst, J., Owen, T.C., Rayman, M.D., Sandel, B.R., Stern, S.A., Thomas, N. and Yelle R.V. (2002) Observations of Comet 19P/Borrelly by the Miniature Integrated Camera and Spectrometer Aboard Deep Space 1. Science, 296, 1087-1091.

Taylor, R., Parsons, J.P., Avent, A.G., Rannard, S.P., Dennis, T.J., Hare, J.P., Kroto, H.W., Walton, D.R.M. (1991) Degradation of C_{60} by light. *Nature*, 351, 277.

Taylor, S. and Brownlee, D.E. (1991) Cosmic spherules in the geological record. *Meteoritics*, 26, 203-211.

Taylor, S., Lever, J.H. and Harvey R.P. (2000) Numbers, types, and compositions if an unbiased collection of cosmic spherules. *Meteorit. Planet. Sci*., 35: 651-666.

Thomas, K.L., Keller, L.P. and McKay, D.S. (1996) A comprehensive study of major, minor, and light element abundances in over 100 interplanetary dust particles. *In Physics, Chemistry and Dynamics of Interplanetary Dust,* B.Å.S. Gustafson and M.S. Hanner, Eds. *Astron. Soc. Pacific Conf. Series*, 104, 283-286, Astron Soc Pacific, San Francisco, California, USA.

Tielens, A.G.G.M., Seab, C.G., Hollenbach, D.J. and McKee, C.F. (1987) Shock processing of interstellar dust: Diamonds in the sky. *Astrophys. J.*, 319, L109-L113.

Tomeoka, K., Kiriyama, K., Nakamura, K., Yamahana, Y. and Sekine, T. (2003) Interplanetary dust from the explosive dispersal of hydrated asteroids by impacts. *Nature*, 423, 60-62.

Westphal, A.J. and Bradley, J.P. (2004) Formation of glass with embedded metal and sulfides from shock-accelerated crystalline dust in superbubbles. *Astrophys. J.*, 617, 1131-1141.

Zolensky, M.E., Wilson, T.L., Rietmeijer, F.J.M. and Flynn, G.J. (Eds.) (1994) *Analysis of Interplanetary Dust, Am Inst. Phys. Conf. Proc.*, 310, 357p., American Institute of Physics Press, Woodbury, NY, USA.

Zwanger, M.S., Banhart, F. and Seeger, A. (1996) Formation and decay of spherical concentric-shell carbon clusters. *J Crystal Growth*, 163, 445-454.

Chapter 8

FULLERENES AND RELATED STRUCTURAL FORMS OF CARBON IN CHONDRITIC METEORITES AND THE MOON

DIETER HEYMANN
Department of Earth Science Rice University, MS 126, Houston, Texas 77251-1892, USA.

FRANCO CATALDO
Soc. Lupi Chemical Research Institute, Via Casilina 1626/A, 00133 Rome, Italy.

MARIE PONTIER-JOHNSON
Carbon Nanotechnology Group, Continental Carbon Company, 333 Cypress Run, Suite 100, Houston, Texas 77094, USA.

FRANS J.M. RIETMEIJER
Department of Earth and Planetary Sciences, MSC03-2040, 1-University of New Mexico, Albuquerque, New Mexico 87131, USA.

Abstract: Reports concerning the presence or absence of fullerenes in chondritic meteorites are reviewed. Structural forms of carbonaceous matter in these meteorites are discussed in the context of fullerene formation.

Key words: Carbon calabashes; carbon cauliflowers; carbon onions; carbon nanoglobules; carbonaceous chondrites; fullerenes; graphite spherules; interstellar dust; meteorites; Moon; nanodiamonds

Frans J.M. Rietmeijer (ed.), Natural Fullerenes and Related Structures of Elemental Carbon, 145–189.

1. INTRODUCTION

When Kroto, Heath, O'Brien, Curl, and Smalley (1985) discovered *Buckminsterfullerene*, they predicted that this surprisingly stable and easily synthesizable carbon molecule should occur abundantly in circumstellar and interstellar media, hence in meteorites (Kroto, 1988; Kroto et al., 1987; Kroto and Walton, 1993). Soon thereafter Heymann (1986) suggested that the so-called *HL* isotopically anomalous noble gases in stony meteorites (Lewis et al., 1975) were actually trapped inside meteoritic fullerene cages. The *HL* designation refers to a mixture of components extracted by acid dissolution methods from a meteorite enriched in the heavy and light isotopes, e.g. Xe-*HL*. The *HL* gases, hence their fullerene "bottles" were thought to be presolar, i.e. had existed as such before the solar system formed at 4.56 Ga. It was later found that these gases were actually contained in presolar meteoritic nanodiamonds (Lewis et al., 1987). Next, it was suggested that certain carbon onions, reported by Smith and Buseck (1981a) in High-Resolution Transmission Electron microscope (HRTEM) images of carbon extracted from the Allende carbonaceous meteorite, were nested fullerene onions as obtained under electron-beam irradiation (Ugarte, 1992), which seemed to imply the possible presence of fullerenes in this meteorite (Becker et al., 1993). Uncontested evidence for the occurrence of interesting interstellar and perhaps interplanetary carbon and carbon compounds in meteorites had already been reported.

HRTEM micrographs of the most abundant elemental carbon in the Allende meteorite (Smith and Buseck, 1980, 1981b; Lumpkin, 1981a, 1981b, 1983a, 1983b) showed a type of pyrocarbon that formed at elevated temperatures. Interstellar graphite in the Allende meteorite (Amari et al., 1990a) occurs as carbon onions and so-called carbon cauliflowers (Bernatowicz et al., 1991). However, silicon carbide was found to be the most abundant interstellar carbon compound in meteorites (Anders and Zinner, 1993). Perhaps the most remarkable fact is that carbon nanotubes were not yet found in meteorites.

While all of this was promising, searches for fullerenes in carbon-rich stony meteorites did not begin until it was discovered that C_{60} and C_{70} are highly soluble in benzene, toluene, and CS_2 (Krätschmer et al., 1990). Hence they might be readily extracted from meteoritic matter and it was then disappointing when early searches failed to find them (Tingle et al., 1991; Ash et al., 1993; Gilmour et al., 1993; De Vries et al., 1993; Oester et al., 1994). Finally, in 1995 fullerenes were reported in the Allende carbonaceous meteorite (Becker et al., 1994) although subsequent searches for fullerenes in this meteorite were negative (Heymann, 1995a, 1995b, 1997).

We present here a review of meteoritic fullerenes in the broader context of other structural forms of elemental, meteoritic carbon forms to highlight evidence which suggests that fullerenes might well occur in Allende and other carbonaceous meteorites and that the failures to find them could have been due to inhomogeneity or to the daunting experimental difficulties of recovering and detecting nanogram quantities of C_{60} and C_{70}. Other elemental carbon forms in meteorites include graphitic carbons, graphite and diamonds that could be interstellar or presolar grains (Rietmeijer, 1988, 1990; Alexander et al., 1990; Anders and Zinner, 1993; Ott, 1993, 2001; Pillinger, 1993; Rotundi et al., 1998; Gilkes and Pillinger, 1999; Buseck, 2002).

Although the chondritic meteorites are not the only stony meteorites that contain substantial amounts of elemental carbon, they are the only class of these meteorites that were searched for fullerenes. The ureilites, a class of achondritic, differentiated, stony meteorites, contain abundant elemental carbons that include carbynes, graphite, and nanodiamonds (e.g. Vdovykin, 1969, 1972; Göbel et al., 1978; Wacker, 1986; Kagi et al., 1991; Goodrich, 1992). Nevertheless, no attempts have been made to detect fullerenes in these meteorites.

Two other, distinct attempts were made to search for fullerenes in extraterrestrial matter. One was a successful discovery of fullerenes associated with a hypervelocity impact crater on the skin of the Long Duration Exposure Facility spacecraft in low-Earth orbit (Radicati di Brozolo et al., 1994). The other was a regrettably unsuccessful search in soil samples from the Moon (Heymann, 1996a, 1996b).

2. METEORITES BRIEFLY

Meteorites are the portions of pre-terrestrial meteoroids that survived passage through the Earth's atmosphere and struck its surface where they were recovered. The velocity of entry into the upper atmosphere is generally between ~10 and 30 km s^{-1}. These meteoroids are slowed down upon entry by collisions with the molecules of the atmosphere, which causes melting and evaporation at the surface, eventually leaving behind a thin, black fusion crust. Frequently the meteoroid breaks up into large numbers of fragments during the passage through the atmosphere. Each fragment commonly has its own fusion crust. Upon impact with the Earth's surface most produced shallow pits. A few meteoroids were sufficiently massive to strike the Earth's surface at supersonic velocities to produce a meteorite impact crater, such as Meteor Crater in Arizona (USA).

Obviously, meteoroids in their last orbits came so close to the orbit of the Earth that a collision was inevitable. However, because the first orbit of a meteoroid may have crossed those of Mars and possibly Earth and Venus, and was, therefore, altered by gravitational interactions with these large planets, the first orbit was almost certainly not the same as the final orbit. Final orbits can sometimes be quite accurately reconstructed from visual and photographic observations of the luminous meteor trail in the atmosphere. Such reconstructed final orbits are thought to be similar to the early orbits for meteorites that reached the Earth very soon after they were broken off from their parent bodies (Heymann and Anders, 1967).

Two recent examples of orbital matching are the Příbram and Neuschwanstein meteorites and their observed fireballs (bolides) (Spurný et al., 2003) and the Park Forest meteorite (Simon et al., 2004). By using reconstructed orbits and transit times from parent body to the Earth, which range from less than 10^6 to 10^8 years, a common origin for meteorites has been deduced. Most meteorites come from parent bodies in the main asteroid belt between Mars and Jupiter, the transitional region between the "once hot" region of the terrestrial planets (Mercury, Venus, Earth, and Mars) and the "eternally cold" region of the giant planets. Some meteorites may come from comets or from the Apollo and Adonis objects whose perihelia are nearer to the Earth's orbit. A few meteorites were ejected from our Moon.

The asteroidal parent bodies themselves were accreted in the nascent solar system. Some of these were never seriously heated, let alone melted. Many parent bodies accreted from early solar-system condensates of inorganic solids, viz. silicates, aluminates, oxides, and sulfides (Grossman, 1972). Melting as revealed by the textures and compositions of chondrules and by the existence of metallic nickel iron meteorites (core-mantle separation) was ubiquitous. Numerous small and large crater-forming impacts on the parent bodies caused local and, occasionally large-scale reheating of surface rocks. One of the greatest challenges in meteorite research is therefore to explain the occurrence in meteorites of very high-temperature minerals side by side with carbonaceous matter that should have been destroyed, become better graphitized, or have lost its trapped noble gases at the inferred high temperatures. We refer the reader who desires details of compositions, textures, and histories, including impact modification, of meteorites to the chapters in Kerridge and Matthews (1981) and in Papike (1998).

Meteorite classification (Table 1) will be incomplete because we have to rely on samples that are randomly delivered and preserved at the Earth's surface that is for 70% covered by water. Thus, some rare, meteorites simply cannot be accommodated by this evolving classification. Nevertheless, a consistent first division of all collected meteorites shows three major categories, viz. iron meteorites, stony meteorites, and stony-iron meteorites. Iron meteorites were once molten nickel-iron metal in which all elemental carbon was dissolved that later during slow, subsolidus cooling exsolved as graphite. No systematic searches for fullerenes in iron and stony-iron meteorites were made so far.

Table 1. Simplified Meteorite Classification

Meteorite Types		**Class (Group)**	**Brief Characteristics**
IRONS			Crystalline nickel-iron. Also graphite
STONY-IRONS			Crystalline nickel-iron. Inorganic minerals
STONY	Chondrites		Inorganic minerals. Chondrules.
		Ordinary Chondrites (H, L, LL)	Inorganic minerals. Chondrules. Generally <1% carbonaceous matter, cf. Table 2
		Carbonaceous Chondrites (CI, CM, CO, CV)	Inorganic minerals. Chondrules. For carbon contents, cf. Table 2
	Achondrites		Inorganic minerals. Differentiated chemical composition
		Ureilites	The only subclass with significant contents of elemental carbon (average 0.65 wt%)
		Others	Incl. lunar and Martian meteorites; not relevant to the present subject

Stony meteorites are made up from silicates, aluminates, oxides, sulfides, carbonates, and halides forming a fine-grained matrix for larger co-accreted chondules (a few millimeters) and Calcium-Aluminum-rich Inclusions (cm-sized). Some contain metallic nickel-iron grains, usually not much larger than 1 mm. Stony meteorites are further subdivided into chondrites and achondrites. Most, but not all, chondrites contain small spherical mineral objects called chondrules and these meteorites show variable degrees of post-formation thermal or aqueous alteration, or both.

Achondrites are differentiated rocks, such as basalts and gabbros, and do not contain chondrules. The abundances of the non-volatile elements in the chondrites are much closer to the solar abundances than they are in achondrites. With few exceptions, the carbonaceous chondrites are the most primitive and therefore the most carbon rich stony meteorites (Table 2) and searches for fullerenes were directed to only these meteorites.

Table 2. Ranges (Vdovykin and Moore, 1960) and averages (McCall, 1973) values of total carbon contents of the LL, L and H ordinary and the CV, CO, CM and CI carbonaceous chondrite groups; few ordinary chondrites contain >1 wt % carbon

Chondrite Groups	**Total Carbon content (weight %)**	
	Range	Average
LL	0.02-0.44	--
L	0.02-0.53	--
H	0.02-0.60	--
CV	0.07-2.5	0.46
CM, CO	1.3-4	2.44
CI2	3.6	--
CI1	2.7-5.0	3.62

Note: Meteorite classification allows adjustments to be made when new meteorites are recovered. The chemical and mineral properties of the Tagish Lake meteorite neither fitted the CI nor CM groups (Brown et al., 2000) but as it was more closely related to the former. The existing CI group became CI1, and Tagish Lake meteorite became a CI2, carbonaceous chondrite

3. ANALYSES OF NATURAL FULLERENES IN METEORITES

When searches for fullerenes in meteorites began, little was known about their geochemical properties. Commonly, searches for fullerenes in chondrites begin with treatments of the powdered meteorite with HF and other acids to dissolve most of the inorganic minerals. The carbon-rich residue is called an Acid Resistant Residue (*ARR*). Since it was not known whether C_{60} and C_{70} could survive these geochemical laboratory procedures, Heymann (1990a, 1990b) and Heymann et al. (1990) proved experimentally that treatment with HF and HCl did not detectably degrade the fullerenes. At the same time, they discovered that C_{60} dissolves in sulfur and that hot fuming nitric and perchloric acids will destroy fullerenes. It seemed therefore safe to make *ARRs* and digest these residues with CS_2, *benzene, toluene, dichlorobenzene* or *trichlorobenzene* to extract fullerenes, possibly including higher fullerenes, but one should refrain from treating samples with strongly oxidizing chemicals. Obviously, soluble organics such as Polycyclic Aromatic Hydrocarbons (PAHs) present in the meteorites are extracted also.

Occasionally, powdered meteorite samples were directly used for fullerene extraction. During these, and subsequent, laboratory bench-procedures, the main risks are the formation of ozonides and oxides by reaction with air-ozone and oxide formation via singlet oxygen as C_{60} and C_{70} in their triplet excited state are quenched by O_2 in the ground state to form singlet O_2. To reduce these risks, extractions were preferably carried out in darkness and, if possible, by purging air from solutions with nitrogen or argon.

The analysis of the extracts for fullerenes was usually done either by High Performance Liquid Chromatography (HPLC) or some variant of mass spectrometry, usually Laser Desorption Ionization Mass Spectrometry (LDI-MS). C_{60} or C_{70} cannot form during HPLC but false positives are possible if extracts of *ARRs* contain organic compounds that have exactly the same retention times and UV-vis absorption spectra as the fullerenes. Taylor and Abdul-Sada (2000), who searched for fullerenes in clays from the Cretaceous-Tertiary Boundary by HPLC, reported a peak at the retention time of C_{60}, which they claimed was actually due to a mixture of hydrocarbons but not C_{60}. However, they did not investigate the UV-vis absorption spectrum of the hydrocarbon mix during the analysis, which would have immediately revealed that the compound was C_{60}. HPLC systems in which the analyzer is a photo-diode array (PDA) are therefore superior to systems in which the detection can be done only at a single wavelength. A peak at the retention time of C_{60} was observed in an extract from the Allende meteorite when a reverse-phase C18-type column was used (Heymann, 1997). With a non-reverse-phase Buckyprep column the retention time of this compound, whose UV-vis spectrum was distressingly similar to that of C_{60}, was now distinctly different from that of C_{60}. Using two types of HPLC columns can evidently help distinguishing between fullerenes and other extracted compounds. Finally, 'double false positives', i.e. the false detection of apparent both C_{60} *and* C_{70} that have the correct, known retention times and absorption spectra of these fullerenes, are essentially ruled out. HPLC analyses can be very easily made quantitative by the calibration with fullerene solutions of known concentrations, or by using the well-known molar extinction coefficients at 336 nm.

When the LDI-MS spectrum of an *ARR* extract contains peaks at masses 720 ($^{12}C_{60}$) and 721 ($^{12}C_{59}{}^{13}C$) with peak-height ratio 721/720 about 0.6, then the extracted sample contained C_{60}. Likewise, if the spectrum contains peaks at masses 840 ($^{12}C_{70}$) and 841 ($^{12}C_{69}{}^{13}C$) with peak-height ratio 840/841 about 0.7, the extracted sample contained C_{70}. A major problem of LDI-MS is that fullerenes can form from carbonaceous or any organic matter in the hot, laser-produced plume. The investigators who have used LDI-MS have carefully studied this issue and have taken precautions to prevent the accidental formation of fullerenes. Also, to the best of our knowledge, LDI-MS analysis of fullerenes was never made truly quantitative, although that would have been, in principle, possible by adding small and known amounts of synthetic C_{76}, C_{78}, or C_{84} to the samples prior to analysis.

Using LDI-MS, C_{60} and C_{70} as well as many PAHs, including *corannulene* ($C_{20}H_{10}$), were reported in the Allende meteorite (Becker et al., 1994). The estimated C_{60} content was 0.1 ppm. The authors suggested that PAHs might have been involved in the fullerene synthesis, perhaps in circumstellar envelopes or in interstellar medium. This suggestion implies dehydrogenation and *zip-up* of the PAHs. The authors further proposed that the fullerenes had formed in the outflows of C-rich stars. Subsequently, using LDI-MS, PAHs, fullerenes and fulleranes ($C_{60}H_2$-$C_{60}H_{60}$) were found in an extract of the Allende meteorite (Becker et al., 1995). Since the Allende meteorite fell in 1969, the very presence in 1995 of these unstable higher fulleranes in the meteorite would be quite unexpected. The reported C_{60} contents in ten Allende samples ranged from less than 3 ppb to 100 ppb (Becker and Bunch, 1997). When using *trichlorobenzene* as extractant, Becker et al. (1999) observed C_{74}^+, C_{76}^+, C_{78}^+, C_{84}^+ ions in the LDI-MS spectrum of the extract and many peaks of carbon clusters between C_{100}^+ and C_{250}^+ that they submitted had formed in the outflows of carbon stars. However, it is very easy to generate fullerenes by laser ablation of sufficient power from practically any carbonaceous substrate that originally does not contain them (Cataldo and Keheyan, 2002).

The search for natural fullerenes took a new direction with a report that approximately 1 in 10^5 of the C_{60} molecules extracted from rocks at the Sudbury impact crater (Canada) contained endogenic ^4He and ^3He (Becker et al., 1996). Traditionally the presence of the noble gas isotopes was revealed by noble gas mass spectrometry (stepwise heating of the sample and ionization by electron bombardment). Because the inferred $^3He/^4He$ ratio was distinctly different from the atmospheric and solar ratios, it was suggested that the He-carrying molecules, i.e. C_{60} fullerene, were presolar (Becker et al., 1996). The $^3He/^4He$ isotope ratio of the noble gas mass spectrometry strongly points to the presence of an extraterrestrial carrier in the sample. However, the LDI mass spectrometry, carried out with a separate split of the same sample, cannot be considered independent confirmation because the alleged peak at m/z = 724 of a $^4He@C_{60}$ ion was actually at m/z = 721 and was due to the $^{13}C^{12}C_{59}^+$ ion. Higher fullerenes C_{100} to C_{400} were reported in the Allende and Murchison meteorites (Becker et al., 2000a, 2000b) but, interestingly, the peak due to the most stable fullerene, C_{240}, was not significantly larger than the peaks for surrounding fullerenes. The noble gas analyses of the extracted fullerenes from Murchison reportedly contained 1.1×10^{-8} cm^3 He per gram of fullerene in the residue and a $^3He/^4He$ ratio of 2.1×10^{-4}. Trapped Argon was also found. It was suggested that fullerenes were carriers of 'planetary noble gases' in the carbonaceous chondrite meteorites (Becker et al., 2000a, 2000b; Pizzarello et al., 2001).

This suggestion has a serious quantitative problem as 10-15 % of the total trapped He content of a chondrite is 'planetary' He (Wieler et al., 1991). The measured ^4He content of an *ARR* of the Murchison meteorite is 4.02×10^{-4} cm^3 per gram of residue (Alaerts et al., 1980). Of this gas, at least 4×10^{-5} cm g^{-1} was therefore planetary He. Essentially all of the reported planetary He in the Murchison *ARR* was

^{4}He (Alaerts et al., 1980), therefore the reported residue must have contained 4 x 10^3 g fullerene per gram of residue. This concentration would be impossibly high. Thus, only a minuscule fraction of the planetary He in the Murchison meteorite could be trapped in fullerene cages. Hence, lacking pertinent quantitative considerations, fullerenes as significant carriers of 'planetary gases' to the Earth still remains uncertain. Every analyzed bulk sample or ARR of a given chondrite, incl. the Murchison and Allende meteorites, contains approximately the same 'planetary noble gas' content (Schultz and Kruse, 1989). However, the reported fullerene contents of the Allende meteorite vary greatly from sample-to-sample (Table 3), which does seems at odds with a claim that the He and Ar noble gasses are uniquely confined to fullerene cages.

Table 3. Reported C_{60}, C_{70} and higher fullerene contents in meteorites

Meteorite	C_{60} (ppm)	C_{70} (ppm)	C>70 atoms	Sources
Allende	100			(a)
Allende	detected	detected	detected	(b)
Allende[1]	all <1			(c), (d)
Allende	detected	detected		(e)
Allende	10			(f)
Allende	5			(f)
Allende[2]	all <3			(f)
Allende	detected	detected	detected	(g), (h)
Murchison	Not found			(i), (j)
Murchison	<2			(k)
Murchison	Not found			(d)
Murray	Not found			(i), (j)
Tagish Lake			detected	(l)

Notes:[1]total of nine distinct samples; [2]total of seven distinct samples

Sources: (a) Becker et al. (1994); (b) Becker et al. (1999); (c) Heymann (1995a, 1995b); (d) Heymann (1997); (e) Becker et al. (1995); (f) Becker and Bunch (1997); (g) Becker et al. (2000a); (h) Becker et al. (2000b); (i) Ash et al. (1993); (j) Gilmour et al. (1993); (k) De Vries et al. (1993); (l) Pizzarello et al. (2001)

Researchers who did not find fullerenes in carbonaceous chondrites included Tingle et al. (1991), Ash et al. (1993), Gilmour et al. (1993), De Vries et al. (1993), Oester et al., (1994), Heymann (1995a; 1995b; 1997) and Buseck (2002). Reasons for these failures are not understood at this time. They could well be due to inhomogeneous fullerene distributions in the meteorites, to experimental difficulties, or both.

4. THE MOON

The unconsolidated, rocky outermost layer of the Moon, called the lunar regolith consists of matter thrown out by the myriads of crater-forming impacts the Moon has experienced during its approximately 4.5 Ga existence. Among the impactors were undoubtedly meteoroids and comet nuclei that contained abundant carbonaceous matter from which fullerenes could have formed during high-energy impact events. A sample of 5.006 g of lunar fines (<1 mm grains) from the Apollo 11 landing site (sample 20084) and 0.972 g of fines from a depth of 7-17 cm (sample 79261) collected in the Van Serg trench at the Apollo 17 site were powdered and extracted with *toluene*. The extracts were analyzed with HPLC. No fullerenes were detected in either sample at the 1 ppb level or higher (Heymann, 1996a, 1996b). The negative result must be appreciated in the context of the absolutely puny fraction of lunar regolith studied. Since the regolith is continuously 'diluted' with fresh matter from the still-ongoing impact cratering and since only a fraction of the impacts might have produced fullerenes, the regolith could be extremely heterogeneous with regards to fullerene content. Furthermore, several processes could actually destroy fullerenes in the course of time. Even relatively small impacts produce molten silicates transiently and it is well known that elemental carbon reacts with molten silicates to form CO and CO_2, which obviously would escape from the Moon. Other potentially destructive processes include thermal decomposition, destruction by electrons and ions from the solar wind and cosmic rays, photo-polymerization and pressure-polymerization. Because of their poor solubility, the fullerene multimers would not have been detected by the experiment.

An additional loss mechanism from the equatorial regions of the Moon, where the Apollo 11 and Apollo 17 sites were located, is due to the relatively short (270 s) estimated sticking time of C_{60} at the equator at lunar noon. That time rapidly lengthens for locations at higher lunar latitudes, which means that fullerenes could have migrated slowly towards the lunar poles. The same migration process can also proceed downwards into the regolith because of the very shallow depth penetration of the diurnal heat wave. Future samples obtained a greater regolith depth and from lunar polar sites should be collected and processed in terrestrial repositories with the survival of possible natural lunar fullerenes in mind.

5. INTERSTELLAR GRAPHITE SPHERULES

Single-crystal graphite is a very stable structural form of elemental carbon outside its thermodynamic stability field in which basal graphene sheets are stacked in an *abab*-sequence. Graphene is defined as six carbon atoms bonded by sp^2 hybridized bonds in a hexagonal arrangement. The length of the C-C bonds in graphite is 0.142 nm, while the inter-sheet distance is 0.335 nm. Well-ordered graphite is exceedingly rare in meteorites but disordered graphitic carbons and diamond are much more common (Rietmeijer, 1988). Although various equilibrium pressure-temperature (*PT*) phase diagrams have been constructed to accommodate at least carbyne, graphite, and diamond, the archetypes of elemental carbon with sp^1, sp^2, and sp^3 hybridization of C-C bonds (e.g. Whittaker, 1978; Grumbach and Martin, 1996), the fundamental problem with all of these diagrams is that they often do not correctly predict which structure(s) of carbon will form by a given process under given *PT* conditions. Obviously the products are seldom determined by equilibrium thermodynamics alone but also by chemical composition, pre-existing grains in the environment, and possibly the degree of ionization. It is also quite apparent that competing reaction rates determine the structure(s) formed in most cases. The long-term survival of diamond and graphite at the surface of the Earth attests to the great reluctance of some metastable structural forms of carbon to transform to the prevailing, thermodynamically stable configuration. That is probably a major reason for the bewildering variety of structural forms of carbon found in meteorites with their complex histories of dust accretion and post-accretion alteration in their parent bodies.

The search for interstellar 'needles' in the meteoritic carbonaceous 'haystack' resulted in the discovery in the Murchison meteorite of graphite grains, 1-4 μm in diameter with a less than 2 ppm abundance (Amari et al., 1990a). It was suggested that the grains had formed in outflows of novae and red giant stars. It became immediately obvious that this particular carbonaceous material was a heterogeneous collection of grains with different densities and structures, viz.

1. Abundant loose aggregates that broke up during ultrasonification,
2. Dense, irregularly shaped grains, and
3. Dense spheres.

The spheres appear to be structurally most closely related to fullerenes. The interstellar origin of these individual grains was firmly established by their $^{13}C/^{12}C$ and $^{22}Ne/^{20}Ne$ isotopic ratios that were both distinctly outside the range of solar system values (Amari et al., 1990b; Hoppe et al., 1992a, 1992b; Lewis and Amari, 1992; Nichols et al., 1992; Travaglio et al., 1999; Croat et al., 2002; Staderman et al., 2002). A salient additional discovery showed that only the spherical grains had isotopically anomalous carbon (Zinner et al., 1990). The few publications with HRTEM images of the internal structures of the interstellar graphite spheres are the only decisive source for deciphering the processes that led to the formation of the onion and cauliflower types of spherules (Bernatowicz et al., 1991, 1996; Bernatowicz and Cowsik, 1996) (Fig. 1).

Figure 1. Transmission electron micrographs showing interstellar graphite spherules in the Murchison carbonaceous chondrite (samples KFC1 and LFC1)

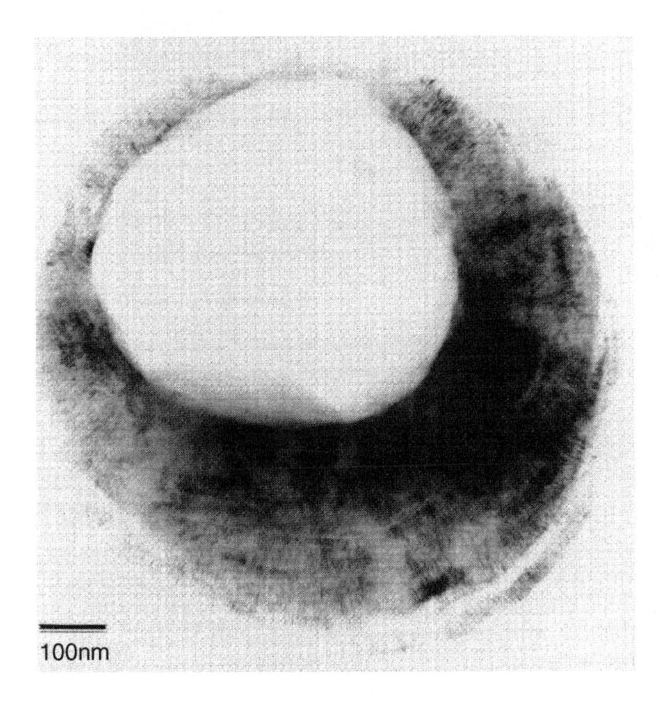

1(a) An onion with nanocrystalline core with a graphitic shell shown dark due to diffraction from (*002*) basal planes in sample KFC1. Scale bar is 100 nm

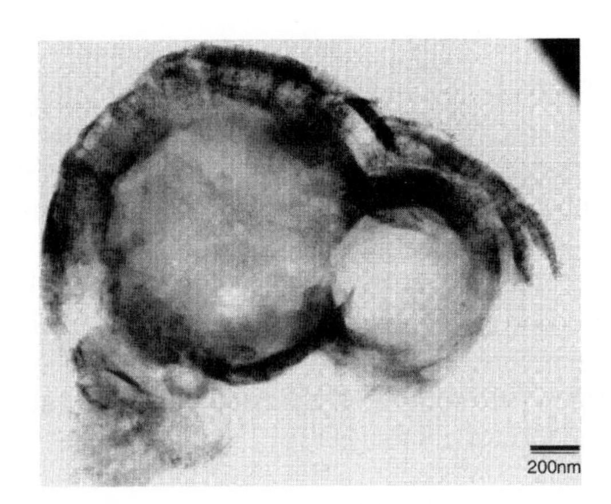

1(b) The double onion dubbed "Mr. Peanut" both with a nanocrystalline core with a well-ordered graphitic shell in sample KFC1. According to Dr. Bernatowicz the cores stuck together and were then covered with graphite. Scale bar is 200 nm

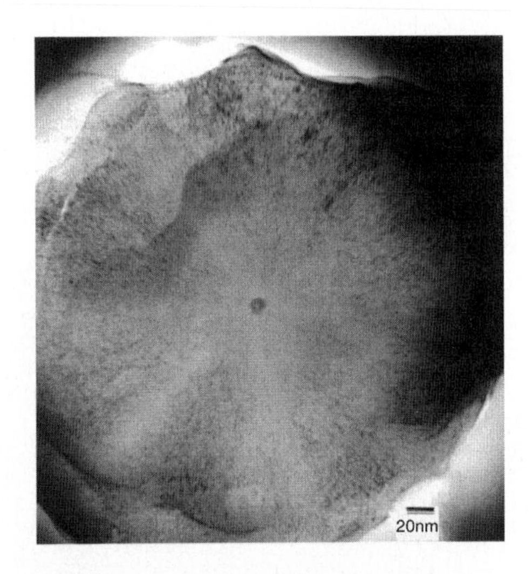

1(c) An onion from sample LFC1 with a central 10 nm TiC crystal (dark central spot) in a nanocrystalline core; the 'radial spokes' are due to a diffraction effect. The onions are spherules with a platy, continuous concentric layering of fairly well developed, parallel graphite (*002*) lattice fringes. Scale bar is 20 nm

For petrographic analyses using a JEOL 2000FX TEM the onions were embedded in LR White Hard resin and sectioned into 70-nm slices with a Reichert-Jung ultra-microtome (Bernatowicz et al., 1991). Unpublished micrographs donated by Prof. T. J. Bernatowicz (Dept. of Physics, Washington University, St. Louis)

These grains most closely resemble the carbon onions produced during electron-beam irradiation in amorphous soot (Ugarte, 1992) but they are completely different from the well-ordered carbon onions found in the Allende meteorites (Smith and Buseck, 1981a).

In 80% of the onions a well-graphitized outer shell encloses an inner core of randomly oriented sheets of PAH-like nanocrystalline matter. The TEM electron diffraction intensity profiles revealed that the core material consisted of 4 nm graphene sheets of several hundreds of carbon atoms in a randomly oriented two-dimensional pattern (Fraundorf and Wackenhut, 2002). This type of shell/core structure is well known from primary soot particles (Hurt et al., 2000; Shim et al., 2000). The cauliflowers consist of corrugated concentric layers of short-wave periodicity with local discontinuity of the graphitic layers (Fig. 2).

The degree of graphitization is low and HRTEM imaging reveals what appears to be turbostratic layering (Bernatowicz et al., 1991, 1996; Bernatowicz and Cowsik, 1996). Some spherules contain much smaller interstellar TiC, Fe, or Fe carbide grains in the carbonaceous matrix either at the center, or dispersed throughout the larger spherule. The onions and cauliflowers are both enigmatic in their own ways. For openers, their sizes are at least one order of magnitude greater than those of typical carbon black particles forming in condensing carbon vapors. While the shell/core structures are well known to occur in soot particles, the dimensions are wholly different between the interstellar grains and soot. In the meteoritic carbon spherules, the cores can be on the order of 100 nm across whereas the diameters of soot cores are typically in the range of 4-6 nm (Hurt et al., 2000). However, the causes for the shell/core structures could be similar. It was suggested that the soot spherules develop at high temperatures through a liquid crystalline phase during which mobile polycyclic aromatic building blocks self-organize either into the isotropic phase of the core or into nematic or columnar phases of the shell (Hurt et al., 2000).

Figure 2. Transmission electron micrographs of two different interstellar carbon 'cauliflower' spherules from the carbonaceous chondrite sample LFC14Murchison

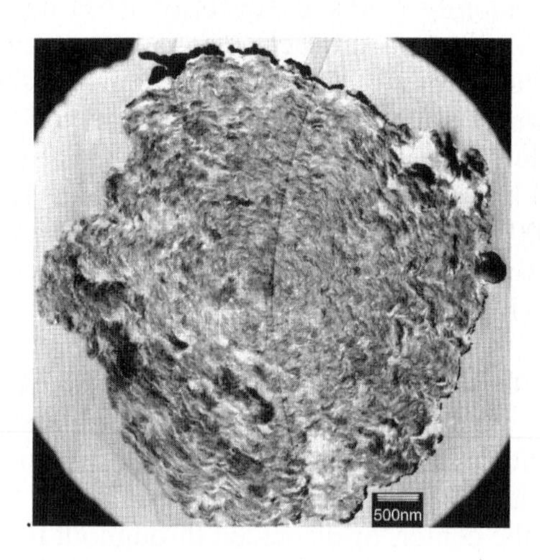

2(a) An entire ultramicrotomed spherule with a corrugated, concentric circular internal texture placed on a holey carbon thin-film (shown in the background on the left) support during TEM analyses. Scale bar is 500 nm

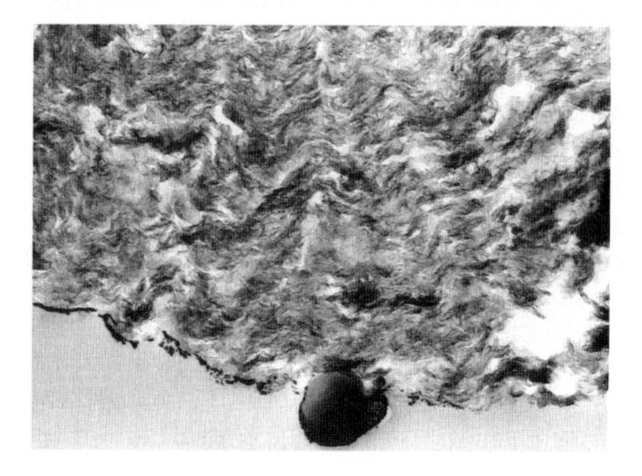

2(b) detail of the typical corrugated texture of carbon sheets in 'cauliflowers' (spherule #84)

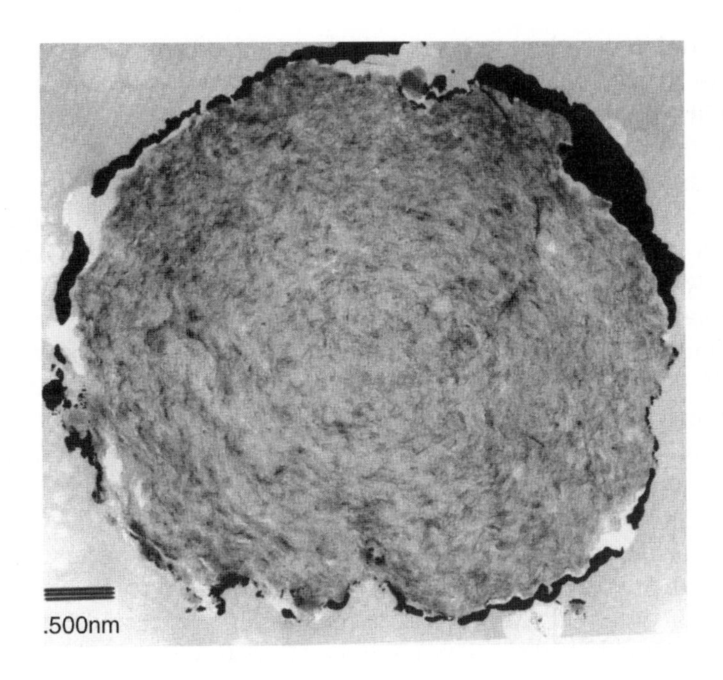

2(c) A faint concentric circular texture. These unpublished micrographs were donated by Prof. T. J. Bernatowicz (Dept. of Physics, Washington University, St. Louis)

The boundary between core and shell is predicted to occur where orientational elastic strain overcomes the free energy difference between bulk isotropic (core) and ordered (shell) phases. Spherules with exactly the same scalloped structure of cauliflower spherules were, to the best of our knowledge, not yet observed among man-made carbons.

The closest analogs were synthesized by condensation, accretion, and coagulation of carbon (Praburam and Goree, 1995) and during carbon graphitization (Oberlin, 1989). The fundamental issue is whether the textures of cauliflower spherules are primary or secondary. Cauliflower grains could form in the interstellar medium by accretion of short stacks of PAHs that were subsequently dehydrogenated (Heymann and Pontier-Johnson, 2002). It is however possible also that the cauliflower spherules began as microcrystalline, poorly ordered carbon spheres that were transformed later into their present structures by shock process, such as hypervelocity shock waves traveling the interstellar medium or were impact-related on the meteorite parent bodies. Some onions and cauliflowers have $^{13}C/^{12}C$ ratios larger, while others have ratios smaller than the solar value to suggest at least two distinctly different environments of formation, perhaps atmospheres of AGB stars and supernova ejecta. There were only few examples of collisions among of meteoritic onions (Fig. 1b).

Perhaps even more enigmatic is the occurrence of smaller crystals (5-200 nm) with compositions ranging from nearly pure TiC to nearly pure Zr-Mo carbide inside approximately one-third of the carbon spherules that were not necessarily nucleation sites for carbon (Bernatowicz et al., 1991; Bernatowicz and Cowsik, 1996). It was suggested that the condensation of carbon probably began in the temperature range about 1775-2275 °C in the atmosphere of a carbon-rich star, that nucleation and growth of TiC crystals occurred during the continued growth of the carbon spherules, and that the growth was quenched by local carbon depletion or by the expulsion of the spherule into the interstellar medium (Bernatowicz et al., 1991). A recent oblique-impact experiment (Eberhardy and Schultz, 2004; Schultz, 1996; Sugita et al., 1998) produced nanoglobules with many of the properties of the interstellar graphite carbon onions (section 9).

The occurrence of essentially monoisotopic ^{22}Ne in interstellar graphite is significant because it is thought to have formed by *in situ* radioactive decay from ^{22}Na. If a high-temperature stage was involved, then ^{22}Na was not quantitatively evaporated from the interstellar graphite grains. Because the ^{22}Na lifetime is only 2.6 years and because ^{22}Ne is highly volatile, a high-temperature stage cannot have lasted for more than a few years and a dehydrogenation and graphitization stage must have started within a few years of complete formation of the spherule and may not have lasted for much longer than a few years.

6. INTERSTELLAR NANODIAMONDS

The first hint of the existence of meteoritic nanodiamonds came from the discovery of a very fine-grained type of carbon that was present in colloidal material from the Allende meteorite (Swart et al., 1982, 1983). It was found that the $^{13}C/^{12}C$ ratio of this component was within the range of terrestrial values and that it was the carrier of what was then called *CCFXe* (Carbonaceous Chondrite Fission Xenon) that is now known to be due to direct nucleosynthesis and is named Xe-*HL*. This carbon was identified as diamond, which became known as nanodiamond because of its very small grain size (Lewis et al., 1987). The nanodiamond grain size-distribution is log-normal with a mass-weighted mean size of 1.6 nm, i.e. median of 1.0 nm (Carey et al., 1987; Fraundorf et al., 1989), later revised to 2.6 nm (Daulton et al., 1996). The bulk density of a Murchison nanodiamond separate was 2.22-2.33 g cm^{-3}, well below the value of 3.51 g cm^{-3} of cubic diamond (Lewis et al., 1987). For the extraction of nanodiamonds from meteorites, the inorganic minerals are dissolved in HF-HCl. Elemental sulfur produced from troilite (FeS) in this step is removed with KOH or CS_2. Non-diamondiferous carbonaceous matter is destroyed with hot HNO_3 and $HClO_4$ (Lewis et al., 1987; Lewis and Anders, 1988; Amari et al., 1994).

We surmise that the formation of these diamonds could have been accompanied by the formation of fullerenes. However, any fullerenes still present after the demineralization would not have survived the oxidative treatment. The reported diamond contents of bulk chondrites range from almost nothing in the Julesburg meteorite to 1436 ppm in the primitive carbonaceous chondrite Orgueil and meteorite diamond contents decrease with increasing metamorphic grade in ordinary chondrites (Lewis et al., 1987; Ming et al., 1987; Ming and Anders, 1988; Huss, 1990; Amari et al., 1994; Arden et al., 1994; Huss and Lewis, 1994, 1995; Russell et al., 1996).

The great majority of the atoms of the nanodiamonds are carbon. Nitrogen atoms can substitute in the lattice for carbon with concentrations in the bulk nanodiamonds phase ranging from 1800 to 13000 ppm (Russell et al., 1991, 1996), i.e. roughly 2 to 15 atoms per 1000 are nitrogen atoms. Hydrogen is abundant in the surface layer (Carey et al., 1987; Virag et al., 1989). Nanodiamonds have generated great scientific interest because they, or else other "hidden" co-recovered phases, are the carrier of significant isotopic anomalies, which is defined as the isotopic composition of an element different from the solar system average. Whereas the $\delta^{13}C$ values of nanodiamond carbon are in the range -5 to -41 per mille, i.e. normal (Swart et al., 1983; Russell et al., 1991, 1992; Fisenko et al., 1992; Arden et al., 1994; Verchovsky et al., 1994), large isotopic anomalies occur in Xe and Kr due to the trapping of noble gases from a source of direct nucleosynthetic matter (Heymann and Dziczkaniec, 1979, 1980). Nitrogen $\delta^{15}N$ values are near -350 per mille (Russell et al., 1991, 1996; Fisenko et al., 1992; Verchovsky et al., 1994).

The properties of the nanodiamonds were essentially obtained by so-called 'blind' statistical techniques which must be complemented by direct imaging, preferably on the atomic scale (HRTEM), because the physiochemical properties of carbonaceous matter depend strongly on texture or microtexture (Oberlin, 1989). Diamond crystals form 2-4 nm clumps with individual crystals that have no epitaxial relationship to one another (Blake et al., 1987, 1988). Individual diamond crystals had irregular and diffuse grain boundaries (structure-less regions) that based on Electron Energy Loss Spectroscope (EELS) analysis are probably an "amorphous" surface component with ample sp^2 bonds. Orgueil nanodiamond diameters were in the 1-6 nm range with roughly 5-10% of the crystals twinned with *(111)* the composition plane (Buseck and Barry, 1988). The twins were about equal in size. The unit cell edge of nanodiamond crystals was that of regular diamond within 0.4%, while EELS showed that no additional carbon component other than diamond and their surfaces was required to provide the observed signal (Bernatowicz et al., 1989, 1990).

In what is arguably the most comprehensive and in-depth HRTEM study of diamonds from Allende and Murchison carbonaceous chondrite meteorites the investigators confirmed that the coherent twin boundaries that are common in meteoritic nanodiamonds are also common in chemical vapor deposition (*CVD*) and detonation-produced synthetic diamonds (Daulton et al., 1996). In addition, linear and non-linear multiple twins and star-twins (Fig. 3) were observed.

The proportions of these twinned types in Allende and Murchison nanodiamonds were more similar to the proportions in *CVD* diamond than in detonation diamond (Daulton et al., 1996). These HRTEM observations led to speculations about the numerous possible processes and environments of formation for these nanodiamonds. The earliest hypothesis was *CVD* formation in the circumstellar atmospheres of red giant stars rich in hydrogen and carbon or in the solar system (Lewis et al., 1987) and in the solar nebula system (Fukunaga et al., 1987). This hypothesis dovetails nicely with ideas about ion-implantation of the isotopically anomalous noble gases (e.g. Jørgensen, 1988; Heymann, 2001). The direct-condensation hypothesis was based on the proposition that diamond is perhaps the thermodynamically stable carbon phase instead of graphite when the condensing grains are smaller than 5 nm (Nuth, 1987).

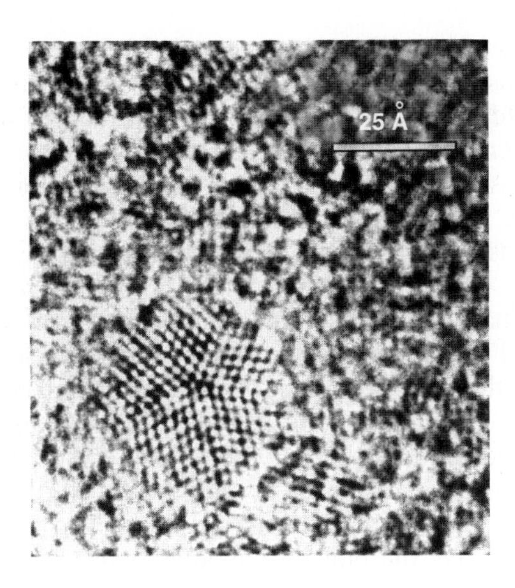

Figure 3. HRTEM electron micrograph showing a pseudo five-fold (star) twinned nanodiamond from the Murchison meteorite oriented along a [011] crystallographic zone axis. Unpublished micrograph donated by Dr. Tyrone L. Daulton (Naval Research Laboratory, NASA Stennis Space Center, MD)

It has long been known that nanometer-size diamonds can be formed by shock compression of almost any carbon-bearing phase, including nm-sized carbon onions and giant fullerenes (among many others, Donnet et al., 2000a, 2000b). Tielens et al. (1987) suggested nanodiamonds would form by transformation of amorphous carbon or graphite grains in supernovae-driven, high-velocity collisions in the interstellar medium. The large surface-to-volume ratio of nano-diamond allows rapid cooling after the shock compression, which greatly reduces the probability of re-graphitization (Tielens et al., 1987). In addition, there are interesting, albeit uncommon syntheses, such as formation of nanodiamonds from carbon onions by electron or ion bombardment (Banhart and Ajayan, 1996; Banhart et al., 1997; Cataldo, 2000, 2001a). A most striking property of nandiamonds is their smallness. For condensation models, which in this case include CVD, this implies a large ratio of the number of seed nuclei to available carbon atoms per unit volume. If the meteoritic nanodiamonds had formed by condensation, the available HRTEM images would indicate homogeneous nucleation of a vast number of critical clusters of about 20 carbon atoms, perhaps still unstructured, which then swiftly depleted the available carbon to accrete into clusters of a few hundred atoms that could then be rearranged to nanodiamonds at still high, ambient temperatures.

The normal carbon isotopic composition of the *bulk* of the nanodiamond carbon is a major conundrum because how can that be consistent with the large isotopic anomalies of nitrogen and the noble gases? While huge numbers of nanodiamond crystals are required for a single $\delta^{13}C$ determination, only about one crystal in 3 x 10^6 of these crystals contains a ^{132}Xe atom of the isotopically anomalous variety. Hence, only these few crystals can be recognized to be truly presolar. The remainder, i.e. the bulk of nanodiamonds, might have formed in the solar system. We think that all nanodiamonds were produced in the star-forming region of the Sun. This environment apparently had atypically large carbon abundances favorable for the formation of carbonaceous matter (Snow and Witt, 1995) but was also seeded with grains from a sufficient variety of astronomical objects to satisfy every proposed formation hypothesis and isotopic composition.

7. PYROCARBONS

Pyrocarbons form by the pyrolitic decomposition of hydrocarbons with deposition of the produced carbon on hot surfaces (Oberlin, 1989, 2002). Carbon black particles form by the pyrolysis of hydrocarbons in the gas phase (Pontier-Johnson, 1998). Glassy polymeric carbon forms when polymerized hydrocarbons are charred (Jenkins and Kawamura, 1976). Although the generic relationship of these forms of carbon and fullerenes is not fully clear, it is conceivable that fullerenes could also form in these processes. Since most of the elemental carbon in the Allende CV carbonaceous chondrite is pyrocarbon, and because this carbon has been most intensely studied, we will focus on this meteorite.

The earliest studies revealed its overwhelmingly elemental carbon matter (Simmonds et al., 1969) instead of the characteristic complex organic polymers of the CI and CM carbonaceous meteorites (Hayatsu et al., 1977). A TEM study of thin Allende foils reported poorly crystalline, disordered graphite that was interstitial to silicate grains

(Green et al., 1971). This Allende carbon was considered amorphous based on X-ray diffraction and TEM evidence (Breger et al., 1972; Dran et al., 1979). Scanning electron microscope analyses showed that most carbonaceous matter occurred as numerous microspherical aggregates of 2 μm in diameter and as surficial films 0.1 μm thick on matrix grains (Bauman and Devaney, 1973; Bauman et al., 1973), often forming carbon micro-mounds concentrated on olivine grain surfaces particularly in the dark haloes surrounding some of the chondrules and other larger aggregates (Bunch and Chang, 1980).

The study of greatly purified carbonaceous matter became the norm with the first preparation of *ARRs* of meteorites (Lewis et al., 1975). Isotopic compositions of *ARR* carbon were reported as $\delta\,^{13}C$ = -16.3 per mille (Chang et al., 1978), $\delta\,^{13}C$ = -19.5 per mille (Grady et al., 1981) and as $\delta\,^{13}C$ = -20 per mille (Swart et al., 1982). All of these values are well within the range of terrestrial carbon values. Apparently the Allende carbon is rich in unpaired electrons, which is common in disordered carbons (Lewis et al., 1982). The *ARR* X-ray photoemission spectra showed "*predominantly carbonaceous matter plus Fe deficient chromite*" (Housley and Clarke, 1980). Moreover, the carbonaceous regions had the appearance of "*crumpled thin film*" (Housley and Clarke, 1980).

Significant progress was by HRTEM studies of *ARR* material in the Allende meteorite that reported "*a tangled aggregate of crystallites*" with a prominent lattice fringe spacing ranging from 0.34 to 0.38 nm (Smith and Buseck, 1980, 1981a, 1981b). This material was named "*a variety of turbostratic carbon*" and described as "*crystallites of this phase contain randomly stacked* sp^2 *hybridized carbon layers and diffraction patterns resemble those from carbon black and glassy carbon*" (Lumpkin, 1981a, 1981b, 1983a, 1983b). The microstructure of this Allende meteorite carbonaceous material appears strikingly similar to that previously described for glassy polymeric carbon, and is interpreted as interwoven ribbon-shaped packets of graphitic layer planes. The HRTEM studies of *ARRs* from the Allende, Vigarano, and Leoville CV carbonaceous meteorites (Harris et al., 2000; Harris and Vis, 2003; Vis et al., 2002) show considerably more microtextural complexity of the *ARR* carbonaceous matter, which in addition to much pyrocarbon material also contain a substantial number of what appeared to be nanometer-sized carbon black particles (Fig. 4).

Figure 4. High-resolution transmission electron micrographs showing curved and tangled graphene sheets typical for pyrolytic carbon nanostructures in bulk elemental carbon *ARRs* of three carbonaceous chondrite meteorites

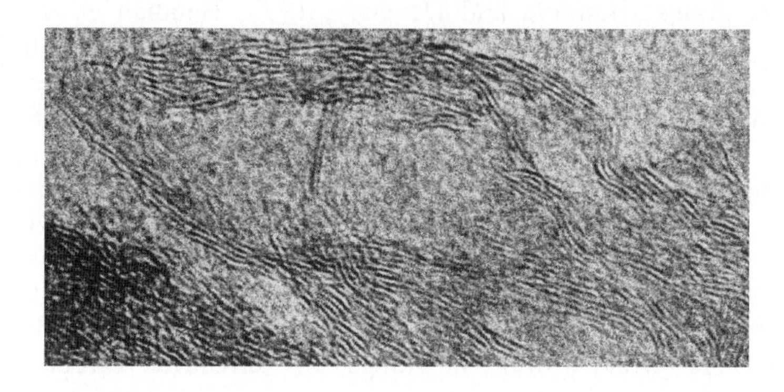

4(a) The Allende meteorite (obtained from the Houston Museum of Natural History)

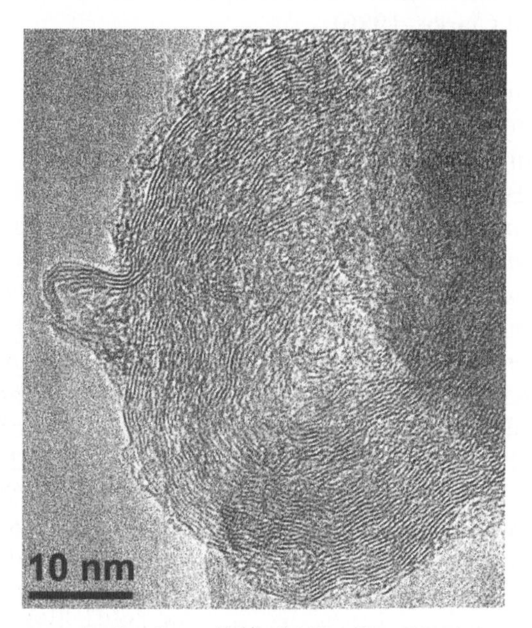

4(b) The Vigarano meteorite (made available by Dr. Gary Huss)

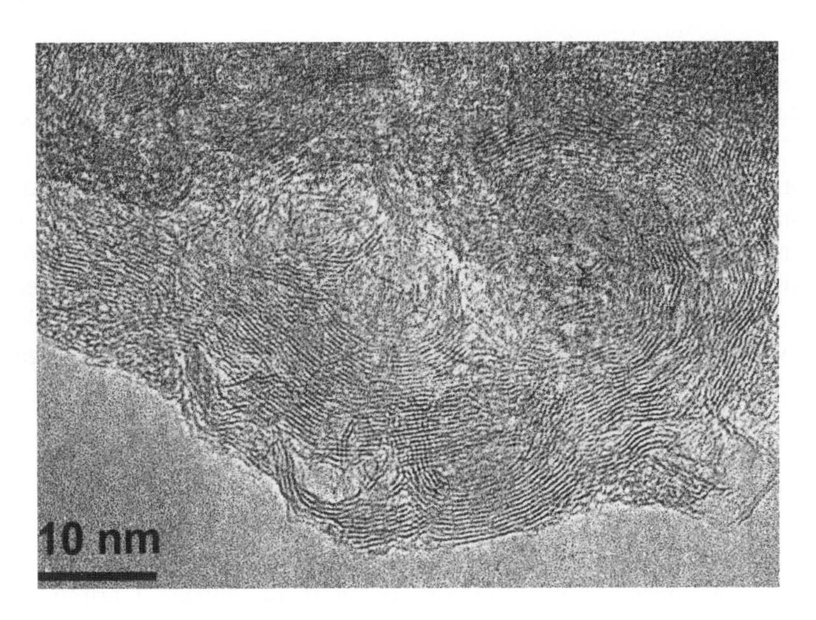

4(c) The Leoville meteorite (made available by Dr. Gary Huss)

The fringe distances range typically from 0.35 nm to 0.42 nm with a most frequently observed value of 0.38 nm indicating turbostratic stacking of graphene sheets. There is a tendency for the fringes to develop closed structures along the periphery, including ring structures (b and c) and apparent internal voids (c).

The images shown in Figure 4 were obtained by Dr. P. Kooyman using a Philips CM 30T TEM operated at 300kV at the National Center for HRTEM, University of Technology, Delft, The Netherlands

Harris et al. (2000), Harris and Vis (2003) and Vis et al. (2002) also observed single-walled and multiple-walled carbon rings, the former similar to such rings reported in condensed soot samples of variable C/H ratios (Rotundi et al., 1998). Structures closely similar to the carbon rings were found in pyrolized polyacetylene (Goto et al., 2001). Such rings in an Allende *ARR* sample could be fullerenes (Harris et al., 2000).

Taken together, all observations strongly suggest that the bulk of *ARR* carbon in CV carbonaceous chondrites is pyrolyzed hydrocarbons and charred organic polymers formed in a gas phase but perhaps also on hot surfaces of inorganic minerals. Because Allende chondrules are often fully jacketed by carbon (Bunch and Chang, 1980), we suggest that hot chondrules may have provided such surfaces. Hewing, as we do, to a close kinship of all carbonaceous chondrites, we further suggest that the parent materials resembled the complex organic polymers that are characteristic of the CI and CM chondrites (Rietmeijer, 2005). Because of the solar isotopic composition of the carbon, we surmise that the pyrolysis occurred in the early solar system in an environment harsher than that postulated (Hayatsu et al., 1977) for the formation of the complex organic polymers.

8. CARBYNES

Another persistent enigma of meteorite research is whether carbyne crystals, a linear carbon allotrope, occur in meteorites. Carbynes are constructed from chains of carbon atoms with *sp*-hybridized carbon bonds. Two major types of chains are possible, the polyyne-type in which the carbon atoms are bound by conjugated triple bonds, and the polycumulene-type in which the bonding occurs by cumulated double bonds. In 1969 Sladkov and Kudryavtsev claimed the discovery of carbynes (see, Kudryavtsev, 1999). A natural crystalline carbyne, called chaoite, was discovered interlayered within graphite at the Ries meteorite impact crater (Germany) (El Goresy and Donnay, 1968).

The first report of the occurrence of carbynes in meteorites was in the Novo Urei and Haverö ureilites based on X-ray diffraction analysis on the powdered bulk meteorite samples (Vdovykin, 1969, 1972). It was claimed that some 80% of the Allende and Murchison *ARRs* were carbynes (Lewis et al., 1980; Whittaker et al., 1980) and that the amorphous carbon in the Allende meteorite was all carbyne material. That particular claim, as it turned out, was premature because the carbyne-hypothesis lost credibility when the first HRTEM analyses of carbons in Allende *ARRs* (Smith and Buseck, 1980, 1981a, 1981b; Lumpkin, 1981a, 1981b, 1983a, 1983b) failed to discover any carbynes but instead identified pyrolitic carbons.

Recent HRTEM studies (Gilkes et al., 1992; Gilkes and Pillinger, 1999) reported the finding of platy, 10-nm carbon grains in *ARR* of the Murchison CM meteorite with the *d*-spacings and electron diffraction pattern of carbyne. At least eleven nanograins with various *d*-spacing values of their lattice fringes that match chaoite were found in *ARRs* of Allende, Leoville, and Vigarano CV chondrites (Vis et al., 2002), but these results must still be checked by electron diffraction analyses. If carbynes do occur in meteorites, they could have formed by condensation of carbon vapor in stellar outflows at high temperatures above 2875 ^{o}C as inferred from many experimental carbon studies (Whittaker and Kintner, 1969, 1976, 1985; Whittaker and Wolten, 1972; Kudryavtsev et al., 1996; Heimann, 1999; Tanuma, 1999; Babaev and Guseva, 1999). The available mineralogical evidence supports the presence of chaoite in meteorites, but none of other suggested forms of carbyne (Rietmeijer and Rotundi, 2005).

The possibility that carbynes could form by shock-transformation of "mainly aromatic hydrocarbons with more or less ordered structures" was first suggested to explain their presence in the differentiated ureilite meteorites (Vdovykin, 1969, 1972). Today, there is a substantial literature on the formation of carbynes by the shock loading of various structural forms of carbon and amorphous acetylene black commonly in the presence of a metal (usually copper) for fast cooling to suppress thermal decomposition of carbynes formed (Setaka and Sekikawa, 1980; Kleiman et al., 1984; Heimann et al., 1984, 1985; Sekine et al., 1987; Yamada et al., 1991, 1994, 2000; Donnet et al., 2000b; Babina et al., 1999; Milyavskiy et al., 2000). The reported peak pressures range from 13.5 to 56 GPa. Diamond was also formed in most of these experiments. The spontaneous transformation of carbyne into carbynoid structures mixed with diamond-like and graphitic carbon is a known process (Cataldo and Capitani, 1999; Cataldo, 2001b).

Judging from the cumulative synthetic evidence, it is apparent that shock-transformation of carbon precursors by strong impacts on meteoritic parent bodies is a viable candidate for the formation of the meteoritic carbynes. However, there are some questions. In several of the laboratory-based shock experiments, the precursor carbons were mixed with metal as heat sinks to avert the potential destruction of carbynes formed by slow cooling from high post-shock temperatures. The Allende, Leoville, and Vigarano meteorites are poor in Fe,Ni-metal and presumably the dust in regions of the solar nebula wherein these meteorite parent bodies accreted 4.56 Ga ago must have been metal-poor also.

9. CARBON CALABASHES AND NANOGLOBULES

9.1 Interrelationships

Hollow carbon calabashes were produced during catalytically supported, using MnO_2 as a catalyst, carbon-rich gas phase deposition between 900-1050 °C (Wang and Yin, 1998). Such a calabash is an enclosed spherical shell that can be either hollow or filled with an amorphous carbon core. Calabashes do not encapsulate the catalyst particles. It was suggested that calabashes could be a new fullerenic carbon nanostructure (Wang and Yin, 1998), which is at odds with observations that with increasing calabash diameter, the initially spherical shell becomes polygonal along with increased crystallographic ordering in the rim (Tracz et al., 1990; Wang and Yin, 1998; Donnet et al., 2000a; Rietmeijer et al., 2003). Hollow calabashes were also, among others, (1) produced in catalytically supported, carbon-rich gas phase deposition between 400-700 °C using a Ni/MgO catalyst (Tracz et al., 1990), (2) by shock-induced transformation of carbon black powder and C_{60} fullerene at 55 GPa using a Co catalyst (Donnet et al., 2000a), and (3) during oblique-impact induced dolomite [$MgCa(CO_3)_2$] decomposition at 2700 to 3700 °C into neutral species of CaI and MgI, CaO, MgO, CO and CO_2 that yielded quenched-liquid CaO, MgO and various carbon phases in an oxygen atmosphere (Rietmeijer et al., 2003). Reported calabash diameters are

1. 30-80 nm with a shell of "a few concentric graphitic layers" formed in an impact-shock experiment at 2.1 km s^{-1} in carbon black mixed with Co powder (Donnet et al., 2000b),
2. ca. 100 nm to ca. 800 nm with a ca.10 to 20 nm thick shell (Tracz et al., 1990; Wang and Yin, 1998), and
3. 45 nm to 308 nm (mean: 114 ± 67.5 nm) with a shell thickness ranging from 2.3 to 6.5 nm (Rietmeijer et al., 2003).

The shell is uniformly thick; its width is invariably a fraction of the shell radius. Parallel lattice fringes in the shells indicate (1) disordered (hkl) 002 graphite (0.36 nm) (Donnet et al., 2000a) and (2) carbyne, a linear carbon polytype, (hkl) 001fringes between 0.9-1.5 nm (mean = 1.1 ± 0.2 nm) (Rietmeijer et al., 2003). The same shock experiments that produced hollow carbon calabashes also yielded carbon nanoglobules (Donnet et al., 2000a, 2000b; Rietmeijer et al., 2003) (Fig. 5).

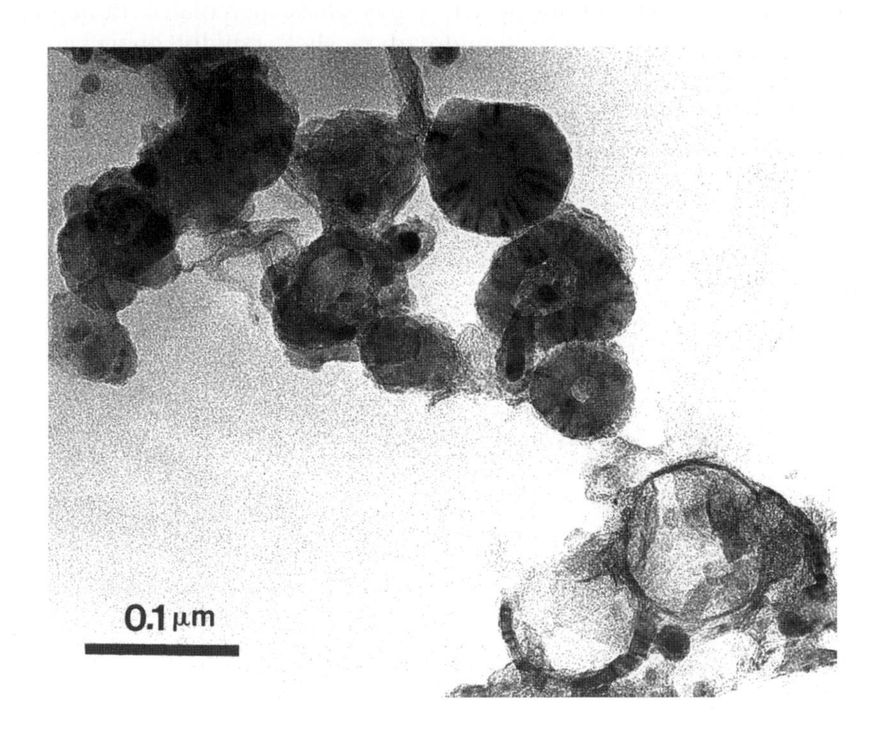

Figure 5. Transmission electron micrograph showing condensed carbons after oblique impact-induced dolomite dissociation and catalytic decomposition of CO_2 vapor. Several grains have an MgO or CaO core (black) surrounded by a graphitic rim (upper left-hand). The sub-spherical grain (top center) has an amorphous carbon core encapsulated by a graphitic rim with radial contrast features due to dislocations in its concentric basal lattice fringes. To the left and below of the filled grain is a hollow fullerenic carbon. Its seemingly dark core is due to an underlying metal-oxide or carbon entity, including an aerodynamically elongated MgO grain encapsulated by a rim. A hollow nanoglobule is associated with these grains. Both objects in the lower left-hand corner are hollow carbon calabashes. The gray background is the material of the ultrathin section. Modified after Rietmeijer et al. (2003)

Their co-occurrence suggests there are no fundamental differences between the physical conditions leading to the formation of these two carbon forms. The oblique impact experiment produced a transient carbon liquid and from this observation the following scenario emerged. The function between the shell width and calabash radius would be consistent with the thinning of a liquid-carbon shell on increasingly larger bubbles when a gas phase percolated through a carbon melt (Fig. 6). The liquid-carbon shell rapidly quenched to amorphous carbon, followed by graphite or carbyne nucleation in thick-walled hollow carbon nanoglobules and calabashes alike (Rietmeijer et al., 2003).

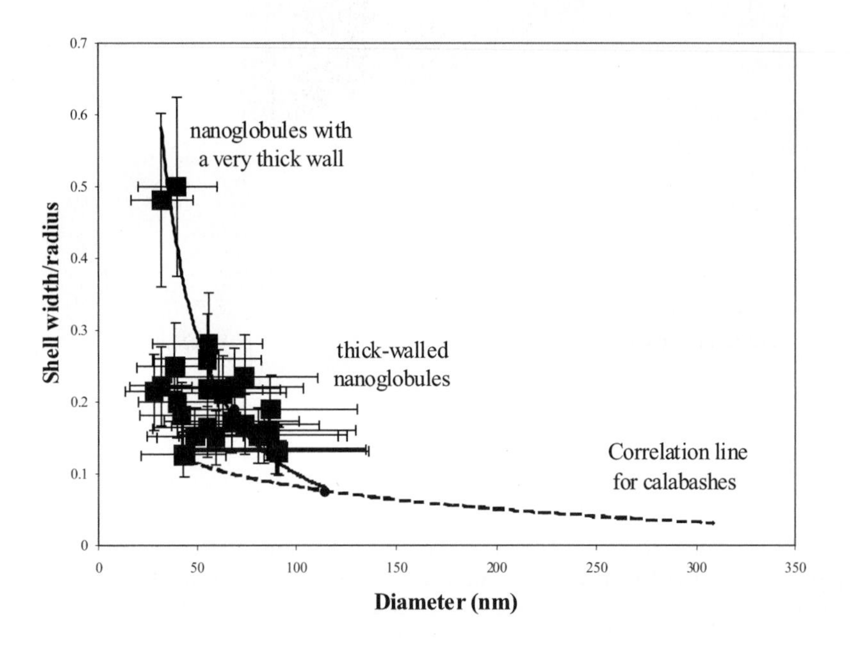

Figure 6. Thick-walled nanoglobule diameters (filled squares) as a function of the shell width to radius ratios in an oblique-impact on a dolomite target. Globules with larger diameters looked to have a thinner wall but the wall thickness for most globules was between 100-200 nm with no apparent correlation. The dashed line indicates the correlation for calabashes (Rietmeijer et al., 2003)

9.2 Hollow Carbon Nanoglobules

Carbon nanoglobules were produced by dynamic impact synthesis of graphite, carbon black, fullerenes and organic substances (Donnet et al., 2000a) and dolomite (Rietmeijer et al., 2003). Their formation is apparently independent of the nature of precursor material as long as a precursor modification produces a transient carbon melt but not necessarily to the exclusion of a carbon vapor phase. The dolomite experiment yielded many nanocarbons in a single, high-temperature and ultra-rapidly quenched thermal event, viz.

1. amorphous balls (<60 nm) encapsulated in a carbyne shell up to 20 nm thick (Fig. 5),
2. hollow nanoglobules (Fig. 5),
3. calabashes (Fig. 5),
4. (rare) fullerenic carbon onions (Fig. 5),
5. quenched-liquid CaO and MgO grains encapsulated in a carbyne shell (Fig. 7),
6. amorphous balls with a distinct tiny CaO or MgO core (Fig. 7), and
7. (rare) same as (2) but encapsulated in a carbyne shell whereby tiny MgO or CaO grains can be found at the interface between the amorphous and crystalline carbon shells.

Infrequent collisions between carbon melt droplets in the oblique impact experiment produced a multiple core surround by a single shell (Fig. 8). The very assemblage of carbon and catalyst/carbon grain morphologies is a strong indicator for catalytic conversion of carbon-bearing or C-rich precursors. It also points to a transient carbon melt. Such a melt is not generally considered in impact-related processes on asteroids or in environments wherein vapor phase condensation was the defining process at high pressures.

9.3 Significance for Presolar Graphite Spherules

1. As was discussed above, the Murchison (section 5) CM carbonaceous meteorite contains graphitic carbon spherules (0.3-9 μm in diameter) with $^{12}C/^{13}C$ ratios that establish a presolar origin in supernovae and asymptotic giant branch (AGB) stars (Bernatowicz et al., 1996). Recapping their petrographic features, these spherules have a nanocrystalline carbon core with a thin shell of well-ordered graphite. In many spherules a Ti-, Zr-, Mo- or Si-

carbide grain is located in the core's center (Bernatowicz et al., 1996). Such grains also occur at the interface of the core and encapsulating graphite shell.

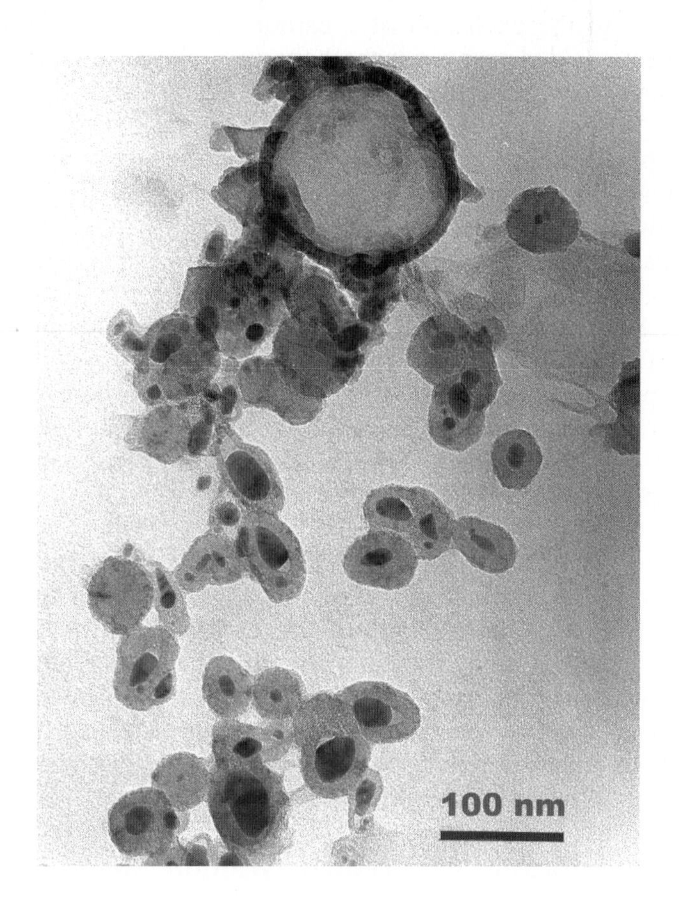

Figure 7. Transmission electron micrograph showing condensed carbons after oblique-impact-induced dissociation of dolomite and catalytic decomposition of the CO_2 vapor. At the top is a large hollow carbon calabash. The rest is carbon-mantled MgO and CaO grains that can have a (sub)-spherical or subhedral shape. The space between core and carbon rim is an aerodynamic feature due to the impact-generated atmosphere. The gray background is the material of the ultrathin section

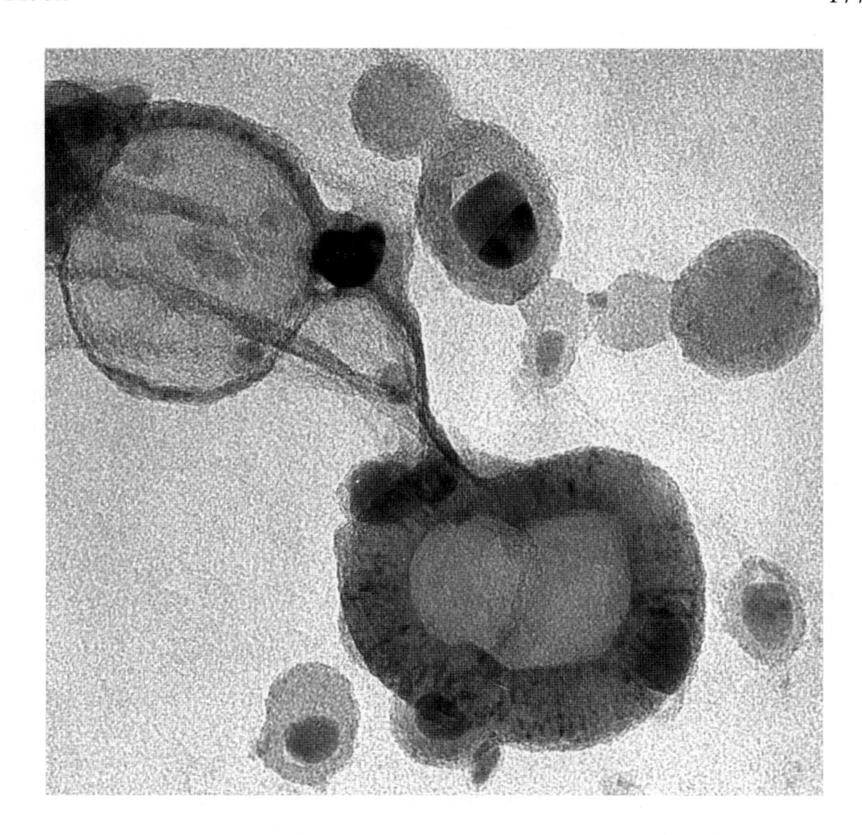

Figure 8. Transmission electron micrograph showing a rare object that formed in the ejecta plume of the oblique impact experiment (Rietmeijer et al., 2003) when two carbon melt droplets collided before they were encapsulated by the well-ordered shell. Three small nanoglobules with a massive catalyst core such as also seen elsewhere in this image are present on the well-ordered shell. The multiple core grain measures 115 nm x 95 nm

The presolar graphite spherules have three different morphologies (Bernatowicz et al., 1996), viz.

2. A nanocrystalline carbon core enclosed by the well-ordered graphite shell,
3. A very small carbide core embedded in a nanocrystalline carbon shell encapsulated in an outer graphite shell, and
4. Spherules with a large Fe or Fe-carbide core surrounded by a graphitic rim encapsulated by a discontinuous outer graphite shell (Bernatowicz et al., 1996; Fig. 7a).

The formation of these interstellar grains involved a two-step process of coagulation of graphene sheets on a carbide grain to form a nanocrystalline carbon core and vapor deposition of a graphite shell (Bernatowicz et al., 1996). The planar graphene sheets could be parallel stacked in larger structures that can occur in random orientations in nanocrystalline carbon. Whether the experimentally produced amorphous carbon (Donnet et al., 2000a; Rietmeijer et al., 2003) contained graphene sheets is unknown. The morphology of first two sphere types closely resembles those produced by catalytically supported decomposition of carbon-rich or light-hydrocarbon gas (Tracz et al., 1990; Wang and Yin, 1998) and dolomite (Figs. 5 and 7). The close fitting "double graphite rims" on a Fe metal or Fe-carbide core were not reproduced in these experiments. The similar morphologies of presolar 'graphite' spherules and experimentally produced spheres support that former are not fullerenic carbon onions.

A first-order implication of these experiments would be that the proposed two-step process for interstellar graphitic carbon spherules is in fact part of a single event of catalytically supported gas phase decomposition that involved a transient liquid carbon phase in a gas-rich atmosphere. If so, could conditions exist to support a transient carbon liquid in supernovae events and in the dust shells around young AGB stars? We don't know the answer to this question.

When the formation of nanoglobules and calabashes requires an oblique impact, it is conceivable that they could have formed on asteroids with a carbonate-rich surface lithology when a fragment (i.e. a meteoroid) was ejected by as such an impact. An example could be the very primitive Tagisch Lake CI2 carbonaceous chondrite meteorite (Brown et al., 2000). Nanoglobules were found in carbonate-poor patches in this meteorite although they were reportedly organic carbon (Nakamura et al., 2002). Carbon calabashes and related nanoglobules are not fullerenic carbon nanostructures such as multi-walled carbon nanotubes and onions formed during carbon-vapor condensation (Dravid et al., 1993; Rotundi et al., 1998; among others) and the well-ordered, concentric circular carbon onions in the Allende CV meteorite (Becker et al., 1993; Smith and Buseck, 1981a), a property that may enhance their survivability in the natural environments of asteroids and on Earth.

10. SUMMARY AND CONCLUSIONS

The great variety of the structural forms of elemental carbon in meteorites implies that these materials were formed in a variety of environments. Judging from the isotopic compositions of the carbon, some environments were definitely presolar whereas others could have been located within the solar system proper. The commonality of the formation processes is that these were 'high-energy' and probably of fairly short duration, including condensation, liquid carbon, shock, high static pressure, or pyrolysis. Hence, one could reasonably expect that fullerene formation, including not only C_{60} and C_{70} but also higher fullerenes, attended these formations of elemental carbon. Why, then were searches for fullerenes, especially for C_{60} and C_{70} so contradictory? It is conceivable that the apparently contradictory results were genuinely due to meteoritic inhomogeneity, either originally during solar nebula dust accretion, during asteroid parent body evolution or both. It is conceivable that the many structures of elemental carbon found in the carbonaceous chondrite meteorites did not form on the parent bodies of these meteorites. They already existed before these bodies accreted and the accretion could have involved material with and without natural fullerenes. Impacts into the surfaces of the parent bodies also could have transiently heated, hence destroyed, fullerenes in random, comparatively small, and quite localized volumes. The meteorites that were studied spent different times exposed to the Earth's atmosphere during which time any existing fullerenes might have been destroyed or severely depleted from their original indigenous abundances. Nearly all searches were done with comparatively small, gram-sized, samples of the meteorites and at that level compositional variations are well known in meteorites.

Acknowledgements: The first author thanks Drs. T. J. Bernatowicz, T. L. Daulton and P. Buseck for information and discussions. FJMR was supported by the National

Aeronautics and Space Administration under Grant NAG5-11762 issued through the Office of Space Science and by RTOPS from the Cosmochemistry and Origins of Solar Systems Research Programs. We are grateful to Peter Schultz (Brown University) and Ted Bunch (Northern Arizona University) who conducted the oblique impact dolomite experiments.

11. REFERENCES

Alaerts, L., Lewis, R.S., Matsuda, J. and Anders, E. (1980) Isotopic anomalies of noble gases in meteorites and their origins-VI. Presolar components in the Murchison C2 chondrite. *Geochim. Cosmochim. Acta*, 44, 189-209.

Alexander, C.M.O'D., Arden, J.W., Ash, R.D. and Pillinger, C.T. (1990) Presolar components in the ordinary chondrites. *Earth. Planet. Sci. Lett.*, 99, 220-229.

Amari, S., Anders, E., Virag, A. and Zinner, E. (1990a) Interstellar graphite in meteorites. *Nature,* 345, 238-240.

Amari, S., Zinner, E. and Lewis, R.S. (1990b) Two types of interstellar carbon grains in the Murchison carbonaceous chondrite (abstract). *Meteoritics*, 25, 348-349.

Amari, S., Lewis, R.S. and Anders, E. (1994) Interstellar grains in meteorites I. Isolation of SiC, graphite, and diamond: Size distributions of SiC and graphite. *Geochim. Cosmochim. Acta*, 58, 459-470.

Anders, E. and Zinner, E. (1993) Interstellar grains in primitive meteorites: Diamond, silicon carbide and graphite. *Meteoritics*, 28, 490-514.

Arden, J.W., Verchovsky, A.B. and Pillinger, C.T. (1994) The abundance of interstellar diamond in meteorites (abstract). *Meteoritics*, 29, 438.

Ash, R.D., Russell, S.S., Wright, I.P. and Pillinger, C.T. (1993) Minor high temperature components confirmed in carbonaceous chondrites by stepped combustion using a new sensitive static mass spectrometer (abstract), *Lunar Planet. Sci.*, 22, 35-36, The Lunar and Planetary Institute, Houston, Texas, USA.

Babaev, V.G. and Guseva, M.B. (1999) Ion-assisted condensation of carbon. *In Carbyne and Carbynoid Structures*, R.B. Heimann, S.E. Evsyukov and L. Kavan, Eds., 159-171, Kluwer Academic Publishers, Dordrecht, the Netherlands.

Babina, V.M., Boustie, M., Guseva, M.B., Zhuk, A.Z., Moigault, A., Milyavskyi, V.V. (1999) Dynamic synthesis of crystalline carbyne from graphite and amorphous carbon. *High Temp. Sci.*, 37, 543-551.

Banhart, F. and Ajayan, P.M. (1996) Carbon onions as nanoscopic pressure cells for diamond formation. *Nature*, 382, 433-435.

Banhart, F., Füller, T., Redlich, Ph. and Ajayan, P.M. (1997) The formation, annealing and self-compression of carbon onions under electron irradiation. *Chem. Phys. Lett.*, 269, 349-355.

Bauman, A.J. and Devaney, J.R. (1973) Allende C-3 chondrite carbonaceous phase: scanning electron morphology, differential thermal analysis, solvent properties, and spark source and electron impact mass spectrometry. *Meteoritics*, 8, 13-14.

Bauman, A.J. and Devaney, J.R. and Bollin, E.M. (1973) Allende meteorite carbonaceous phase: intractable nature and scanning electron morphology. *Science*, 241, 264-267.

Becker, L. and Bunch, T.E. (1997) Fullerenes, fulleranes and polycyclic aromatic hydrocarbons in the Allende meteorite. *Meteorit. Planet. Sci.*, 32, 479-487.

Becker, L., McDonnald, G.D. and Bada, J.L. (1993) Carbon onions in meteorites. *Nature*, 361, 595.

Becker, L., Bada, J.L., Winans, R.E. and Bunch, T.E. (1994) Fullerenes in Allende meteorite. *Nature*, 372, 507.

Becker, L., Bada, J.L. and Bunch, T.E. (1995) PAH's, fullerenes and fulleranes in the Allende meteorite (abstract). *Lunar Planet. Sci.*, 26, 87-88, The Lunar and Planetary Institute, Houston Texas, USA.

Becker, L., Poreda, R.J. and Bada, J.L. (1996) Extraterrestrial helium trapped in fullerenes in the Sudbury Impact Structure. *Science*, 272, 249-252.

Becker, L., Bunch, T.E. and Allamandola, L.J. (1999) Higher fullerenes in the Allende meteorite. *Nature*, 400, 227-228.

Becker, L., Poreda, R.J. and Bunch, T.E. (2000a) Fullerenes and noble gases in the Murchison and Allende meteorites. *Lunar Planet. Sci.*, 31, abstract #1803, The Lunar and Planetary Institute, Houston, Texas, USA (CD ROM).

Becker, L., Poreda, R.J. and Bunch, T.E. (2000b) Fullerenes: An extraterrestrial carbon carrier phase for noble gases. *Proc. Natl. Acad. Sci.*, 97, 2997-2983.

Bernatowicz, T.J. and Cowsik, R. (1996) Conditions in stellar outflows from laboratory studies of presolar grains. *In Astrophysical Implications of the Laboratory Study of Presolar Materials*, T. J. Bernatowicz and E. Zinner, Eds., *Am. Inst. Physics Conf. Proc.*, 402, 451-476, AIP Press, Woodbury, New York.

Bernatowicz, T.J., Gibbons, P.C. and Lewis, R.S. (1989) Meteoritic diamonds: Nature of the amorphous component (abstract). *Lunar Planet. Sci.*, 20, 65-66, The Lunar and Planetary Institute, Houston, Texas, USA.

Bernatowicz, T.J., Gibbons, P.C. and Lewis, R.S. (1990) Electron energy loss spectrometry of interstellar diamonds. *Astrophys. J.*, 359, 246-255.

Bernatowicz, T.J., Amari, S., Zinner, E.K. and Lewis, R.S. (1991) Interstellar grains within interstellar grains. *Astrophys. J.*, 373, L73-L76.

Bernatowicz T.J., Cowsik R., Gibbons P.C., Lodders K., Fegley, Jr. B., Amari S. and Lewis R.S. (1996) Constraints on stellar grain formation from presolar graphite in the Murchison meteorite. *Astrophys. J.*, 472, 760-782.

Blake, D.F., Freund, F., Shipp, R., Bunch, T.E., Flores, J., Chang, S., Krishnan, K.M., Echer, C. and Ackland, D. (1987) Analytical electron microscopy of interstellar diamond from Allende (abstract). *Meteoritics*, 22, 329-330.

Blake, D.F., Freund, F., Krishnan, K.F.M., Echer, C.J., Shipp, R., Bunch, T.E., Tielens, A.G., Lipari, R.J., Hetherington, C.J.D. and Chang, S. (1988) The nature and origin of interstellar diamond. *Nature*, 332, 611-613.

Breger, I.A., Zubovic, P., Chandler, R.S. and Clarke, R.S. (1972) Occurrence and significance of formaldehyde in the Allende carbonaceous chondrite. *Nature*, 236, 155-158.

Brown, P.G., Hildebrand, A.R., Zolensky, M.E., Grady, M., Clayton, R.N., Mayeda, T.K., Tagliaferri, E., Spalding, R., MacRea, N.D., Hoffman, E.L., Mittlefehldt, D.W., Wacker, J.F., Bird, J.A., Campbell, M.D., Carpenter, R., Gingerich, H., Glatiotis, M., Greiner, E., Mazur, M.J., McCausland, P. JA., Plotkin, H. and Mazur, T.R. (2000) The fall, recovery, orbit, and composition of the Tagish Lake meteorite: a new type of carbonaceous chondrite. *Science*, 90, 320-325.

Bunch, T.E. and Chang, S. (1980) Carbonaceous chondrites-II. Carbonaceous chondrite phyllosilicates and light element geochemistry as indicators of parent body processes and surface conditions. *Geochim. Cosmochim. Acta*, 44, 1543-1577.

Buseck, P. (2002) Geologic fullerenes: review and analysis. *Earth Planet. Sci. Lett.*, 203, 781-792.

Buseck, P. and Barry, J.C. (1988) Twinned diamonds in the Orgueil carbonaceous chondrite (abstract). *Meteoritics*, 23, 261-262.

Carey, W., Zinner, E., Fraundorf, P. and Lewis, R.S. (1987) Ion probe and TEM studies of a diamond bearing Allende residue (abstract). *Meteoritics*, 22, 349-350.

Cataldo, F. (2000) Raman spectra of radiation-damaged graphite. *Carbon*, 38, 634-636.

Cataldo, F. (2001a) Some implications of the radiation treatment of graphite and carbon black. *Fullerene Sci. Tech.*, 9, 409-424.

Cataldo, F. (2001b) Raman scattering investigation of carbynoid and diamond-like carbon. *Fullerene Sci. Tech.*, 9, 153-160.

Cataldo, F. and Capitani, D. (1999) Preparation and characterization of carbonaceous matter rich in diamond-like carbon and carbyne. *Mater. Chem. Phys.*, 59, 225-231.

Cataldo, F. and Keheyan, Y. (2002) On the mechanism of carbon clusters formation under laser irradiation. The case of diamond grains and solid C_{60} fullerene. *Fullerenes, Nanotubes, and Carbon Nanostructures*, 10, 313-332.

Chang, S., Mack, R. and Lennon, K. (1978) Carbon chemistry of separated phases of Murchison and Allende (abstract). *Lunar Planet. Sci.*, 9, 157-158, The Lunar and Planetary Institute, Houston, Texas, USA.

Croat, K., Bernatowicz, T.J., Stadermann, F.J., Messenger, S. and Amari, S. (2002) Coordinated isotopic and TEM studies of a supernova graphite. *Lunar Planet. Sci.*, 33, abstract #1315, The Lunar and Planetary Institute, Houston, Texas, USA (CD-ROM).

Daulton, T.L., Eisenhour, D.D., Bernatowicz, T.J., Lewis, R.S. and Buseck, P.R. (1996) Genesis of presolar diamonds: Comparative high-resolution transmission electron microscopy study of meteoritic and terrestrial nano-diamonds. *Geochim. Cosmochim. Acta*, 60, 4853-4872.

De Vries, M.S., Reihs, K., Wendt, H.R., Golden, W.G., Hunziker, H.E., Fleming, R., Peterson, E. and Chang, S. (1993) A search for C_{60} in carbonaceous chondrites. *Geochim. Cosmochim. Acta*, 57, 933-935.

Donnet, J.B., Fousson, E., Wang, T.K., Samirant, M., Baras, C. and Pontier-Johnson, M. (2000a) Dynamic synthesis of diamonds. *Diamond Rel. Mat.*, 9, 887-892.

Donnet, J-B., Fousson, E., Samirant, M., Wang, T.K., Pontier-Johnson, M.. and Eckhardt, A. (2000b) Shock synthesis of nanodiamonds from carbon precursors: identification of carbynes. *Comptes Rendues Acad. Sci,. Paris, Series IIc, Chimie/Chemistry*, 3, 359-364.

Dran, J.C., Klossa, J. and Maurette, M. (1979) The predicted irradiation record of asteroidal regoliths and the origin of gas-rich meteorites (abstract). *Lunar Planet. Sci.*, 10, 312-315, The Lunar and Planetary Institute, Houston, Texas, USA.

Dravid, V.P., Lin, X., Wang, Y., Wang, X.K., Yee, A., Ketterson, J.B. and Chang, R.P.H. (1993) Buckytubes and derivatives: Their growth and implications for buckyball formation. *Nature*, 259, 1601-1604.

Eberhardy, C.A. and Schultz, P.H. (2004) Probing impact-generated vapor plumes. *Lunar Planet. Sci.*, 35, abstract #1855, Lunar and Planetary Institute, Houston, Texas, USA (CD-ROM).

El Goresy, A. and Donnay, G. (1968) A new form of allotropic carbon from the Ries crater. *Science*, 161, 363-364.

Fisenko, A.V., Russell, S.S., Ash, R.D., Semjenova, L.F., Verchovsky, A.B. and Pillinger, C.T. (1992) Isotopic composition of carbon and nitrogen in the diamonds from the unequilibrated ordinary chondrite Krymka LL3.0 (abstract). *Lunar. Planet. Sci.*, 23, 365-366, The Lunar and Planetary Institute, Houston, Texas, USA.

Fraundorf, P. and Wackenhut, M. (2002) The core structure of presolar graphitic onions. *Astrophys. J.*, 578, L153-L156.

Fraundorf, P., Fraundorf, G., Bernatowicz, T.J., Lewis, R.S. and Ming, T. (1989) Stardust in the TEM. *Ultramicroscopy*, 27, 401-412.

Fukunaga, K., Matsuda, J., Ito, K., Nagao, K. and Miyamoto, M. (1987) Chemical vapor-deposition of diamonds from CH_4-H_2 gas mixtures and the origin of diamonds in meteorites (abstract). *Meteoritics*, 22, 381-382.

Gilkes, K.W.R., Gaskell, P.H., Russell, S.S., Arden, J.W. and Pillinger, C.T. (1992) Do carbynes exist as interstellar material after all? (abstract). *Meteoritics*, 27, 224.

Gilkes, K.W.R. and Pillinger, C.T. (1999) Carbon-How many allotropes associated with meteorites and impact phenomena? *In Carbyne and Carbynoid Structures*, R.B. Heimann, S.E. Evsyokov and L. Kavan, Eds., 17-30, Kluwer Academic Publishers, Dordrecht, the Netherlands.

Gilmour, I., Russell, S.S., Newton, J., Pillinger, C.T., Arden, J.W., Dennis, T.J., Hare J.P., Kroto, H.W., Taylor, R. and Walton, D.R.M. (1993) A search for the presence of C_{60} as an interstellar grain in meteorites (abstract). *Lunar. Planet. Sci.*, 22, 445-446, The Lunar and Planetary Institute, Houston, Texas, USA.

Göbel, R., Ott, U. and Begemann, F. (1978) On trapped noble gases in ureilites. *J. Geophys. Res.*, 83, 855-867.

Goodrich, C. (1992) Ureilites: A critical review. *Meteoritics*, 27, 327-352.

Goto, A., Kyotani, M., Tsugawa, K., Piao, G., Akagi K. and Koga, Y. (2001) Structure of pyrolytic carbon from polyacetylene. *Carbon*, 39, 2082-2086.

Grady, M.M., Swart, P.K. and Pillinger, C.T. (1981) Stable carbon isotopic measurements of ordinary chondrites (abstract). *Meteoritics*, 16, 319.

Green, H.W., Radcliffe, S.V. and Heuer, A.H. (1971) Allende meteorite: A high-voltage electron petrographic study. *Science*, 172, 936-939.

Grossman, L. (1972) Condensation in the primitive solar nebula. *Geochim. Cosmochim. Acta*, 49, 2433-2444.

Grumbach, M.P and Martin, R.M. (1996) Phase diagram of carbon at high pressures and temperatures. *Phys. Rev. B*, 54, 15730-15741.

Harris, P.J.F. and Vis, R.D. (2003) High resolution electron microscopy of carbon and nanocrystals in the Allende meteorite. *Proc. Roy. Soc. London A.*, 459, 2069-2076.

Harris, P.J.F., Vis, R.D. and Heymann, D. (2000) Fullerene-like carbon nanostructures in the Allende meteorite. *Earth Planet. Sci. Lett.*, 183, 355-359.

Hayatsu, R., Matsuoka, S., Scott, R.G., Studier, M.H. and Anders, E. (1977) Origin of organic matter in the early solar system-VII. The organic polymer in carbonaceous chondrites. *Geochim. Cosmochim. Acta*, 41, 1325-1339.

Heimann, R.B. (1999) Resistive heating and laser irradiation. *In Carbyne and Carbynoid Structures*, R.B. Heimann, S.E. Evsyukov, L. Kavan, Eds., 139-148, Kluwer Academic Publishers Dordrecht, the Netherlands.

Heimann, R.B., Kleiman, J. and Salansky, N.M. (1984) Structural aspects and conformation of linear carbon polytypes. *Carbon*, 22, 147-156.

Heimann, R.B., Fujiwara, S., Kakudate, Y., Koga, Y., Komatsu, T. and Nomura, M. (1985) A new carbon form obtained by weak shock compression of carbyne. *Carbon*, 33, 859-861.

Heymann, D. (1986) Buckminsterfullerene C_{60} and siblings: Their deduced properties as traps for inert gas atoms (abstract). *Lunar Planet. Sci.*, 17, 337-338, The Lunar and Planetary Institute, Houston, Texas, USA.

Heymann, D. (1990a) On the chemical attack of fullerene, soot, graphite, and sulfur with hot perchloric acid. *Carbon*, 29, 684-685.

Heymann, D. (1990b) The geochemistry of Buckminsterfullerene (C_{60}) I: Solid solutions with sulfur and oxidation with perchloric acid (abstract). *Lunar Planet. Sci.*, 22, 569-570, The Lunar and Planetary Institute, Houston, Texas, USA.

Heymann, D. (1995a) Search for extractable fullerenes in the Allende meteorite (abstract). *Lunar Planet. Sci.*, 25, 595-596, The Lunar and Planetary Institute Houston, Texas, USA.

Heymann, D. (1995b) Search for extractable fullerenes in the Allende meteorite. *Meteoritics*, 30, 436-438.

Heymann, D. (1996a) Fullerenes were not found in lunar samples 20084 and 79261 (abstract). *Lunar Planet. Sci.*, 27, 541-542, The Lunar and Planetary Institute, Houston, Texas, USA.

Heymann, D. (1996b) Search for fullerenes in lunar fines 10084 and 79261. *Meteorit. Planet. Sci.*, 31, 362-364.

Heymann, D. (1997) Fullerenes and fulleranes in meteorites revisited. *Astrophys. J.*, 489, L111-L114.

Heymann, D. (2001) 'Isotopically strange xenon' in meteoritic nanodiamonds: Implantation by stellar winds? *Astrophys. Space Sci.*, 275, 415-423.

Heymann, D. and Anders, E. (1967) Meteorites with short cosmic-ray exposure ages, as determined from their Al^{26} contents. *Geochim. Cosmochim. Acta*, 31, 1793-1809.

Heymann, D. and Dziczkaniec, M. (1979) Xenon from intermediate zones of supernovae. *Proc. Lunar Planet. Sci. Conf.*, 10, 1943-1960, The Lunar and Planetary Institute, Houston, Texas, USA.

Heymann, D. and Dziczkaniec, M. (1980) A first roadmap for kryptology. *Proc. Lunar Planet. Sci. Conf.*, 11, 1179-1213, The Lunar and Planetary Institute, Houston, Texas, USA.

Heymann, D. and Pontier-Johnson, M.A. (2002) New prescriptions for growing interstellar carbonaceous cauliflowers *Astrophys. J. Lett.*, 574, L91-L94.

Heymann, D., Stormer, Jr., J.C. and Pierson, M. (1990) Buckminsterfullerene (C_{60}) dissolves in molten and solid sulfur. *Carbon*, 29, 1053-1055.

Hoppe, P., Amari, S., Zinner, E. and Lewis, R.S. (1992a) Large oxygen isotopic anomalies in graphite grains from the Murchison meteorite (abstract). *Meteoritics*, 27, 235.

Hoppe, P., Amari, S., Zinner, E. and Lewis, R.S. (1992b) Just how many types of interstellar carbon? (abstract). *Lunar Planet. Sci.*, 23, 553-554, The Lunar and Planetary Institute, Houston Texas, USA.

Housley, R.M. and Clarke, D.R. (1980) XPS and STEM studies of Allende acid insoluble residues. *Proc. Lunar Planet. Sci. Conf.*, 11, 945-958, The Lunar and Planetary Institute, Houston, Texas, USA.

Hurt, R.H., Crawford, G.P. and Shim, H.S. (2000) Equilibrium nanostructure of primary soot particles. *Proc. Combustion Inst.*, 28, 2539-2546.

Huss, G.R. (1990) Ubiquitous interstellar diamond and silicon carbide in primitive chondrites: abundances reflect metamorphism. *Nature*, 347, 159-162.

Huss, G.R. and Lewis, R.S. (1994) Noble gases in presolar diamonds II: Component abundances reflect thermal processing. *Meteoritics*, 29, 811-829.

Huss, G.R. and Lewis, R.S. (1995) Presolar diamond, SiC, and graphite in primitive chondrites: Abundances as a function of meteorite class and petrologic type. *Geochim. Cosmochim. Acta*, 59, 115-160.

Jenkins, G.M. and Kawamura, K. (1976) Polymeric carbons-carbon fibre, glass and char, 178p., Cambridge University Press, Cambridge, London, New York, Melbourne.

Jørgensen, U.G. (1988) Formation of Xe-HL-enriched diamond grains in stellar environments. *Nature,* 332, 702-705.

Kagi, H., Takahashi, K., Shimizu, H., Kitajima, F. and Masuda, A. (1991) *In-situ* micro raman studies on graphitic carbon in some Antarctic ureilites. *Proc. NIPR Symp. Antarct. Meteorites*, 4, 371-383.

Kerridge J.F. and Matthews M.S., Eds. (1988) *Meteorites and the Early Solar System*, 1269p., The University of Arizona Press, Tucson, Arizona, USA.

Kleiman, J., Heimann, R.B., Hawken, D. and Salansky, N.M. (1984) Shock compression and flash heating of graphite/metal at temperatures up to 3200 K and pressures up to 25 GPa. *J. Appl. Phys.*, 56, 1440-1454.

Krätschmer, W., Lamb, L.D., Fostiropoulos, K. and Huffman, D. (1990) Solid C_{60}: A new form of carbon. *Nature*, 347, 354-358.

Kroto, H.W. (1988) Space, stars, C_{60} and soot. *Science*, 242, 1139-1145.

Kroto, H.W. and Walton, D.R.M. (1993) Polyynes and the formation of fullerene. *Phil. Trans. R. Soc. Lond. A*, 343, 103-112.

Kroto, H.W., Heath, J.R., O'Brien, S.C., Curl, R.F. and Smalley, R.E. (1985) C_{60}: Buckminsterfullerene. *Nature*, 318, 162-163.

Kroto, H.W., Heath, J.R., O'Brien, S.C., Curl, R.F. and Smalley, R.E. (1987) Long carbon chain molecules in circumstellar shells. *Astrophys. J.*, 314, 352-355.

Kudryavtsev, Y.P., Heimann, R.B. and Evsyukov, S.E. (1996) Carbynes: Advances in the field of linear carbon chain compounds. *J. Mater. Sci.*, 31, 5557-5571.

Kudryavtsev, Y.P. (1999) The discovery of carbyne. *In Carbyne and Carbynoid Structures*, R.B. Heimann, S.E. Evsyukov, L. Kavan, Eds., 1-6, Kluwer Academic Publishers, Dordrecht, the Netherlands.

Lewis, R.S. and Amari, S. (1992) Interstellar Murchison graphite: How many noble gas components? (abstract). *Lunar Planet. Sci.*, 23, 775-776, The Lunar and Planetary Institute, Houston, Texas, USA.

Lewis, R.S. and Anders, E. (1988) Xenon-HL in diamonds from the Allende meteorite-composite nature (abstract), *Lunar Planet. Sci.*, 19, 679-680, The Lunar and Planetary Institute, Houston, Texas, USA.

Lewis, R.S., Srinivasan, B. and Anders, E. (1975) Host phase of a strange Xe component in Allende. *Science*, 190, 1251-1262.

Lewis, R.S., Matsuda, J.-I., Whittaker, A.G., Watts, E.J. and Anders, E. (1980) Carbynes: Carriers of primordial noble gases in meteorites (abstract). *Lunar Planet. Sci.*, 11, 624-625, The Lunar and Planetary Institute, Houston, Texas, USA.

Lewis, R.S., Ebihara, M. and Anders, E. (1982) Unpaired electrons: An association with primordial gases in meteorites (abstract). *Meteoritics*, 17, 244-245.

Lewis, R.S., Ming, T., Wacker, J.F., Anders, E. and Steel, E. (1987) Interstellar diamonds in meteorites. *Nature*, 326,160-162.

Lumpkin, G.R. (1981a) Electron microscopy of carbon in Allende acid residues (abstract). *Lunar Planet. Sci.*, 12, 631-633, The Lunar and Planetary Institute, Houston, Texas, USA.

Lumpkin, G.R. (1981b) Electron microscopy of carbonaceous matter in Allende acid residues. *Proc. Lunar Planet. Sci. Conf.*, 12, 1153-1166, The Lunar and Planetary Institute, Houston, Texas, USA.

Lumpkin, G.R. (1983a) Electron microscopy of carbonaceous matter in acid residues from the Orgueil (C1) and Cold Bokkeveld (C2) meteorites (abstract). *Lunar Planet. Sci.*, 14, 450-451, The Lunar and Planetary Institute, Houston, Texas, USA.

Lumpkin, G.R. (1983b) Microstructural variations in Allende carbonaceous matter (abstract). *Lunar Planet. Sci.*, 14, 452-453, The Lunar and Planetary Institute, Houston, Texas, USA.

McCall, G.J.H. (1973) *Meteorites and their origin*, 352p., Davis and Charles, Newton Abbott, Australia.

Milyavskiy, V.V., Borodina, T.I., Zhuk, A.Z. and Fortov, V.E. (2000) Shock-wave-induced transformation of graphite to carbyne. *Mol. Mater.*, 13, 361-366.

Ming, T. and Anders, E. (1988) Isotopic anomalies of Ne, Xe and C in meteorites. II. Interstellar diamond and SiC. Carriers of exotic noble gases. *Geochim. Cosmochim. Acta*, 52, 1235-1244.

Ming, T., Lewis, R.S. and Anders, E. (1987) Diamond and silicon carbide: carriers of presolar noble gases in carbonaceous chondrites. *Meteoritics*, 22, 462-463.

Nakamura, K, Zolensky, M.E., Tomita, S., Nakashima, S. and Tomeoka, K. (2002) Hollow organic globules in the Tagish Lake meteorite as possible products of primitive organic reactions. *Intl. J. Astrobiology*, 1, 179-189.

Nichols, Jr, R.H., Hohenberg, C.M., Hoppe, P., Amari, S. and Lewis, R.S. (1992) ^{22}Ne-E(H) and ^{4}He in single SiC grains and ^{22}Ne-E(L) in single C_{α} grains of known C-isotopic composition (abstract). *Lunar Planet. Sci.*, 23, 989-990, The Lunar and Planetary Institute, Houston, Texas, USA.

Nuth, III, J.A. (1987) Small-particle physics and interstellar diamonds. *Nature*, 329, 589.

Oberlin, A. (1989) High-resolution TEM studies of carbonization and graphitization. *In Chemistry and Physics of Carbon*, P.A. Thrower, Ed., 22, 1-144, Marcel Dekker, Inc., New York.

Oberlin, A. (2002) Pyrocarbons. *Carbon*, 40, 7-24.

Oester, M.Y., Kuechl, D., Sipiera, P.P. and Welch, C.J. (1994) Search for fullerenes in stone meteorites (abstract). *Meteoritics*, 29, 513.

Ott, U. (1993) Interstellar grains in meteorites. *Nature*, 364, 25-33.

Ott, U. (2001) Presolar grains in meteorites: an overview and some implications. *Planet. Space Sci.*, 49, 763-767.

Papike, J.J., Ed. (1998) *Interplanetary Dust Particles, Planetary Materials, Revs. Mineral.*, 36, 1052p., The Mineralogical Society of America, Washington, DC, USA.

Pillinger, C.T. (1993) Elemental carbon as interstellar dust. *Phil. Trans. R. Soc. Lond. A*, 343, 73-86.

Pizzarello, S., Huang, Y.S., Becke, L., Poreda, R.J., Nieman, R.A., Cooper, G. and Williams, M. (2001) The organic content of the Tagish Lake meteorite. *Science*, 293, 2236-2239.

Pontiert-Johnson, M.A. (1998) Noirs de carbone au four: méchanisme de formation des particules (in French). Ph. D. Thesis, 185p., Université de Haute-Alsace, Mulhouse, France.

Praburam, G. and Goree, J. (1995) Cosmic dust synthesis by accretion and coagulation. *Astrophys. J.*, 441, 830-838.

Radicati di Brozolo, F., Bunch, T.E., Fleming, R.H. and Macklin, J. (1994) Observation of fullerenes in an LDEF impact crater. *Nature*, 369, 37-40.

Rietmeijer, F.J.M. (1988) On graphite in primitive meteorites, chondritic interplanetary dust, and interstellar dust. *Icarus*, 74, 446-453.

Rietmeijer, F.J.M. (1990) On diamond, graphite and amorphous carbons in primitive extraterrestrial solar system materials. *In Carbon in the galaxy: Studies from Earth and Space*, J.C. Tarter, S. Chang and D.J. DeFrees, Eds., *NASA Conf. Publ.*, 3061, 339-340, NASA Scientific and Technical Information Division.

Rietmeijer, F.J.M. (2005) Iron-sulfides and layer silicates: A new approach to aqueous processing of organics in interplanetary dust particles, CI and CM meteorites. *Adv. Space Res.*, in press.

Rietmeijer, F.J.M. and Rotundi, A. (2005) Natural carbynes, including chaoite, on Earth, in meteorites, comets, circumstellar and interstellar dust. *In Polyynes: Synthesis, Properties, and Application*, F. Cataldo, Ed., 339-370, CRC Press, Taylor & Francis Publishing Group, Boca Raton, Florida, USA.

Rietmeijer, F.J.M., Schultz, P.H. and Bunch, T.E. (2003) Carbon calabashes in a shock-produced carbon melt. *Chem. Phys. Lett.*, 374, 464-470.

Rotundi, A., Rietmeijer, F.J.M., Colangeli, L., Mennella, V., Palumbo, P. and Bussoletti, E. (1998) Identification of carbon forms in soot materials of astrophysical interest. *Astron. Astrophys.*, 329, 1087-1096.

Russell, S.S., Arden, J.W. and Pillinger, C.T. (1991) Evidence for multiple sources of diamond from primitive chondrites. *Science*, 254, 1188-1191.

Russell, S.S., Pillinger, C.T., Arden, J.W., Lee, M.R. and Ott, U. (1992) A new type of meteoritic diamond in the enstatite chondrite Abee. *Science*, 256, 206-209.

Russell, S.S., Arden, J.W. and Pillinger, C.T. (1996) A carbon and nitrogen isotope study of diamonds from primitive chondrites. *Meteorit. Planet. Sci.*, 31, 343-355.

Schultz, L. and Kruse, H. (1989) Helium, neon, and argon in meteorites-A data compilation. *Meteoritics*, 24, 155-172.

Schultz, P.H. (1996) Effect of impact angle on vaporization. *J. Geophys Res.*, 101(E9), 21117-21136.

Sekine, T., Akaishi, M., Setaka, N. and Kondo, K.I. (1987) Diamond synthesis by weak shock loading. *J. Mat. Sci.*, 22, 3615-3619.

Setaka, N. and Sekikawa, Y. (1980) Chaoite: a new allotropic form of carbon produced by shock compression. *J. Am. Ceram. Soc.*, 63, 238-239.

Shim, H.S., Hurt, R.H. and Yang, N.Y.C. (2000) A methodology for analysis of 002 lattice fringe images and its application to combustion-derived carbons. *Carbon*, 38, 29-45.

Simmonds, P.G., Bauman, A.J., Bollin, E.M., Gelpi, E. and Oro, J. (1969) The unextractable organic fraction of the Pueblito de Allende meteorite: evidence for its indigenous nature. *Proc. Nat. Acad. Sci.*, 64, 1927-1934.

Simon, S.B., Grossman, L., Clayton, R.N., Mayeda, T.K., Schwade, J.R., Sipiera, P.P., Wacker, J.F. and Wadhwa, M. (2004) The fall, recovery, and classification of the Park Forest meteorite. *Meteorit. Planet. Sci.*, 39, 625-634.

Smith, P.P.K and Buseck, P. (1980) High resolution transmission electron microscopy of an Allende acid residue (abstract). *Meteoritics*, 15, 368-369.

Smith P.P.K. and Buseck P.R. (1981a) Graphitic carbon in the Allende meteorite: a microstructural study. *Science*, 212, 322-324.

Smith, PP.K. and Buseck, P.R. (1981b) Carbon in the Allende meteorite: Evidence for poorly graphitized carbon rather than carbyne. *Proc. Lunar Planet. Sci. Conf.*, 12, 1167-1175, The Lunar and Planetary Institute, Houston, Texas, USA.

Snow, T.P. and Witt, A.N. (1995) The interstellar carbon budget and the role of carbon in dust and large molecules. *Science*, 270, 1455-1460.

Spurný, P., Oberst, J. and Heinlein, D. (2003) Photographic observations of Neuschwanstein, a second meteorite from the orbit of the Příbram chondrite. *Nature*, 423, 151-153.

Stadermann, F.J., Bernatowicz, T.J., Croat, T.K., Zinner, E., Messenger, S. and Amari, S. (2002) Presolar graphite in the nanosims: A detailed look at the isotopic makeup of the spherule and its sub-components. *Lunar Planet. Sci.*, 33, abstract #1796, The Lunar and Planetary Institute, Houston, Texas, USA (CD-ROM).

Sugita, S., Schultz, P.H. and Adams, M.A. (1998) Spectroscopic measurements of vapor clouds due to oblique impacts. *J. Geophys Res.*, 103(E8), 19427-19441.

Swart, P.K., Grady, M.M., Wright, I.P. and Pillinger, C.T. (1982) Carbon components and their isotopic compositions in the Allende meteorite. *Proc. Lunar. Planet. Sci. Conf.*, 13, A283-A288, The Lunar and Planetary Institute, Houston, Texas, USA.

Swart, P.K., Grady, M.M., Pillinger, C.T., Lewis, R.S. and Anders, E. (1983) Interstellar carbon in meteorites. *Science*, 220, 406-410.

Tanuma, S. (1999) Condensation of carbon vapor obtained by electrical discharge. *In Carbyne and Carbynoid Structures*, R.B. Heimann, S.E. Evsyukov and L. Kavan, Eds., 149-158, Kluwer Academic Publishers, Dordrecht, the Netherlands.

Taylor, R. and Abdul-Sada, A.K. (2000) There are no fullerenes at the K-T boundary. *Fullerene Sci. Tech.*, 8, 47-54.

Tielens, A.G.G.M., Seab, C.G., Hollenbach, D.J. and McKee, C.F. (1987) Shock processing of interstellar dust: Diamonds in the sky. *Astrophys. J.*, 319, L109-L113.

Tingle, T.N., Becker, C.H. and Malhotra, R. (1991) Organic compounds in the Murchison and Allende carbonaceous chondrites studied by photoionization mass spectrometry. *Meteoritics*, 26, 117-127.

Tracz, E., Borowiecki, T. and Scholz, R. (1990) TEM study of carbon shell formation on Ni/MgO catalyst. *Proc. XIIth Intl Congr. Electron Microscopy*, 302-303, San Francisco Press Inc., San Francisco, California, USA.

Travaglio, C., Gallino, R., Amari, S., Zinner, E., Woosley, S. and Lewis, R.S. (1999) Low-density graphite grains and mixing in Type II supernovae. *Astrophys. J.*, 510, 325-354.

Ugarte, D. (1992) Curling and closure of graphitic networks under electron-beam irradiation. *Nature*, 359, 707-709.

Verchovsky, A.B., Huss, G.R. and Pillinger, C.T. (1994) Nitrogen and carbon isotopes in presolar diamond samples with known noble gas isotopic signatures (abstract). *Meteoritics*, 29, 544-545.

Vdovykin, G.P. (1969) New hexagonal modification of carbon in meteorites. *Geochem. Int.*, 6, 915-918.

Vdovykin, G.P. (1972) Forms of carbon in the new Haverö ureilite of Finland. *Meteoritics*, 7, 547-552.

Vdovykin, G.P. and Moore, C.B. (1960) Carbon. *In Handbook of Elemental Abundances*, B. Mason, Ed., 81-91, Gordon and Breach, New York, USA.

Virag, A., Zinner, E., Lewis, R.S. and Ming, T. (1989) Isotopic compositions of H, C, and N in C_α diamonds from the Allende and Murray carbonaceous chondrites (abstract). *Lunar Planet. Sci.*, 20, 1158-1160, The Lunar and Planetary Institute, Houston, Texas, USA.

Vis, R.D., Mrowiec, A., Kooyman, P.J., Matsubara, K. and Heymann, D. (2002) Microscopic search for the carrier phase Q of the trapped planetary noble gases in Allende, Leoville, and Vigarano. *Meteorit. Planet. Sci.*, 37, 1391-1400.

Wacker, J.F. (1986) Noble gases in the diamond-free ureilite, ALHA 78019: The roles of shock and nebular processes. *Geochim. Cosmochim. Acta*, 50, 633-642.

Wang, Z.L. and Yin, J.S. (1998) Graphitic hollow carbon calabashes. *Chem. Phys. Lett.*, 289, 189-192.

Wieler, R., Anders, E., Baur, H., Lewis, R.S. and Signer, P. (1991) Noble gases in "phase Q": Closed-system etching of an Allende residue. *Geochim. Cosmochim. Acta*, 55, 1709-1722.

Whittaker, A.G. (1978) Carbon: A new view of its high-temperature behavior. *Science*, 200, 763-764.

Whittaker, A.G. and Kintner, P.L. (1969) Carbon: Observations on the new allotropic form. *Science*, 165, 589-591.

Whittaker, A.G. and Kintner, P.L. (1976) Particle emission and related morphological changes occurring during the sublimation of graphitic carbons. *Carbon*, 14, 257-265.

Whittaker, A.G. and Kintner, P.L. (1985) Carbon: Analysis of spherules and splats formed from the liquid state and of the forms produced by quenching gas and solid. *Carbon*, 23, 255-262.

Whittaker, A.G. and Wolten, G.M. (1972) Carbon: a suggested new hexagonal crystal form. *Science*, 178, 54-56.

Whittaker, A.G., and Watts, E.J., Lewis, R.S. and Anders, E. (1980) Carbynes: carriers of primordial noble gases in meteorites. *Science*, 209, 1512-1514.

Yamada, K., Kunishige, H. and Sawaoka, A.B. (1991) Formation process of carbyne produced by shock compression. *Naturwissenschaften*, 78, 450-452.

Yamada, K., Burkhard, G., Dan, K., Tanabe, Y. and Sawaoka, A.B. (1994) Microstructures of carbon polymorphs formed in shocked compressed diamond powder utilizing an interaction of oblique shock-waves. *Carbon*, 32, 1197-1213.

Yamada, K., Tanabe, Y. and Sawaoka, A.B. (2000) Allotropes of carbon shock synthesized at pressures up to 15 Gpa. *Phil. Mag. A*, 80, 1811-1828.

Zinner, E., Wopenka, B., Amari, S. and Anders, E. (1990) Interstellar graphite and other carbonaceous grains from the Murchison meteorite: structure, composition, and isotopes of C, N, and Ne (abstract). *Lunar Planet. Sci.*, 21, 1379-1380, The Lunar and Planetary Institute, Houston, Texas, USA.

Chapter 9

FULLERENES IN THE CRETACEOUS-TERTIARY BOUNDARY

DIETER HEYMANN
Rice University. Department of Earth Science, MS 126, Houston, Texas 77251-1892, USA.

WENDY S. WOLBACH
DePaul University. Department of Chemistry, 1036 W Belden Ave, Chicago, Illinois 60614, USA.

Abstract: Fullerenes C_{60} and C_{70} were found in the thin clay seams of nine worldwide locations of the geologic boundary between the Cretaceous and Tertiary periods. These clays are also rich in soot. One hypothesis suggests that the fullerenes were synthesized in global wildfires that followed the meteorite impact 65 Ma ago. An alternative hypothesis suggests that the fullerenes predate the formation of the solar system because they contain isotopically anomalous helium in their carbon cages.

Key words: Cretaceous-Tertiary Boundary (KTB); fullerenes; wildfires

1. INTRODUCTION

The first searches for fullerenes in natural materials were carried out in meteorites because the discoverers of C_{60} had suggested that this molecule might occur quite abundantly in circumstellar media, interstellar media, the nascent solar system, *ergo* in meteorites (Kroto et al., 1985). The first quests for C_{60} in stony meteorites, which were first reported at the Lunar and Planetary Science Conference of 1991 (de Vries et al., 1991), failed to find any fullerenes (Tingle et al.,

Frans J.M. Rietmeijer (ed.), Natural Fullerenes and Related Structures of Elemental Carbon, 191–212.

1991). Subsequent studies also yielded negative results (Ash et al., 1993; de Vries et al., 1993; Gilmour et al., 1993; Oester et al., 1994). Despite these early failures, the question arose whether favorable conditions for the formation of fullerenes might have existed on the Earth in which case abundant material might be potentially available. Discussions of that topic engendered searches for terrestrial C_{60} along different strategies. One idea was to investigate rocks that are rich in carbonaceous matter and that had been heated by metamorphic events, or shock-metamorphosed by large meteorite impacts, or both, because significant energy seemed to be required to transform carbonaceous matter into fullerenes. Along these lines, Buseck et al. (1992) searched for and found C_{60} by high-resolution transmission electron microscopy (HRTEM) in shungite, a metamorphosed carbon-rich meta-sedimentary rock of Lower Proterozoic age (2.1-2.0 Ga) on the Kola Peninsula (Russia). Theirs was the first report of a "*Buckyball in Nature*". Another hypothesis assumed that the energy of lightning striking the Earth's surface in the presence of dry plant remains might be sufficient to transform these remains to form C_{60} in the resulting fulgurite, which is typically a hollow tubular structure of molten surface material. Daly et al. (1993) reported the finding of fullerenes in such a fulgurite, but Heymann (1998) found no fullerenes in the char produced by a lightning strike in a Norway spruce. However, the potential fullerene-synthesis-by-lightning differed fundamentally from that of the fulgurites because most of the energy of the lightning strike into this living and wet tree produced steam which kept the maximum temperature reached well below that which would have been produced by lightning-strike ignition of dry wood. The failure to distinguish between the consequences of the burning of wet versus dry biota eventually caused much confusion about fullerene contents in soot from modern wildfires, wherein none were found, compared to fullerenes in soot proposed to have been produced by the global wildfires at the Cretaceous-Tertiary Boundary.

2. THE CRETACEOUS-TERTIARY BOUNDARY (*KTB*)

The *KTB* defines the end of the Cretaceous Period and the beginning of the Tertiary 65 Ma ago. It is widely known, if not famous for the catastrophic worldwide extinction of many species at that time, including the dinosaurs. In 1980, Alvarez and his collaborators discovered a thin layer of clay-like geologic deposit that was greatly enriched in iridium at exposed *KTB* locations distributed on global scale (Alvarez et al., 1980). At most of the known *KTB* (Fig. 1) locations the Ir-rich seams are a few millimeters to several centimeters thick.

Figure 1. World Map showing the locations of the **Cretaceous-Tertiary Boundary** (*KTB*) and the Permian-Triassic Boundary (*PTB*) sites samples listed in the Tables

1 = Woodside Creek, New Zealand. 2 = Flaxbourne River, New Zealand. 3 = Chancet Rocks, New Zealand. 4 = Stevns Klint, Denmark. 5 = Caravaca, Spain. 6 = Agosto, Spain. 7 = Gubbio, Italy. 8 = Raton Basin, USA. 9 = El Kef, Tunisia. 10 = Elendgraben, Austria. 11 = Sumbar, Turkmenistan. 12 = Koshak, Turkmenistan. 13 = Malyi Balkhan, Turkmenistan. 14 = Brazos River, USA. 15 = Deep Sea Drilling Project site 465, Pacific Ocean. 16 = Tetri Tskaro, Republic of Georgia.

The **Permian-Triassic Boundary** sites (*italics*): 17 = Meishan, People's Republic of China. 18 = Sasayama, Japan. 19 = 1.85 Ga-old Sudbury Impact Feature.
The Allende meteorite fell in Mexico. The Murchison meteorite fell in Australia.

At a few locations, such as at the Brazos River site in Texas (USA), the entire *KTB* sequence of sediments is approximately one meter thick, but the iridium-rich layer is only about 0.4-0.5 cm thick (Smit et al., 1996). Because the iridium contents of essentially all terrestrial crustal rocks are orders of magnitude smaller than those of the *KTB* clays, Alvarez et al. (1980) concluded that a kilometer-sized extraterrestrial, iridium-rich body had collided with the Earth and that the aftermath of this event had triggered the catastrophic biologic extinctions.

This hypothesis raised the question where on Earth the corresponding and presumably very large crater might be. The existence of a large buried circular structure at the tip of the Yucatán Peninsula (Mexico) had been known since 1950 (Cornejo Toledo and Hernandez Osuna, 1950). This structure was ignored as a candidate with the right geological age until 1991 when that crater, now known as Chicxulub, was firmly identified as the Cretaceous-Tertiary impact site (Hildebrand et al., 1991). The age of the crater is now determined as 65 Ma. The target at Chicxulub was a thick carbonate rock platform with sulfate seams (Sharpton et al., 1996). The rocks contain little, if any carbonaceous matter today, but, when one considers the extent of the currently known Mexican oil and gas deposits, one may speculate that the formations could have contained Permian oil and natural gas at the time of the impact (Dr. Albert Bally, pers. comm., 2003).

3. WILDFIRES

Geochemical investigations of the *KTB* sediments revealed that these were uncommonly rich in elemental carbon and contained abundant soot, allegedly due to global wildfires triggered by the impact (Wolbach et al., 1985, 1988, 1990a, 1990b, 2003; Anders et al., 1986, 1991; Argyle et al., 1986; Gilmour et al., 1990, Heymann et al., 1997, 1998). These investigators defined the *KTB* elemental carbon phases as consisting of three components,

1. Fusinite (natural charcoal),
2. Submicrometer sized spherules of amorphous carbon welded into clusters and chains like a bunch-of-grapes (aciniform carbon), and
3. Carbonaceous microgel of spheroidal carbon particles embedded in a carbonaceous matrix.

Soot was defined in the aciniform component only and it was thought to form above a flame from direct condensation from the carbon gas to a solid carbon phase (see also, Kroto, 1988). In the early studies, the *KTB* clay samples were demineralized with HF and HCl. This procedure left too much kerogen that is insoluble organic matter derived from plants behind, making it difficult to identify elemental carbon morphologies in the Scanning Electron Microscope (SEM). In subsequent work the demineralization was therefore followed by dichromate oxidation under controlled conditions, which removed most of the kerogen. The mass fraction of soot was determined by morphological recognition and planimetric analysis in the SEM. No Transmission Electron Microscope (TEM) or HRTEM studies of this extracted *KTB* carbon were reported to date.

The most detailed geochemical study of carbonaceous matter was made at the *KTB* outcrop of Woodside Creek (New Zealand) where the Cretaceous rocks below and the Tertiary rocks above the *KTB* clay layer are massive, that is hard, low-porosity limestone. Iridium, elemental carbon, and soot contents were determined at 22 locations straddling the boundary from about 40 cm below to about 250 cm above the *KTB* clay layer. The contents of Ir, C and soot rise steeply from the Cretaceous rock below the boundary into the *KTB* clay by factors of 1400, 210, and 3600, respectively. Above the clay layer, the contents decrease again. The maximum iridium content is 114 ppb and the maximum content of carbon is 0.462%. Soot is a major component of the carbon phases with a maximum content of about 70%. Single maxima in the concentrations of iridium, carbon and soot are by no means a typical feature in a *KTB* layer. For example, at the nearby Flaxbourne River site and at the Brazos River site (Texas), iridium contents show two or more maxima. Kyte et al. (1985) suggested that only the lowest maximum of such cases is due to primary, that is worldwide, fallout and that higher layers are due to later deposits delivered from local rivers, creeks, and streams. A summary of all carbon data is presented in Table 1.

Anders et al. (1991) remarked "*No charcoal layers approaching 11 mg cm^{-2} have been found anywhere* [in the geologic record], *let alone layers that are (1) global, (2) isochronous, and (3) enriched in soot. The K/T soot layer, coinciding with a very sharp, worldwide marker [iridium], appears to be unique*." Note that at the Brazos River site the carbon and soot deposits are more than double the average worldwide values.

Table 1. Carbon abundance and soot concentrations at the Cretaceous-Tertiary Boundary sites

Site	Carbon abundance (mg cm^2)	Soot %
Woodside Creek, NZ[1]	4.8	69
Chancet Rocks, NZ	35	68
Stevns Klint, DK	11	27
Caravaca, E	10	14
Gubbio, I	13	17
Agost, E	>3.6	11
Raton, USA	4.0	13
El Kef, TN	15	19
Flaxbourne River, NZ	<1.3	55
Elendgraben, A	>0.8	12
Sumbar, T	11	15
Brazos River, USA[2]	28	64
Mean[3]	11	32
	Carbon abundance (ppm)	**Soot %**
DSDP site 465[4]		
465-3-3,129-131	3,600 ± 400	50
465A-3-3,130-132	1800 ± 200	30

Notes:
[1] From Anders et al. (1991) but without uncertainties.
[2] From Heymann et al. (1998).
[3] The Brazos River site is well preserved, but extremely complex because of thick deposits from either storm-driven waves or a tsunami. For this reason, the Brazos River data were not used in the calculation of the mean value.
[4] From Wolbach et al. (2003). Data are from the levels of the highest iridium concentrations. Soot percentages calculated for this paper.

County Key to Tables

NZ	New Zealand
DK	Denmark
E	Spain (España)
I	Italy
USA	United States of America
TN	Tunisia
A	Austria
T	Turkmenistan
A	Austria
K	Kazakhstan
G	Republic of Georgia
C	Canada
J	Japan

Of all *KTB* locations, this site was the closest to the Chicxulub impact at the time of the event but the significance of this geographic fact is not clear. Were potential combustibles more thoroughly dried here than elsewhere on Earth? Or was the burnable biomass here much larger than elsewhere? For completeness we note that the carbon abundance and soot concentration at the Deep Sea Drilling Project (DSDP) sites in the Pacific Ocean are originally given in ppm concentrations, hence they cannot be directly compared to values reported at the other *KTB* sites.

The fundamental issue was "*where did this mass of carbon come from, especially the spheroidal clusters of submicrometer sized particles that show the morphology of carbon formed in flames?*" Wolbach et al. (1985) considered three possible hypotheses to explain the high *KTB* carbon content.

1. An impacting, asteroidal (Kyte, 1998), meteorite with sufficient mass estimated to have caused the Chicxulub crater and having a carbon content of typical carbonaceous chondrites would have provided almost half of the *KTB* carbon needed worldwide. However, the resulting fireball from the impact would have quantitatively incinerated this carbon to CO_2. Moreover, the aciniform carbon structures of the *KTB* carbon are totally different from those found so far in carbonaceous chondrites (Smith and Buseck, 1980, 1981a, 1981b; Lumpkin, 1981a, 1981b, 1983a, 1983b; Amari et al., 1990; Bernatowicz et al., 1991, 1996; Bernatowicz and Cowsik, 1996; Daulton et al., 1996; Harris et al., 2000; Harris and Vis, 2002; Vis et al., 2002).
2. On the same grounds, they dismissed that carbon had formed from fossil carbon (oil, coal, etc.) from the impact crater site that was initially explored for the presence of fossil fuels (Cornejo Toledo and Hernandez Osuna, 1950). Still, the impact may have been a sort of natural "super-oil-drill", oil and gas to escape locally into the atmosphere where it was ignited.
3. The impact of a meteor with estimated 10-km diameter would have generated a huge fireball that ignited wildfires on continents more than 1000 km from the point of impact with ensuing winds dispersing the soot worldwide. The scale of the wildfires must have been staggering as approximately 10% of the then-present total biomass on Earth must had to have become ignited to account for the estimated total carbon at the *KTB*. The occurrence of the polynuclear aromatic hydrocarbon retene (*1-methyl, 7-isopropyl*

phenanthrene) in several *KTB* clays is apparently diagnostic of resinous wood fires (Venkatesan and Dahl, 1989).

The wildfire theory was met with skepticism (Scott, 2000; Scott et al., 2000) but also with constructive criticism, such as that by Argyle et al. (1986). A serious problem was that the essentially simultaneous deposition of excess (relative to indigenous terrestrial) iridium and high amounts of soot demanded that the fires had started very soon after the impact. As Argyle at al. (1986) pointed out, 'green', living biota is not conducive for the development of large wildfires. Anders et al. (1986) therefore suggested that trees and other combustibles had been freeze-dried during the 'global winter' following the impact. Later, Anders et al. (1991) suggested that it was the thermal radiation from incandescent ejecta reentering the atmosphere that would have dried most wood on Earth very swiftly. Melosh et al. (1990) and Kring and Durda (2002) have since proposed satisfactory models for igniting of even wet biomass based on the total amount of energy and its dissipation in the Chicxulub impact event. As was pointed out above, it is unclear whether such fires could have produced fullerenes.

4. FULLERENE PRODUCTION IN COMBUSTION

Although fullerene production in wildfires has not been demonstrated to occur naturally, it has long been known that soot produced by oxygen-starved combustion of fuels does contain fullerenes (Gerhardt et al., 1987, 1988; Howard et al., 1991, 1992, 1994; Anacleto et al., 1992; Baum et al., 1992; Howard, 1992; Pope et al., 1993; Bachmann et al., 1995). It is conceivable that copious local production of soot and fullerenes from the ignition and burning of natural gas and crude oil after the impact could have happened, especially in the ancient Gulf of Mexico region. When taken into account the occurrence of the Ir- and soot-rich *KTB* seams, the wildfire-hypothesis for the origin of the soot, and the known occurrence of fullerenes in soot produced by oxygen-starved combustion, the authors of this chapter decided that a search for C_{60} and C_{70} in *KTB* clay seams might have a reasonable chance of success.

5. FULLERENE GEOCHEMISTRY

The search for fullerenes in rocks is a comparatively simple geochemical procedure. Both C_{60} and C_{70} dissolve well in CS_2 and aromatic organic solvents, hence most analytical procedures begin with solvent-extraction of either a powdered rock sample or its acid-resistant residue (*ARR*). This carbon-enriched residue obtained by the dissolution of inorganic minerals usually with HF-HCl. CS_2 was seldom used because of its flammability and tendency to produce elemental sulfur. As the number of carbon atoms in the fullerene molecules increases, their solubility decreases and other solvents are used for extraction. One assumes that the fullerenes are not too strongly locked up in the powdered samples or *ARRs* and that the chemical treatments do not destroy them. Studies have shown that the HF-HCl treatments do not degrade fullerenes but that very strongly oxidizing chemicals, such as hot perchloric acid, will do so (Heymann, 1991). In darkness or even in the moderate light of laboratory benches, C_{60} does not react with molecular oxygen. Ozone in air is only dangerous compound that reacts very swiftly with fullerenes in solution (Heymann and Chibante, 1993).

Essentially no dedicated research was done on the issue of survival of fullerenes in sediments such as the *KTB* clays. By extrapolation of the solubility of C_{60} in alcohols, Heymann (1996) estimated the solubility of non-clustered C_{60} and C_{70} molecules in water to be 1.3×10^{-11} and 1.3×10^{-10} ng mL^{-1}, with an one order-of-magnitude uncertainty. Given the current volume of the oceans of 1.3×10^{9} km^{3}, it was estimated that 1 kg to 100 kg of fullerenes might dissolve in the worldwide ocean. However, it was discovered more recently that C_{60}, in contact with water might form water-soluble "nano-C_{60}" clusters that have dramatically increased solubility up to 100 ppm concentrations (Simonin, 1991; Bensasson et al., 1994; Scrivens et al., 1994; Ying et al., 1994; Andrievsky et al., 1999, 2002; Schuster et al., 2004). Deguchi et al. (2001) found that C_{70} also forms colloidal solutions in which, however, the maximum concentrations of the fullerenes decrease significantly when NaCl was added to the solutions. The colloidal fullerene solutions in water are extremely toxic (Sayes et al., 2004). As such, fullerenes in solution could have been potentially detrimental to non-marine aquatic life at the end of the Cretaceous. These are intriguing results, but it is not very likely that fullerene solution in natural waters was relevant in the case of combustion-produced fullerenes of the global wildfires where fullerene molecules would find themselves packaged inside insoluble soot. Once such fullerene-soot packages are ensconced in sediments, the fullerenes are, most likely, in protective microenvironments.

6. SEARCHES FOR *KTB* FULLERENES WITH HIGH PERFORMANCE LIQID CHROMATOGRAPHY

In this section we describe and discuss the results of searches for *KTB* fullerenes extracted from *KTB* clays with High Performance Liquid Chromatography (HPLC). This search began jointly at Rice University and Illinois Wesleyan University where a small sample of clay (~10 g) from Woodside Creek (New Zealand) was available. Subsequently, we collected much larger fresh samples at Woodside Creek and Flaxbourne River (New Zealand) as well as at the Brazos River site (Texas, USA). Dr. Brooks donated additional samples from New Zealand. Dr. Smit provided samples from the Caravaca (Spain) and Stevns Klint (Denmark) locations while Drs. Nazarov and Korochantsev provided samples from *KTB* sites in the (former) USSR. Samples from Austria were received from Dr. Preisinger.

The powdered samples of boundary clays were either directly extracted with toluene or were extracted after destruction of most inorganic minerals with HF-HCl. Limestone and other carbon-poor rocks from below and above the clays were always directly extracted. Only new glassware was used. Procedural blanks that were identical to actual runs, but in which no sample was added, were done. The filtered solutions were analyzed with HPLC. The *KTB* samples were not injected until at least ten injections of toluene (instrumental blanks) showed no detectable presence of C_{60} and C_{70}. The sensitivity of the HPLC analysis was calibrated with injections of known, but very small amounts of the fullerenes dissolved in toluene. Amounts of 1 ng C_{60} could be detected. The identification of fullerenes was made on the basis of retention times and the well-known UV-vis spectra of C_{60} and C_{70} in the range 300 nm to 800 nm. The stronger absorption lines below 300 nm could not be used because of the opacity of the solvent below 300 nm.

Most of the *KTB* samples were rich in toluene-soluble organic compounds, including PAH's. Hence the fullerene peaks in HPLC were "riding on the tails" of large organic peaks. This problem was significantly alleviated by pre-separation of much, but not all, organic compounds from fullerenes with a large semi-preparative HPLC column followed by volume reduction with N_2 (Heymann et al., 1995a). Heymann et al. (1994a) reported the first hint that C_{60} was present in the small sample from the *KTB* site at Woodside Creek (New Zealand) and that no fullerenes were found in the massive limestone deposits above and below the *KTB* clay at this site. Subsequently, fullerenes were studied in much larger samples and were found to be present or absent at thirteen different *KTB* sites around the world (Heymann et al., 1994b, 1994c, 1995a, 1995b, 1996a, 1996b, 1997, 1998) (Fig. 1). The C_{60} contents range from essentially zero to 12 ppb; C_{70}, when present, was typically between 0.2 and 0.4 times C_{60} (Table 2).

Given that fullerenes were found in so many samples at different *KTB* sites around the world, makes this unique fullerene occurrence the best-documented case of natural fullerenes. Nevertheless, there are still unanswered issues, controversies and serious skepticism that need to be considered. The fullerene contents worldwide vary greatly (Table 2) but so do the contents of iridium, carbon, and soot. Numerous factors may have contributed to these variations such as local ease of ignition, the kinds of biota, degree of oxygen starvation, winds, rain, and near-shore ocean currents following the impact. In addition, there are issues of fullerene in soot and soot preservation.

Fortunately, as mentioned above the solubility of fullerenes in water is exceedingly small hence water percolating through the *KTB* clays was not likely to have removed detectable amounts of fullerenes. The reaction with molecular oxygen in darkness (the "*nuclear winter scenario*") is apparently very slow but it is fast in the presence of light owing to the formation of chemically reactive singlet oxygen. Fullerene molecules deposited on land were therefore more vulnerable to oxidation than molecules deposited in water. Ozone swiftly attacks fullerenes to form ozonides that lose O_2 to form oxides, which, in turn, have a tendency to polymerize to compounds that are very poorly soluble in organic solvents (Heymann et al., 2000; Weisman et al., 2001). Of course, the ozone-peril also threatens fullerenes in *KTB* samples that were stored for long periods in laboratories unless preventive steps were taken. Taken together, these factors could explain some of the variations in fullerene contents in *KTB* samples, especially local variations such as between Woodside Creek, Chancet Rocks, and Flaxbourne River sites (New Zealand) (see, Fig. 1) that are only separated by a few miles.

Table 2. Fullerenes in samples from Cretaceous-Tertiary Boundary Sites

Site	Country	C_{60} (ppb)	C_{70}/C_{60}	Sources
Woodside Creek[1]	NZ	2.17	[8]	(a)
Woodside Creek[1,2]	NZ	2.74	[8]	(a)
Woodside Creek[1]	NZ	2.5	<0.002	(a)
Woodside Creek[1]	NZ	4.7	<0.002	(a)
Woodside Creek[1]	NZ	5.4	<0.002	(a)
Woodside Creek[3]	NZ	0.325	<0.002	(a)
Woodside Creek[1,2]	NZ	2.48	0.34	(a)
Woodside Creek[3]	NZ	0	0	(b)
Flaxbourne River[4]	NZ	0.556	[8]	(a)
Flaxbourne River[4,2]	NZ	1.14	[8]	(a)
Flaxbourne River[4]	NZ	0.0058	[8]	(a)
Flaxbourne River[4]	NZ	2.3	0.28	(a)
Flaxbourne River[4]	NZ	1.84	0.28	(a)
Flaxbourne River[4,2]	NZ	11.9	0.36	(a)
Flaxbourne River[4,2]	NZ	12.2	0.29	(a)
Caravaca	E	2.7	0.22	(d)
Caravaca[2]	E	11.9	0.26	(d)
Sumbar-4[2]	T	4.0	0.21	(d)
Sumbar-5[2]	T	1.2	[8]	(d)
Koshak[2]	K	0	0	(d)
Malyi Balkhan	K	1.8	0.36	(d)
Stevns Klint	DK	0	0	(d)
Stevns Klint	DK	0	0	(d)
Elendgraben	A	0	0	(d)
Tetri Tskaro	G	0	0	(d)
Brazos River[5]	USA	0	0	(e)
Gubbio[6]	I	detected		(f)
Raton Basin	USA	detected		(f)
Stevns Klint[7]	DK	detected		(g)

Notes:

[1] This material came from high on a hillside at Woodside Creek at which no sampling had ever been done before.
[2] Demineralized samples.
[3] This sample came from the northern bank of the creek, at river level.
[4] This material came from a location at Flaxbourne River where no sampling had been done before.
[5] No fullerenes were found in any of the 55 samples studied.
[6] Becker et al. (1994a; ref. #13).
[7] Higher fullerenes were also reported.
[8] C_{70} was not detected.

Sources: (a) Heymann et al. (1994a); (b) Taylor and Abdul-Sada (2000); (c) Heymann et al. (1994b); (d) Heymann et al. (1995b, 1996a, 1996b); (e) Heymann et al. (1997, 1998), (f) Becker et al. (1994a); (g) Becker et al. (2000a)

A second concern is that no C_{70} was detected in most of the Woodside Creek samples (Table 2). Perhaps that was largely due to the greater difficulty of detecting this fullerene than C_{60} or perhaps to variable local conditions of combustion.

Overall, a major unresolved issue surrounding the findings of natural fullerene is centered on the question of their survivability in natural terrestrial settings during their original emplacement in the environment, subsequent geological processing and during exposure to the atmosphere-soil-rock interactions (i.e. weathering) after exhumation of the very sites wherein they were found. It is not an uncommon experience that seemingly fresh looking in hand specimen upon closer inspection in the laboratory may show evidence of weathering that is a highly unpredictable phase of bio-geological modification.

Taylor and Abdul-Sada (2000), who reported no traces of C_{60} or C_{70} in a single sample from the Woodside Creek site, raised a serious objection to the validity of claims to the contrary and suggested that earlier HPLC results measured only "a mixture of hydrocarbons". However, there are questions about their study also. Considering the natural variability in surviving fullerene abundances in *KTB* samples, a single measurement on a small, 20 g sample from the southern exposure of the Woodside Creek location by Taylor and Abdul-Sada (2000) is not convincing evidence for claiming there are "no fullerenes in the K-T boundary layer". These authors may not have appreciated that, even for small fullerene concentrations, the C_{60} peak in the HPLC chromatograms can indeed stick out above a tail caused by a mixture of hydrocarbons in the sample (Heymann et al., 1994b). The obvious reason is that C_{60} has an exceptionally large molar extinction at 336 nm that is much larger than the extinctions due to hydrocarbons. Taylor and Abdul-Sada (2000) offered no plausible explanation why several of all the *KTB* extracts studied showed not only C_{60} at its expected retention time, and with its known UV-vis spectrum, but also C_{70} at the expected retention time and with its known UV-vis spectrum.

Becker et al. (1995a) questioned reports of fullerenes from *KTB* sites on the grounds that the atmospheric oxygen content at the time of the Chicxulub event was too high, that is higher than during wood burning associated with the 1993 Malibu Fires wherein no fullerenes were found in a collected sooty sample. Gilmour et al. (1990) have argued that the atmospheric oxygen concentration at the end of the Cretaceous could not have been significantly larger than what it is today. Clearly, oxygen is needed to remove hydrogen from the fullerene precursors in flame-produced fullerenes. The 1993 Malibu Fire consumed live trees as opposed to assumed dead trees at the *KTB*. It was also a very small fire, 14,000 acres of brush, compared to the deduced extent of the Cretaceous-Tertiary wildfires (Kring and Durda, 2003; Durda and Kring, 2004).

7. SEARCHES FOR *KTB* FULLERENES WITH LASER-DESORPTION MASS SPECTROMETRY

Early searches for fullerenes in *KTB* samples using laser-desorption mass spectrometry (LDMS) were also undertaken on a sample from the Gubbio *KTB* site, Italy (Becker et al., 1994a). It is widely recognized that LDMS is a potentially problematic analytical method in the search for indigenous fullerenes because fullerenes can actually be synthesized in the laser-produced plume material. A more serious shortcoming of LDMS is the deduction of the fullerene content of a sample from the strength of the mass spectrometric signal alone. Becker at al. (1994a) reported the finding of C_{60} in the Gubbio (Italy) *KTB* sample along with preliminary data for fullerenes in the Black Member sediments of the Onaping Formation at the giant (180 x 250 km) Sudbury Impact Feature (Canada) was formed by a meteorite impacting Earth at 1.85 Ga (Fig. 1). Becker et al. (1994a) suggested that trace quantities of fullerenes might be common in sedimentary deposits associated with giant meteorite impact events such as the Cretaceous-Tertiary impact event. Quoting Heymann et al. (1994c) they submitted that fullerenes at these sites were possibly produced during a global conflagration but Becker et al. (1995a) expressed doubts of the wildfire origin of fullerenes, and asked if it is possible that fullerenes in *KTB* clays might be produced during the impact events at Sudbury and Chicxulub from processing indigenous carbonaceous matter in the bolide.

If we assume, for example, that the bolide was 10 km in diameter (Hildebrand et al., 1991) and contained 1 weight % of organic carbon, that is the organic carbon content of an average carbonaceous chondrite meteorite, then the total organic carbon present in the bolide was about 10^{16} g. If we further assume a worldwide fullerene concentration of 1 to 5 ppb for the *KTB* clays (Table 2), then roughly 10^{10} g of C_{60} were deposited globally at the time of the Chicxulub impact event. Therefore, the conversion of only 10^{-4} % of the total bolide carbon into C_{60} could account for C_{60} fullerene found in the clay layers. This would be the best-case scenario because the carbon content of a bolide of ordinary chondrite meteorite, representing ca. 85% of all meteorites, would be much lower. The strength of hypotheses based on meteorite impacts is that quantitative estimates of this kind are easy to make because the unique nature of such an event. The wildfire-based hypothesis is hampered by the fact that global biomass burning was not a single event, but stretched over a period of up to several years

Becker et al. (1996) reported the finding of fullerenes in the Black Member sediments of the Onaping Formation at the Sudbury impact structure that contained isotopically greatly anomalous Helium. From this observation they advanced the new hypothesis that since the He isotopes were clearly consistent with an extraterrestrial source, the C_{60} molecules had to be extraterrestrial. Ergo, the fullerene had come to the Earth locked inside the bolide. The next logical step was to search for extraterrestrial He@C_{60} in the collected meteorites, in particular in the most primitive carbonaceous chondrites. Becker et al. (1994b, 1995b) and Becker and Bunch (1997) had already reported the occurrence of fullerenes in the Allende CV carbonaceous chondrite and eventually He@C_{60} was reported in the Allende meteorite and in the Murchison CM carbonaceous chondrite (Becker et al., 2000a, 2000b) and in *KTB* clays (Becker et al., 2000a, 2000c). The $^{3}He/^{4}He$ ratios reported for He@C_{60} fullerene in *KTB* samples are shown in Table 3 along with this ratio in Sudbury samples and at the Permian-Triassic boundary where at the Chinese location essentially no fullerene was reported (Farley and Mukhopadhyay, 2001).

Nevertheless, this bold and intriguing hypothesis has inherent problems. The Cretaceous-Tertiary bolide was probably a typical carbonaceous chondrite (Kyte, 1998). Fullerene contents of numerous Allende samples, a CV carbonaceous chondrite, are greatly variable: from 0 to 10 ppm. If one assumes 1 ppm, this bolide contained on the order of 10^{12} g C_{60}, or about 100 times the estimated *KTB* amount. Is a "survival rate" of 1% reasonable? We really do not know. The fact that the typical carbon structures of carbonaceous chondrite meteorites such as Allende are not found at *KTB* sites is not so serious because other, CI and CM, carbonaceous chondrites are known to have different types of carbonaceous matter.

Table 3. Fullerenes with anomalous $^3He/^4He$ ratios at the Cretaceous-Tertiary Boundary (*KTB*) and Permian-Triassic Boundary (*PTB*)

Sample Sites	$^3He/^4He$	Sources
Sudbury (C); Capreol[1]	5.88×10^{-4}	(a)
Sudbury (C); Dowling[1]	5.46×10^{-4}	(a)
Raton Basin (USA); *KTB*	100 times air (1.4×10^{-4}) [2]	(b)
Stevns Klint [3] (DK); *KTB*	$>4 \times 10^{-3}$	(b)
Woodside Creek (Z); *KTB*	400 times air (5.56×10^{-4}) [2]	(c)
Meishan (China); *PTB*	1.6 to 1.9×10^{-4} [4]	(d)
Sasayama (J); *PTB*	1.6 to 1.9×10^{-4} [4]	(d)
Meishan (China); *PTB*	Essentially no helium	(e)
Murchison Meteorite	2.1×10^{-4}	(b)
Allende Meteorite	See, [5]	(b)

Notes:
[1] The $^3He/^4He$ ratios were given in units of "the atmospheric ratio", which is 1.39×10^{-6}. We have recalculated these to proper isotopic ratios.
[2] Calculated for this table.
[3] In high temperature fraction.
[4] The precise quotation: "The helium isotopic compositions from both the Meishan and Sasayama boundary sites are within the range reported for the "planetary" component in meteorites (1.6 to 1.9×10^{-4})".
[5] The precise quotation is: "The isotopic composition of the He contained within the Allende fullerene extract is more than one order of magnitude greater than the values found in Murchison"

Sources: (a) Becker et al. (1996); (b) Becker et al. (2000b); (c) Becker et al. (2000a); (d) Becker et al. (2001); (e) Farley and Mukhopadhyay (2001)

8. SUMMARY AND CONCLUSIONS

Results of searches for C_{60} and C_{70} in *KTB* clay layers have established a convincing case for the occurrence of natural fullerenes because the findings were confirmed by studies of two independently working research groups that employed fundamentally different analytical techniques. These findings were only questioned by the result of a single measurement from a third group. The principal debate was not about the veracity of the discoveries but about the source of the fullerenes, that is global wildfires or an extraterrestrial source. The wildfire hypothesis cannot account for the fullerene molecules that carry the isotopically anomalous helium but it is conceivable that one fraction of the fullerenes comes from wildfires and another from extraterrestrial sources. This dual origin would obviously increase the already alarmingly high He concentration in the extraterrestrial component. The extraterrestrial hypothesis has some difficulty accounting for the source of the isotopically anomalous helium and the process of its implantation inside the fullerene cages. Large $^{3}He/^{4}He$ ratios occur in the cosmic ray produced helium atoms in meteorites. These atoms then had to be trapped in fullerenes during their recoil phase, which, given the huge abundance of other potentially available solids that could trap helium atoms in the meteorites had to be a greatly inefficient process to favor fullerenes. The Sun, and presumably other stars as well, occasionally emit ^{3}He-rich flares in which the isotopic ratios are close to unity. This ^{3}He-rich gas would have to be trapped by condensing or condensed fullerenes before they accumulated in meteorites. If there exists, or existed a galactic or solar system reservoir with the enlarged $^{3}He/^{4}He$ ratio, why do we see this gas only in fullerenes and not somewhere else in the bulk meteorites? Was the inner solar system around KT time transiently invaded by matter that contained this isotopically anomalous He, for example by passage through a small interstellar dark cloud? Lastly, one should be mindful of Edward Anders's quip about the 'Russian Doll Effect' by which he meant that new carrier phases of isotopically anomalous noble gases were almost endlessly uncovered in meteorites.

The proof that fullerenes are present in *KTB* clays rests only on the reliability of the geochemical analytical procedures. If these are to be trusted and fullerenes show up in the analyses, then they present in the *KTB* clays. It is more difficult to prove that helium atoms occur inside the natural meteoritic fullerene cages and is not present in some other accompanying rare and hard to detect phase. The proof based on an LDMS spectrum with a substantial peak at 724 amu where $He@C_{60}$ would appear (Becker et al., 1996) was not entirely convincing when it was not demonstrated this peak was not due to $^{13}C^{12}C_{59}$.

Acknowledgements: We thank T. J. Bernatowicz, T. Bunch, T. L. Daulton, P. Buseck, I. Gilmour, A. C. Scott, and R. E. Smalley for providing additional information, correcting misunderstandings, and advice on contested issues.

9. REFERENCES

Alvarez, L., Alvarez, W., Asaro, F. and Michel, H.V. (1980) Extraterrestrial cause for the Cretaceous/Tertiary extinction. *Science*, 208, 1095-1108.

Amari, S., Anders, E., Virag, A. and Zinner, E. (1990) Interstellar graphite in meteorites. *Nature*, 345, 238-240.

Anacleto, J.F., Boyd, R.K., Pleasance, J.B., Quilliam, M.A., Howard, J.B., Lafleur, A.L. and Makarovsky, Y. (1992) Analysis of minor constituents in fullerene soots by LC-MS using a heated pneumatic nebulizer interface with atmospheric pressure chemical ionization. *Can. J. Chem.*, 70, 2558-2568.

Anders, E., Wolbach, W.S. and Lewis R.S. (1986) Cretaceous extinctions and wildfires. *Science*, 234, 261-264.

Anders, E., Wolbach, W.S. and Gilmour, I. (1991) Major wildfires at the Cretaceous/Tertiary boundary. *In Global Biomass Burning*, J.S. Levine, Ed., 489-492, The MIT Press, Cambridge, Massachusetts, USA.

Andrievsky, G.V., Klochkov, V.K., Karyakina, E.L. and Mchedlov-Petrossyan, N.O. (1999) Studies of aqueous colloidal solutions of fullerene C_{60} by electron microscopy. *Chem. Phys. Lett.*, 300, 392-396.

Andrievsky, G.V., Klochkov, V.K., Bordyuh, A.B. and Dovbeshko, G.I. (2002) Comparative analysis of two aqueous-colloidal solutions of C_{60} fullerenes with help of FTIR reflectance and UV-vis spectroscopy. *Chem. Phys. Lett.*, 364, 8-17.

Argyle, E., Cisowski, S.M., Fuller, M., Officer, C.B. and Ekdale, A.A. (1986) Cretaceous extinctions and wildfires. *Science*, 234, 261-264.

Ash, R.D., Russell, S.S., Wright, I.P. and Pillinger, C.T. (1993) Minor high temperature components confirmed in carbonaceous chondrites by stepped combustion using a new sensitive static mass spectrometer (abstract). *Lunar Planet. Sci.*, 22, 35-36, Lunar and Planetary Institute, Houston, Texas, USA.

Bachmann, M., Wiese, W. and Homann, K.H. (1995) Fullerenes versus soot in benzene flames. *Combustion and Flame*, 101, 548-550.

Baum, T.H., Löffler, S., Löffler Ph., Weilmünster, P. and Homann, K.H. (1992) Fullerene ions and their relation to PAH and soot in low-pressure hydrocarbon flames. *Ber. Bunsenges. Phys. Chem.*, 96, 841-857.

Becker, L. and Bunch, T.E. (1997) Fullerenes, fulleranes and polycyclic aromatic hydrocarbons in the Allende meteorite. *Meteorit. Planet. Sci.*, 32, 479-487.

Becker, L., Bada, J.L., Winans, R.E., Hunt, J.E., Bunch, T.E. and French, B.E. (1994a) Fullerenes in the 1.85-billion-year-old Sudbury impact structure. *Science*, 256, 642-644.

Becker, L., Bada, J.L., Winans, R.E. and Bunch T.E. (1994b) Fullerenes in Allende meteorite. *Nature*, 372, 507.

Becker, L., Bada, J.L. and Bunch, T.E. (1995a) Fullerenes in the K/T boundary: Are they a result of global wildfires? (abstract). *Lunar. Planet. Sci.*, 26, 85-86, Lunar and Planetary Institute, Houston, Texas, USA.

Becker, L., Bada, J.L., Bunch, T.E. (1995b) PAH's, fullerenes and fulleranes in the Allende meteorite (abstract). *Lunar. Planet. Sci.*, 26, 87-88. Lunar and Planetary Institute, Houston, Texas, USA.

Becker, L., Poreda, R.J. and Bada, J.L. (1996) Extraterrestrial helium trapped in fullerenes in the Sudbury impact structure. *Science*, 272, 249-252.

Becker, L., Poreda, R.J. and Bunch, T.E. (2000a) Fullerenes: An extraterrestrial carbon carrier phase for noble gases. *Proc. Natl. Acad. Sci.*, 97, 2979-2983.

Becker, L., Poreda, R.J. and Bunch, T.E. (2000b) Fullerenes and noble gases in the Murchison and Allende meteorites. *Lunar Planet. Sci.*. 31, abstract #1803, Lunar and Planetary Institute, Houston, Texas, USA (CD-ROM).

Becker, L., Poreda, R.J. and Bunch, T.E. (2000c) The origin of fullerenes in the 65 myr old Cretaceous/Tertiary 'K/T' boundary. *Lunar Planet. Sci.*, 31, abstract #1832, Lunar and Planetary Institute, Houston, Texas, USA (CD-ROM).

Becker, L., Poreda, R.J., Hunt, A.G., Bunch, T.E. and Rampino, M. (2001) Impact event at the Permian-Triassic boundary: Evidence from extraterrestrial noble gases in fullerenes. *Science*, 291, 1530-1533.

Bensasson, R.V., Bienvenue, E., Dellinger, M., Leach, S. and Seta, P. (1994) C_{60} in model biological systems. A visible-UV absorption study of solvent-dependent parameters and solute aggregates. *J. Phys. Chem.*, 98, 3492-3500.

Bernatowicz, T.J. and Cowsik, R. (1996) Conditions in stellar outflows from laboratory studies of presolar grains. *In Astrophysical Implications of the Laboratory Study of Presolar Materials*, T.J. Bernatowicz and E. Zinner, Eds., *Am.. Inst. Physics Conf. Proc.*, 402, 451-476, American Institute of Physics Woodbury, New York, USA.

Bernatowicz, T.J., Amari, S., Zinner, E.K. and Lewis, R.S. (1991) Interstellar grains within interstellar grains. *Astrophys. J.*, 373, L73-L76.

Bernatowicz, T.J., Cowsik, R., Gibbons, P.C., Lodders, K., Fegley, Jr., B., Amari, S. and Lewis, R.S. (1996) Constraints on stellar grain formation from presolar graphite in the Murchison meteorite. *Astrophys. J.*, 472, 760-782.

Buseck, P.R., Tsipurski, S.J. and Hettich, R. (1992) Fullerenes from the geological environment. *Nature*, 247, 215-217.

Cornejo Toledo, A. and Hernandez Osuna, A. (1950) Las anomalias gravimetricas en la cuenca salina del istmo, planicie costera de Tabasco, Campeche y Peninsula de Yucatan (in Spanish). *Boletín de la Asociación Mexicana de Geólogos Petroleros*. 2, 435-460.

Daly, T.K., Buseck, P.R., Williams, P. and Lewis, C.F. (1993) Fullerenes from a fulgurite. *Science*, 259, 1599-1601.

Daulton, T.L., Eisenhour, D.D., Bernatowicz, T.J., Lewis, R.S. and Buseck, P.R. (1996) Genesis of presolar diamonds: Comparative high-resolution transmission electron microscopy study of meteoritic and terrestrial nano-diamonds. *Geochim. Cosmochim. Acta*, 60, 4853-4872.

Deguchi, S., Alargova, R.G. and Tsujii, K. (2001) Stable dispersions of fullerenes, C_{60} and C_{70} in water, preparation and characterization. *Langmuir*, 17, 6013-6017.

de Vries, M.S., Wendt, H.R., Hunziker, H., Peterson, E. and Chang, S. (1991) Search for high molecular weight polycyclic aromatic hydrocarbons and fullerenes in carbonaceous meteorites (abstract). *Lunar Planet. Sci.*, 22, 315-316, Lunar and Planetary Institute, Houston, Texas, USA.

de Vries, M.S., Reihs, K., Wendt, H.R., Golden, W.G., Hunziker, H.E., Fleming, R., Peterson, E. and Chang, S. (1993) A search for C_{60} in carbonaceous chondrites. *Geochim. Cosmochim. Acta*, 57, 933-935.

Durda, D.D. and Kring, D.A. (2004) Ignition threshold for impact-generated fires. *J. Geophys. Res., 109*, E08004, doi:10.1029/2004JE002279, 14p.

Farley, K.A. and Mukhopadhyay, S. (2001) An extraterrestrial impact at the Permian-Triassic boundary? *Science*, 293, 2343.

Gerhardt, Ph., Löffler, S. and Homann, K.H. (1987) Polyhedral carbon ions in hydrocarbon flames. *Chem. Phys. Lett.*, 137, 306-310.

Gerhardt, Ph., Löffler, S. and Homann, K.H. (1988) The formation of polyhedral carbon ions in fuel-rich acetylene and benzene flames. *Twenty-second Internl. Symp. Combustion*, 395-401, The Combustion Institute, Philadelphia, Pennsylvania, USA.

Gilmour, I., Wolbach, W.S. and Anders, E. (1990) Major wildfires at the Cretaceous/Tertiary boundary. *In Catastrophes and Evolution: Astronomical Foundations*, S.V.M. Clube, Ed., 195-213, Cambridge University Press, Cambridge, Great Britain.

Gilmour, I., Russell, S.S., Newton, J., Pillinger, C.T., Arden, J.W., Dennis, T.J., Hare, J.P., Kroto, H.W., Taylor, R. and Walton, D.R.M. (1993) A search for the presence of C_{60} as an interstellar grain in meteorites (abstract). *Lunar. Planet. Sci.*, 22, 445-446, The Lunar and Planetary Institute, Houston, Texas, USA.

Harris, P.J.F. and Vis, R.D. (2003) High resolution electron microscopy of carbon and nanocrystals in the Allende meteorite. *Proc. Roy. Soc. London A.*, 459, 2069-2076.

Harris, P.J.F., Vis, R.D. and Heymann, D. (2000) Fullerene-like carbon nanostructures in the Allende meteorite. *Earth Planet. Sci. Lett.*, 183, 355-359.

Heymann, D. (1991) On the chemical attack of fullerene, soot, graphite, and sulfur with hot perchloric acid. *Carbon*, 29, 684-685.

Heymann, D. (1996) Solubility of C_{60} and C_{70} in seven normal alcohols and their deduced solubility in water. *Fullerene Sci. Techn.*, 4, 509-515.

Heymann, D. (1998) Search for C_{60} fullerene in char produced on a Norway spruce by lightning. *Fullerene Sci. Techn.*, 6, 1079-1086.

Heymann, D. and Chibante, F. (1993) Reaction of C_{60}, C_{70}, C_{76}, C_{78} and C_{84} with ozone at 23.5 ^{o}C. *Recl. Trav. Chim. Pays-Bas*, 112, 639-642.

Heymann, D., Wolbach, W.S., Chibante, L.P.F. and Smalley, R.E. (1994a) Search for extractable fullerenes in clays from the KT boundary of the Woodside Creek and Flaxbourne River sites, New Zealand (abstract). *In New Developments Regarding the KT Event and Other Catastrophes in Earth History, LPI Contribution*, # 825, 47-48, Lunar and Planetary Institute, Houston, Texas, USA.

Heymann, D., Chibante, L.P.F., Brooks, R.R., Wolbach, W.S. and Smalley, R.E. (1994b) Fullerenes in the K/T boundary layer. *Science*, 265, 645-647.

Heymann, D., Wolbach, W.S., Chibante, L.P.F., Brooks, R.R. and Smalley, R.E. (1994c) Search for extractable fullerenes in clays from the Cretaceous/Tertiary boundary of the Woodside Creek and Flaxbourne River sites, New Zealand. *Geochim. Cosmochim. Acta*, 58, 3531-3534.

Heymann, D., Chibante, L.P.F. and Smalley, R.E. (1995a) Determination of C_{60} and C_{70} fullerenes in geologic materials by high performance liquid chromatography. *J. Chrom. A*, 689, 157-163.

Heymann, D., Nazarov M.A., Korochantsev, A. and Smit, J. (1995b) The Chicxulub event: did it produce a global layer of fullerene-bearing sediments? (abstract). *Lunar Planet. Sci.*, 26, 597-598, The Lunar and Planetary Institute, Houston, Texas, USA.

Heymann, D., Chibante, L.P.F., Brooks, R.R., Wolbach, W.S., Smit, J., Korochantsev, A., Nazarov, M.A. and Smalley, R.E. (1996a) Fullerenes of possible wildfire origin in Cretaceous-Tertiary boundary sediments. *In The Cretaceous-Tertiary event and other catastrophes in Earth history*, G. Ryder, D. Fastovsky and S. Gartner, Eds., *Geol Soc. Am. Spec. Paper*, 307, 453-464.

Heymann, D., Korochantsev, A., Nazarov, M.A. and Smit, J. (1996b) Search for fullerenes C_{60} and C_{70} in Cretaceous-Tertiary boundary sediments from Turkmenistan, Kazakhstan, Georgia, Austria, and Denmark. *Cretaceous Geol.*, 17, 367-380.

Heymann, D., Yancey, T.E., Wolbach, W.S., Johnson, E.A., Roach, D. and Moecker, S. (1997) Carbon in sediments at the KT boundary site of the Brazos River, Texas (abstract). *Lunar Planet. Sci.*, 28, 567-568, Lunar and Planetary Institute, Houston, Texas, USA.

Heymann, D., Yancey, T.E., Wolbach, W.S., Thiemens, M.H., Johnson, E.A., Roach, D. and Moecker, S. (1998) Geochemical markers of the Cretaceous-Tertiary boundary event at Brazos river, Texas, USA. *Geochim. Cosmochim. Acta*, 62, 173-181.

Heymann, D., Bachilo, S.M., Weisman, R.B., Cataldo, F., Fokkens, R.H., Nibbering, N.M.M., Vis, R.D. and Chibante, L.P.F. (2000) $C_{60}O_3$, a fullerene ozonide: Synthesis and dissociation to $C_{60}O$ and O_2. *J. Am. Chem. Soc.*, 122, 11473-11479.

Hildebrand, A.R., Penfield, G.T., Kring, D.A., Pilkington, M., Jacobsen, S. and Boynton, W.V. (1991) The Chicxulub crater: A possible Cretaceous-Tertiary boundary impact crater on the Yucatán Peninsula, Mexico. *Geology*, 19, 867-871.

Howard, J.B. (1992) Fullerenes formation in flames. *Twenty-forth Internl. Symp.Combustion*, 933-946, The Combustion Institute, Philadelphia, Pennsylvania, USA.

Howard, J.B., McKinnon, J.Th., Makarovsky, Y., Lafleur, A.L. and Johnson, M.E. (1991) Fullerenes C_{60} and C_{70} in flames. *Nature*, 352, 139-141.

Howard, J.B., McKinnon, J.Th., Johnson, M.E. Makarovsky, Y. and Lafleur, A.L. (1992) Production of C_{60} and C_{70} in benzene-oxygen flames. *J. Phys. Chem.*, 96, 6657-6662.

Howard, J.B., Chowdhury, K.D. and van der Sande, J.B. (1994) Carbon shells in flames. *Nature*, 370, 603.

Kroto, H.W. (1988) Space, Stars, C_{60}, and Soot. *Science*, 242, 1139-1145.

Kroto, H.W., Heath, J.R., O'Brien, S.C., Curl, R.F. and Smalley, R.E. (1985) C_{60}: Buckminsterfullerene. *Nature*, 318, 162-163.

Kring, D.A. and Durda, D.D. (2002) Trajectories and distribution of material ejected from the Chicxulub impact crater: Implications for postimpact wildfires. *J. Geophys. Res.*, 107, 6/1-6/22.

Kyte, F.T. (1998) A meteorite from the Cretaceous/Tertiary boundary. *Nature*, 396, 237-239.

Kyte, F.T., Smit, J. and Wasson, J.T. (1985) Siderophile interelement variations in the Cretaceous-Tertiary boundary sediments from Caravaca, Spain. *Earth Planet. Sci. Lett.*, 73, 183-195.

Lumpkin, G.R. (1981a) Electron microscopy of carbon in Allende acid residues (abstract). *Lunar Planet. Sci.*, 12, 631-633,.The Lunar and Planetary Institute, Houston. Texas, USA.

Lumpkin, G.R. (1981b) Electron microscopy of carbonaceous matter in Allende acid residues. *Proc. Lunar Planet. Sci. Conf.*, 12, 1153-1166.

Lumpkin, G.R. (1983a) Electron microscopy of carbonaceous matter in acid residues from the Orgueil (C1) and Cold Bokkeveld (C2) meteorites (abstract). *Lunar Planet. Sci.*, 14, 450-451, The Lunar and Planetary Institute, Houston, Texas, USA.

Lumpkin, G.R. (1983b) Microstructural variations in Allende carbonaceous matter (abstract). *Lunar Planet. Sci.*, 14, 452-453, The Lunar and Planetary Institute, Houston, Texas, USA.

Melosh, H.J., Schneider, N.M., Zahnle, K.J. and Latham, D. (1990) Ignition of global wildfires at the Cretaceous-Tertiary boundary. *Nature*, 343, 251-254.

Oester, M.Y., Kuechl, D., Sipiera, P.P. and Welch, C.J. (1994) Search for fullerenes in stone meteorites (abstract). *Meteoritics*, 29, 513.

Pope, C.J., Marr, J.A. and Howard, J.B. (1993) Chemistry of fullerenes C_{60} and C_{70} formation in flames. *J. Phys. Chem.*, 97, 11001-11013.

Sayes, C.M., Fortner, J.D., Guo, W., Lyon, D., Boyd, A.M., Ausman, K.D., Tao, Y.J., Sitharaman, B., Wilson, L.J., Hughes, J.B., West, J.L. and Colvin, V.L. (2004) The differential cytotoxicity of water-soluble fullerenes. *Nanoletters*, 4, 1881-1887.

Schuster, D.I., Cheng, P., Jarowski, P.D., Guldi, D.M., Luo, C., Echegoyen, L., Pyo, S., Holzwarth, A.R., Braslavsky, S.E., Williams, R.M. and Klihm, G. (2004) Design, synthesis and photophysical studies of a porphyrin-fullerene dyad with parachute topology; charge recombination in the Marcus Inverted Region. *J. Am. Chem. Soc.*, 126, 7257-7270.

Scott, A.C. (2000) The pre-Quaternary history of fire. *Paleogeogr. Paleoclimatol. Paleoecol.*, 164, 281-329.

Scott, A.C., Lomax, B.H., Collinson, M.E., Upchurch, G.R. and Beerling, D.J. (2000) Fire across the K-T boundary: initial results from the Sugarite coal, New Mexico, USA. *Paleoclimatol. Paleoecol. Paleoecol.*, 164, 381-395.

Scrivens, W.A., Tour, J.M., Creek, K.E. and Pirisi, L. (1994) Synthesis of ^{14}C-labeled C_{60}, its suspension in water, and its uptake by human keratinocytes. *J. Am. Chem. Soc.*, 116, 4517-4518.

Sharpton, V.L., Marín, L.E., Carney, J.L., Lee, S., Ryder, G., Schuraytz, B.C., Sikora, P. and Spudis, P.D. (1996) A model of the Chicxulub impact basin based on evaluation of geophysical data, well logs, and drill core samples. *In The Cretaceous-Tertiary Event and Other Catastrophies in Earth History*, G. Ryder, D. Fastovsky and S. Gartner, Eds., *Geol. Soc. Am. Spec. Paper*, 307, 55-74.

Simonin, J.-P. (1991) Solvent effects on osmotic second virial coefficient studied using analytical molecular models. Application to solutions of C_{60} fullerene. *J. Phys. Chem. B*, 105, 5262-5270.

Smit, J., Roep, Th.B., Alvarez, W., Montanari, A., Claeys, P., Grajales-Nishimura, J.M. and Bermudez, J. (1996) Coarse-grained clastic sandstone complex at the K/T boundary around the Gulf of Mexico: Deposition by tsunami waves induced by the Chicxulub impact? In *The Cretaceous-Tertiary Event and Other Catastrophies in Earth History*, G. Ryder, D. Fastovsky and S. Gartner, Eds., *Geol. Soc. Am. Spec. Paper*, 307, 151-182.

Smith, P.P.K. and Buseck, P.R. (1980) High resolution transmission electron microscopy of an Allende acid residue (abstract) *Meteoritics*, 15, 368-369.

Smith, P.P.K. and Buseck, P.R. (1981a) Carbon in the Allende meteorite: Evidence for poorly graphitized carbon rather than carbyne. *Proc. Lunar Planet. Sci. Conf.*, 12, 1167-1175.

Smith, P.P.K. and Buseck, P.R. (1981b) Graphitic carbon in the Allende meteorite: a microstructural study. *Science*, 212, 322-324.

Taylor, R. and Abdul-Sada, A.K. (2000) There are no fullerenes in the K-T boundary layer. *Fullerene Sci. Techn.*, 8, 47-54.

Tingle, T.N., Becker, C.H. and Malhotra, R. (1991) Organic compounds in the Murchison and Allende carbonaceous chondrites studied by photoionization mass spectrometry. *Meteoritics*, 26, 117-127.

Venkatesan, M.I. and Dahl, J. (1998) Organic geochemical evidence for global fires at the Cretaceous/Tertiary boundary. *Nature*, 338, 57-60.

Vis, R.D., Mrowiec, A., Kooyman, P.J., Matsubara, K. and Heymann, D. (2002) Microscopic search for the carrier phase Q of the trapped planetary noble gases in Allende, Leoville, and Vigarano. *Meteorit. Planet. Sci.*, 37, 1391-1400.

Weisman, R.B., Heymann, D. and Bachilo, S.M. (2001) Synthesis and characterization of the 'missing' oxide of C_{60}: [5,6]-open $C_{60}O$. *J. Am. Chem. Soc.*, 123, 9720-9721.

Wolbach, W.S., Lewis, R.S. and Anders, E. (1985) Cretaceous extinctions: Evidence for wildfires and search for meteoritic material. *Science*, 230, 167-170.

Wolbach, W.S., Gilmour, I., Anders, E., Orth, C.J. and Brooks, R.R. (1988) Global fire at the Cretaceous/Tertiary boundary. *Nature*, 334, 665-669.

Wolbach, W.S., Gilmour, I., Anders, E. (1990a) Major wildfires at the Cretaceous/Tertiary boundary. *In Global catastrophes in Earth history; an interdisciplinary conference on impacts, volcanism, and mass mortality*, V.L. Sharpton and P.D. Ward, Eds., *Geol. Soc. Am. Spec. Paper*, 247, 391-400.

Wolbach, W.S., Anders, E. and Nazarov, M.A. (1990b) Fires at the K/T boundary: Carbon at the Sumbar, Turkmenia, site. *Geochim. Cosmochim. Acta*, 54, 1133-1146.

Wolbach, W.S., Widicus, S. and Kyte, F.T. (2003) A search for soot from global wildfires in Central Pacific Cretaceous-Tertiary Boundary and other extinction and impact horizon sediments. *Astrobiology*, 3, 91-97.

Ying, Q.C., Marecek, J. and Chu, B. (1994) Solution behavior of buckminsterfullerene (C_{60}) in benzene. *J. Chem. Phys.*, 101, 2665-2672.

Chapter 10

FULLERENE C_{60} IN SOLID BITUMEN ACCUMULATIONS IN NEO-PROTEROZOIC PILLOW-LAVAS AT MÍTOV (BOHEMIAN MASSIF)

JAN JEHLIČKA
Institute of Geochemistry, Mineralogy and Mineral Resources, Faculty of Science, Charles University in Prague, Albertov 6, 128 43 Prague 2, Czech Republic.

OTAKAR FRANK
Institute of Geochemistry, Mineralogy and Mineral Resources, Faculty of Science, Charles University in Prague, Albertov 6, 128 43 Prague 2, Czech Republic.

Abstract: Fullerene C_{60} occurs at extremely low concentration (0.2-0.3 ppm) in hard solid bitumen that was accumulated in the pillow lava bodies exposed at Mítov (Teplá-Barrandian Neo-proterozoic of the Bohemian Massif). No higher fullerenes were documented. The C_{60} fullerene and its precursors are present in non-graphitic and non-graphitizable carbonaceous matter. Inclusions of fullerene are found embedded within glass-like carbons that are found within the pillow lava bodies. Rare fullerenes in Mítov were probably preserved in the closed micropores and mesopores of the glass-like, solid bitumen. Pyrolysis of biogenic PAHs precursors seems a plausible way to form fullerenes in geological environments.

Key words: Algal precursors; Bohemian Massif; C_{60}; electron impact mass spectrometry (EIMS); fullerene formation; fullerene preservation; glass-like carbon; high-performance liquid chromatography (HPLC); metamorphic rocks; Neo-Proterozoic rocks; organic matter; pillow-lavas; polycyclic aromatic hydrocarbons; pyrolysis; solid bitumen

Frans J.M. Rietmeijer (ed.), Natural Fullerenes and Related Structures of Elemental Carbon, 213–240.

1. INTRODUCTION

For about ten years after the discovery of fullerenes, the presence of these strange cage molecules was reported in several terrestrial rock types and in meteorites (for reviews see, Buseck, 2002, Heymann et al., 2003a). Natural C_{60} and C_{70} fullerenes were first reported in rocks that had experienced unique but atypical geological events defined by high-energy conditions such as in a lightning strike that produced a silica-rich glassy fulgurite (Daly et al., 1993). Mass spectrometry with electron ionization confirmed the results obtained by time-of-flight mass spectrometry that the fulgurite at Sheep Mountain (Colorado) contained C_{60} fullerene as $^{12}C_{60}^{+}$, $^{12}C_{59}\,^{13}C^{+}$, and $^{12}C_{58}\,^{13}C_2^{+}$ (Daly et al., 1993).

Shallow-water deposits that formed near landmasses and are related to wildfires at the Cretaceous-Tertiary boundary (KTB) contain fullerene at many KTB sites around the world: Woodside Creek and Flaxbourne River (New Zealand), Caravaca (Spain), Sumbar and Malyi Balkan (Turkmenistan), Stevens Klint (Denmark) and the Brazos River (Texas) (Heymann et al., 1994a, 1994b, 1996, 1998). These KTB rocks are claystones, occasionally enriched in organic carbon and soot, with fullerene concentrations from about 0.1 to 11 ppb. It was suggested that the KTB fullerenes were formed by oxygen-starved global wildfires following a major meteoroid impact event 65-Ma ago that produced the Chicxulub meteorite crater on the Yucatan peninsula (Mexico). Fullerenes in sediments at the Permian-Triassic boundary (Becker et al., 2001) might also be associated with a meteoroid impact event. Some of the rocks associated with the meteorite impact at the Sudbury impact structure (Canada) also contain fullerenes (Becker et al., 1994; Mosmann et al., 2003). Becker et al. (1996) has proposed an extraterrestrial origin of the fullerenes at Sudbury based on the isotopically anomalous composition of helium trapped in the fullerene cages.

Fullerenes are also present in Precambrian carbonaceous rocks called shungite wherein they were initially described in a sample from the Shunga location (Buseck et al., 1992). Shungite is solid carbonaceous matter forming a black, amorphous carbon mass with traces of N, O, S, and H. Different types of this rock ranging from shungitic shale to pure bright shungite occur in the 2.1-2.0 Ga-old, 600-m thick Zaonezhskaya Formation near Lake Onega, Karelia (Russia). In this region, rocks rich in organic carbon are exposed over an area of 9000 km^2 with an estimated total carbon reserve exceeding 25 x 10^{16} g (Galdobina, 1993). Shungite layers and veins of brittle shungite that both contain about 95-99% C_{org} are considered to be allochthonous, which is migrated, bitumen from organic carbon-rich strata, initially deposited elsewhere, that was transported to its present geological setting. Due to only very low-grade regional metamorphic processes (Melezhik et al., 1999) the bitumen was transformed to chemically stable polymeric, solid bitumen (Jehlička and Rouzaud, 1993a, 1993b; Buseck et al., 1997; Kovalevski et al., 2001).

Fullerenes were first detected in situ in a sample from the shungite type locality Shunga by high-resolution transmission electron microscopy (HRTEM) and laser desorption-Fourier transform mass spectrometry (LD-FTMS) (Buseck et al., 1992). Using electron-impact mass spectrometry (EIMS), powder X-ray diffraction and ^{13}C-NMR spectrometry, Parthasarathy et al. (1998) detected fullerenes in shungite from Kondopoga, Karelia (Table 1).

No fullerenes were detected in other samples of shungite, and in samples of anthraxolite and thucholite, by high-performance liquid chromatography (HPLC) and electron impact mass spectrometry (EIMS) analyses (Ebbesen et al., 1995; Heymann, 1995; Gu et al., 1995). Anthraxolite is highly carbonified, infusible, solid bitumen found in many hydrothermal mineral deposits. When this solid bitumen contains uranium and thorium, it is designated as thucholite. Ebbessen et al. (1995) concluded that natural fullerenes in the type-locality of Shunga are the products of localized high-energy events such as lightnings. We note that the composition and structure of shungite and the solid bitumen from Mítov are very similar (Jehlička and Rouzaud, 1993a, 1993b).

2. FULLERENE ANALYSES TECHNIQUES: ADVANCES AND LIMITATIONS

The analytical procedures in use to detect natural fullerenes can be critical for a definitive detection as some of the mass spectrometric methods were described as being “problematic”. For example, some results are based on laser desorption mass spectrometry (LDMS) analysis of toluene extracts but this method can cause in situ production of fullerenes during the analysis, specifically by interactions of the laser beam with the carbonaceous matter (Ebbesen et al., 1995; Gu et al., 1995). When the laser energy is suitably

Table 1. Positive reported detections of natural fullerenes in rocks

Locality	Samples	Age (Ga)	Fullerene Content	Analytical Methods	Sources
Shunga, Karelia, Russia	shungite	2.0-2.1	0.1 ppm	HRTEM, TD/EC MS HPLC, MS	(1) (2), (3)
Kondopoga, Karelia, Russia	shungite	2.0-2.1		EIMS, XRD, NMR	(4)
Mítov, Czech Republic	solid bitumen	0.6	0.2 ppm 0.3 ppm	HPLC, HPLC, EIMS	(5) (6)
Sudbury, Canada	black tuff	2	1 – 10 ppm 5 ng/g	LDI, EIMS, TOF MS	(7) (8)
Woodside Creek, Colorado, USA Flaxbourne River, New Zealand Stevns Klint, Denmark Caravaca, Spain Sunbar, Turkmenistan	KTB clays	0.065	0.3 – 2.7 ppb 0.006 – 16.0 ppb 0.9 ppb 2.0 – 12.0 ppb 1.2 – 4.0 ppb	HPLC	(9) (10) (11) (12)
Inuyama, Japan Meishan, China Sasayama, Japan	PTB clays	0.251		HPLC	(13), (14)

Sources: (1) Buseck et al. (1992); (2) Konkov et al. (1994); (3) Masterov et al. (1994); (4) Parthasarathy et al. (1998); (5) Jehlička et al. (2000); (6) Jehlička et al. (2003); (7) Becker et al. (1994); (8) Mossman et al. (2003); (9) Heymann et al. (1994a); (10) Heymann et al. (1994b); (11) Heymann et al. (1994c); (12) Heymann et al. (1996); (13) Chijiwa et al. (1999); (14) Becker et al. (2001)

Note: KTB (Cretaceous-Tertiary Boundary); PTB (Permian-Triassic Boundary)

Note: HRTEM (high-resolution transmission electron microscopy); TD/EC MS (thermal desorption/electron-capture mass spectroscopy); MS (mass spectroscopy); HPLC (high performance liquid chromatography), EIMS (electron impact mass spectrometry); XRD (X-ray diffraction); NMR (nuclear magnetic resonance); LDI (laser desorption ionization); TOF MS (time-of-flight MS)

reduced, it is possible to carefully use this technique for fullerene detection (Buseck, 2002). The more recent mass spectrometric methods seem well suited for fullerene detections of low fullerene concentrations in rocks and minerals (Mossmann et al., 2003). With the use of HPLC and suitable columns (e.g. Cosmosil Buckyprep) fullerenes can be detected by their retention times. The subsequent characterization of these fractions by mass spectrometry is required for unambiguous fullerene identification. Taylor and Abdul-Sada

(2000) recommend EIMS as an efficient detection tool that would not generate fullerene artifacts when correctly employed.

Several isolation (Soxhlet extraction, under-critical extraction) and in situ fullerene detection techniques, e.g. Raman microspectroscopy and transmission electron microscopy (TEM), were tested to estimate the lower detection limit of C_{60} in carbonaceous geological samples. Raman microspectroscopy can be used to detect fullerene but only in natural samples with a relatively high (>100 ppm) fullerene content as was demonstrated using prepared mixtures of coal, kerogen or graphite powders with pure C_{60} in various concentrations (Jehlička et al., 2005a).

TEM analyses successfully identified the presence of very small concentrations of C_{60} and higher fullerenes in condensed amorphous soot produced in an arc-discharge experiment that also yielded co-condensed fullerenic carbon onions and nanotubes (Rietmeijer et al., 2004). Raman microspectroscopy of this soot did not show evidence for C_{60} or other fullerenes but the separation of the ca. 1350 cm^{-1} '*D*' ("disorder") and ca. 1580 cm^{-1} '*G*' ("graphite") peaks was poorly-defined with regard to their separation in a typical Raman spectrum of disordered, pre-graphic carbon, and the intensity ratio of the '*D*/*G*' peaks was equal to, or larger than, unity for mature pre-graphitic carbons. These Raman spectral features might be caused by small quantities of C_{60} (Rietmeijer et al., 2005)

Searches for natural fullerene could also be biased when fullerene is strongly bound in the matrix of carbonaceous matter. This spurious effect was investigated by mixing the powders of natural carbonaceous matter (graphite, shungite, bituminous coal) with known amounts of fullerene C_{60} in solution. Very low fullerene recovery, occasionally only 5% of the original fullerene content, was observed pointing to the possibly strong adsorption or fullerene destruction during the extraction procedure, or both (Jehlička et al., 2005b).

We will review the characteristics of the carbonaceous matter in black shales of the Neo-proterozoic Teplá Barrandian Unit and of the glass-like carbonaceous accumulations with fullerene at Mítov. We will also discuss the formation processes of this natural fullerene and its preservation.

3. GEOLOGICAL SETTING

The Bohemian Massif forms the easternmost part of the Variscan branch of the European Hercynian Complex. The Teplá-Barrandian Neo-proterozoic rocks form a part of the so-called Bohemicum where in the southwestern part Neo-proterozoic rocks form the basis of the large Teplá-Barrandien Unit (*TBU*) (Fig. 1).

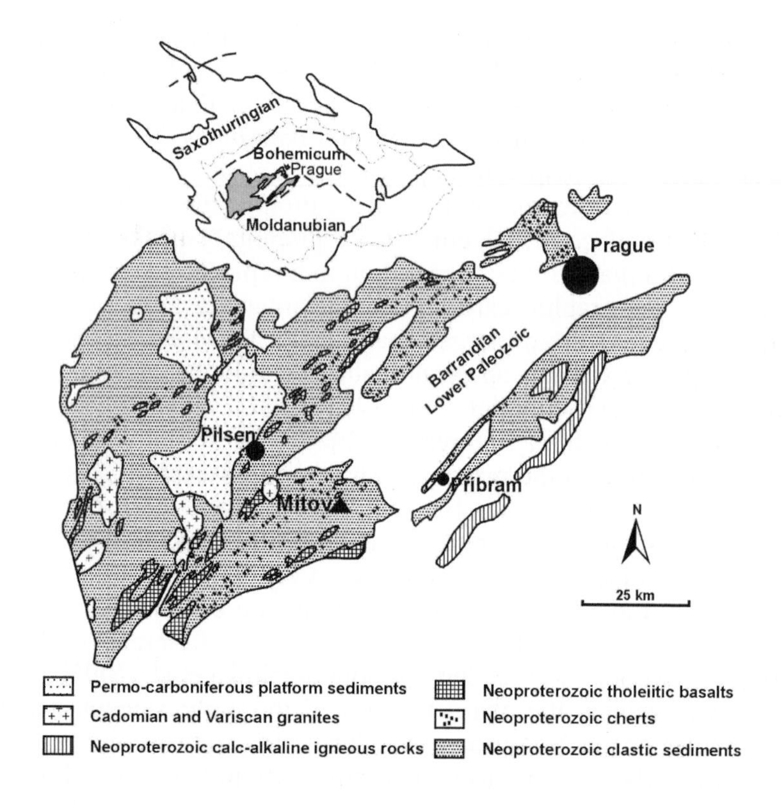

Figure 1. Simplified geological map of the Bohemian massif (modified after Jehlička et al., 2003)

The *TBU* consists of the Neo-proterozoic and Lower Paleozoic sedimentary and volcanic-sedimentary sequences, igneous rocks of Cambrian, Ordovician and Devonian ages, Permian-Carboniferous continental molasse, *i.e.* soft sedimentary deposits, and Cretaceous and Tertiary platform sediments (Zoubek et al., 1988). Neo-proterozoic formations of the *TBU* comprise tholeitic and calc-alkaline volcanic fallout deposits dated between about 700 Ma and 540 Ma (Mašek, 2000), volcaniclastic rocks, greywackes, black shales and black cherts.

This lithology reflects periods of clastic sediment deposition alternating with periods of low tectonic activity characterized by the deposition of black shales, cherts and, less frequent, carbonates (Zoubek et al., 1988, Kříbek et al., 2000). The deposition of black shales and cherts mostly corresponds to periods of increased submarine hydrothermal activity, when the hot water in these emanations may have significantly contributed to locally increased bioactivity.

3.1 Pillow Lavas, Metamorphic Facies, and the Mítov Quarry

Pillow lavas are formed when an extruding basalt lava flow enters a body of water where it breaks up and is rapidly quenched in rounded, pillow-like boulders. The pillow lavas of the *TBU* unit that, at Mítov site contain rare, hard solid bitumen are part of the Neo-proterozoic volcano-sedimentary complex. All these volcanic rocks are metamorphosed under mostly prehnite-pympellyite and greenschist facies conditions (ca. 250 °C; 0.2 GPa), and locally, amphibolite facies conditions (ca. 450 to 500 °C, up to 1 GPa) (Cháb and Suk, 1977).

Mineral assemblages in sedimentary, volcanic, and igneous rocks will change when their host rock is submitted to increasing temperatures and pressures during burial following a path along the geothermal gradient. During this process, thermodynamically stable mineral associations are formed, replacing pre-existing ones. The diagnostic, stable mineral associations defining a metamorphic facies of regional metamorphism are constrained as a function of temperature and lithostatic pressure (Fig. 2). These temperature-pressure regimes can be quantitatively defined by high-pressure/high-temperature experiments. The diagnostic minerals are readily identified using a light-optical petrographic microscope, which in turn then allows pressure and temperature constraints to be assessed.

The geochemical evolution of the volcanic suite points to the continuing maturation of an originally primitive, oceanic-arc, crust terrain. The lithofacies characteristic for this southeastern part of the Barrandian Neo-proterozoic includes stromatolitic cherts (silicified remnants of biomats and colonies of blue-green algae), volcanic and volcaniclastic rocks, dark grey to black cherts, and lastly, shales and black shales (Pouba et al., 2000). This lithologic facies is often developed in the periphery of large bodies of andesitic pillow-lavas. The Mítov quarry is situated ca. 2 km east of the village Nové Mitrovice that is located 25 km southeast from Pilsen (Fig. 1).

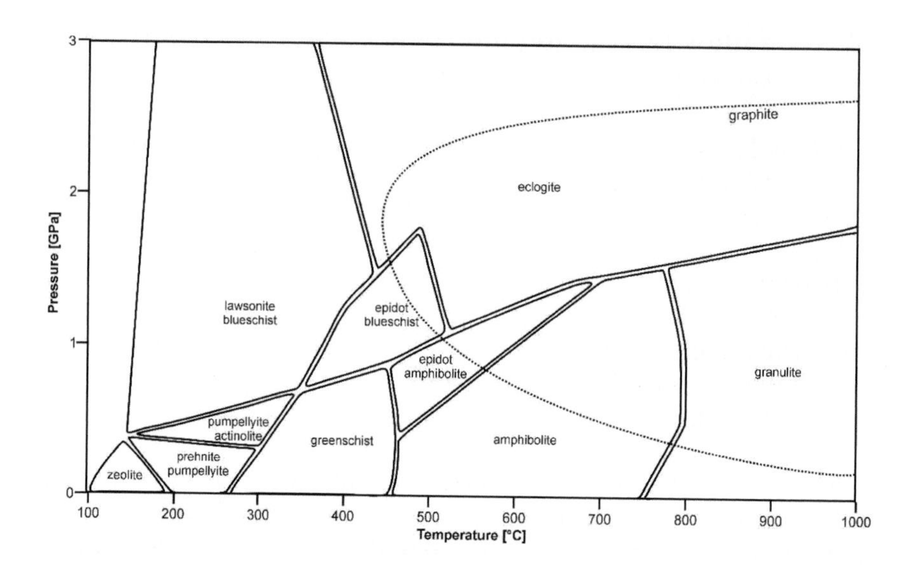

Figure 2. The binary temperature (°C) and lithostatic pressure (GPa) diagram showing the distribution of the metamorphic facies (modified after Yardley, 1989) and also showing the graphite stability field from data by Jehlička and Bény (1992), Jehlička and Rouzaud (1990), Wopenka and Pasteris (1993), Beyssac et al. (2002), Křibek et al. (1994)

At the Mítov quarry, a one meter-thick horizon of stromatolitic siliceous breccia with solid bitumen is present between two lava flows of amygdaloidal basaltic andesite (Fiala, 1977) (Fig. 3). Solid bitumen forms centimeter to decimeter size aggregates also directly enclosed within the basalts. Solid bitumen enclosed in siliceous matter also occurs in inter-pillow spaces in the andesitic basalts.

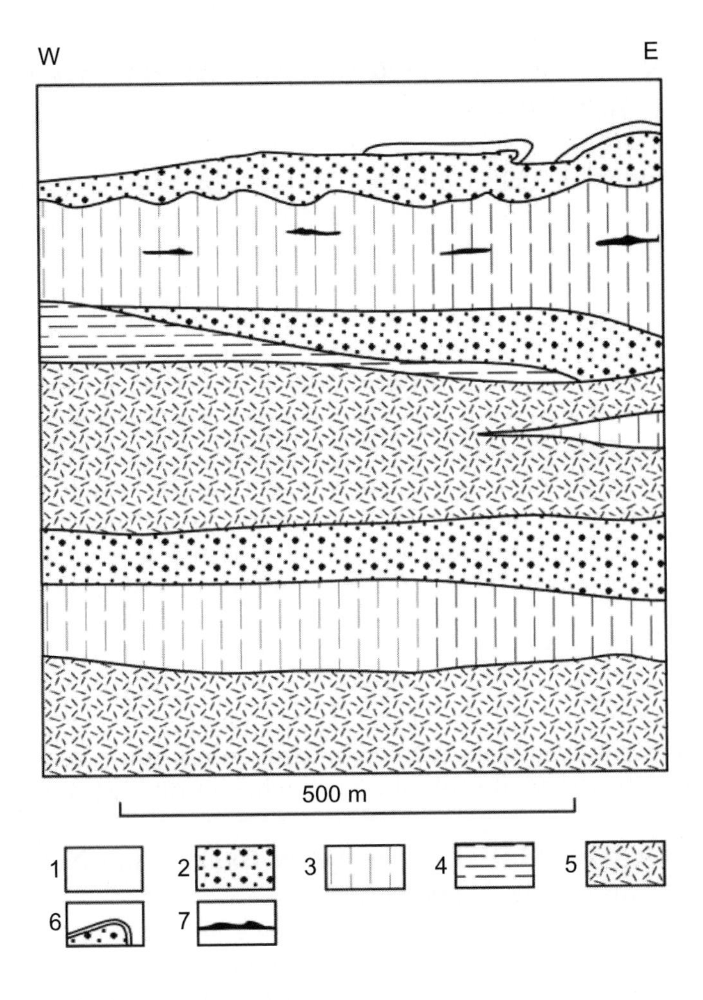

Figure 3. Schematic reconstructed cross-section of the geological relationships at Mítov (modified after Křibek et al., 1993). 1 seawater; 2 basalt-andesite pyroclastics; 3 pillow lavas (andesite basalts); 4 shales, siltstones, and cherts; 5 graywackes and pyroclastics; 6 stromatolitic siliceous mats; 7 anthraxolites

The basalts enriched in solid bitumens are underlain by a 20-meter thick pillow lava complex and are overlain by a 5-10 meter thick layer of basaltic andesite (Fig. 3). It passes upwards into dark cherts or black shales with intercalated chert seams or silicified stromatolites (Pouba, 1988). Samples were taken directly from the fresh, unweathered andesitic basalts and from isolated, hard and black, brittle solid bitumen accumulations of 5 mm in diameter.

4. KEROGEN MATURATION AND BITUMEN FORMATION

The formation and evolution of kerogen can be traced from sedimentation, through diagenesis and up to complete graphitization during high-grade, regional metamorphism at high tectonic stresses. The transformation of carbonaceous matter to graphite at the low pressures, which are typical for the majority of contact-metamorphic events, is much more exotic. During diagenesis of kerogen, various functional groups from the original organic materials are continuously eliminated and numerous new compounds are generated, forming mixtures of low to medium molecular-weight hydrocarbons. This initially liquid bitumen can solidify under the influence of various alteration processes such as fluid migration, oxidation, and thermal alteration. Bitumen accumulations may occur in the cracks and fissures in minerals or rocks, or form vein-type deposits. Bitumens can be also transformed to very specific types of carbon such as the Karelian shungite and Mitov solid bitumen as a function of a number of different geological processes, e.g. regional or contact metamorphism and by oxidation during interactions with aqueous fluids. Solid bitumen when soluble in carbon disulfide is designated as bitumen *sensu stricto*; it is pyrobitumen when insoluble (Jacob, 1993).

4.1 Organic Matter of the Teplá-Barrandian Neo-Proterozoic Black Shales

Black shales in the *TBU* appear individually or associated with volcanic rocks. The content of organic carbon varies between 1-6 wt %; the amount of sulfur from sulfides ranges from 1 to 16 wt %. Organic matter thought to be of algal origin (Pouba and Kříbek, 1986; Jehlička, 1994) is finely dispersed in the mineral matrix of the shale. Diagenetic mobilization of this organic matter along small joints and bedding planes and bitumen migration are relatively common. Traces of aliphatic and aromatic hydrocarbons were found in the very-low grade metamorphosed parts of the black shales (Jehlička et al., 1984). In the easternmost area the black shales were only slightly metamorphosed (prehnite-pumpellyite zone) but regional metamorphism in the central and western zones had raised temperatures to about 250-450 °C. The degree of carbonization and graphitization of organic matter generally increases with the increasing metamorphic temperatures from east to the west and southwest (Jehlička, 1984; Kříbek and Jehlička, 1986).

4.1.1 Transformation of Dispersed Organic Matter

HCl and HF insoluble carbonaceous residue – *kerogen* – isolated from the low-grade metamorphic black shales is highly carbonified with a carbon content >95%. Its isotopic composition ranges between $\delta^{13}C$ values of –38‰ and –25‰, which are typical for Neo-proterozoic algal organic matter. With increasing metamorphic grade, the carbon isotopic composition shifts to higher values of about –23‰ (Jehlička, 1984; Kříbek et al., 1999). HRTEM analyses of the 002-lattice fringe images show that kerogen of the prehnite-pumpellyite facies shales has a turbostratic structure wherein aromatic carbon layers are piled up in small individual stacks that are rotated randomly in three dimensions. Thus, the bulk crystalline ordering is bi-periodic in contrast to the tri-periodic crystalline ordering of graphite wherein infinite strata of hexagonal, aromatic carbon rings (called graphenes) are perfectly parallel stacked. The HRTEM images also show the low degree of organization due to sets of discontinuous lattice fringes (Jehlička and Rouzaud, 1990).

In the black schists from the western *TBU* (Fig. 1) the finely dispersed organic matter (micrometer size) was transformed to graphite at metamorphic temperatures reaching up to 350 °C at about 0.5-1 GPa. Graphite crystallites range from a few hundreds of nanometers to micrometers in size. The HRTEM 002-lattice fringe images show the typical, perfectly straight, parallel arrangement of graphite basal planes with a characteristic 0.3354-nm spacing (Jehlička and Rouzaud, 1990). This transformation to graphite was facilitated by the presence of anisotropic tectonic stresses and the enhancement of anisotropic strain in the interstratified carbonaceous layers. During sedimentation the organic matter was initially finely dispersed within the mineral matrix of anisometric, platy, mixed-layer clay minerals. Regional metamorphism modified this mixture of original layer silicates and organic matter into composite assemblage of interstratified organic and mica layers up to micrometers in size. Mica formed during metamorphosis of the shales. This interstratified, lamellar texture can enhance the graphitizability of organic carbon resulting in the formation of single-crystal graphite plates between two mica layers (or other layer silicates) at temperatures outside the graphite stability field (see, Fig. 2).

The Raman spectra of the regionally metamorphosed, dispersed carbonaceous matter show a progressive increase of structural organization. That is, the gradual disappearance of the '*D*' band at 1350 cm^{-1}, the reduction of the peak width of the E_{2g2} '*G*' band and its progressively decreasing frequency from ca. 1590 cm^{-1} for the poorly-organized kerogens of the prehnite pumpellyite facies to 1575-1580 cm^{-1} for graphite (Jehlička and Bény, 1992). Similar textural ordering was documented in the Raman spectra of samples from other metamorphosed rocks (e.g. Wopenka and Pasteris, 1993, Beyssac et al., 2002; Rantitsch et al., 2004) and in heat-treated graphitizing carbons (Bény-Bassez and Rouzaud, 1985).

5. SOLID BITUMEN FROM BASALTIC ROCKS AT MÍTOV

At the Mítov site, solid bitumen appears as accumulations or 'nests' up to a few centimeters in diameter directly in (1) pillows, (2) in the inter-pillow fillings of pillow lavas, and (3) inside the quartz filled cracks in the inter-pillow matter. The primary hydrocarbons originated from kerogen evolution in the black shales when this bitumen was later mobilized and then interacted with submarine volcanic effusions (Kříbek et al., 1993; Jehlička, 1994). Only solid bitumen thus produced contains C_{60} fullerene (Jehlička et al., 2003).

Solid bitumen is black and lustrous with conchoidal fracture surface. Its measured bi-reflectance is 6.64 - 3.62%; its reflectance is 6.46%, consistent with the typical value of 6.46% (546 nm) for solid bitumen (Kříbek et al., 1993). It is optically anisotropic showing a mosaic texture. It is highly aromatic and highly carbonized as revealed by its elemental composition, viz. H/C = 0.11 and O/C = 0.067, and results of ^{13}C Nuclear Magnetic Resonance spectroscopy (Jehlička, 1994). A biogenic source of solid bitumen is generally agreed upon, because its carbon is isotopically light with δ-values ranging from -29.0‰ to -31.0‰ (Kříbek et al., 1993; Jehlička, 1994).

The X-ray diffraction patterns of different samples of solid bitumen correspond to non-graphitized carbonaceous matter with a mean interlayer spacing, d_{002} = 0.347 nm, compared to 0.336 nm for fully-ordered, single-crystal graphite. The crystallite thickness (L_c) is ca. 3.5 nm that corresponds to an average of ten, stacked carbon layers in a '*basic structural unit*' (Jehlička, 1994). The *hk*-bands are not modulated at the *hkl*-positions demonstrating its completely turbostratic character. The first-order Raman spectra show the two main characteristic Raman bands for carbonaceous matter at 1350 cm^{-1} and 1575 cm^{-1}. The band surface ratio is almost unity and both bands have a relatively small half-width (ca. 60 cm^{-1}) in each of the bitumen samples investigated. Well-resolved broad bands at 2680 cm^{-1} and 2930 cm^{-1} appear in the second-order spectrum. The splitting of the Raman band at 2720 cm^{-1} was not observed, which is consistent with the turbostratic nature of this solid bitumen (Jehlička et al., 1997) that is reminiscent of glass-like carbon (Nakamizo et al., 1974). The HRTEM analyses (002-lattice fringe images) reveal considerable micro-textural heterogeneity (Jehlička, 1994; Jehlička and Rouzaud, 2000) (Fig. 4). The microtextural heterogeneity includes:

1. Turbostratic carbonaceous matter with no obviously pronounced parallel orientation of the aromatic lattice planes (Fig. 4a) forming ~70% of the bitumen sample,
2. Aggregates of ~100 nm consisting of quasi-parallel sets of wrinkled graphene layers representing ~20% of the sample, and
3. Microporous glass-like carbon (Fig. 4b) forming 10% of the sample.

5.1 Solid Bitumen Evolution

The accumulated solid bitumen is non-graphitic, i.e. non-crystalline carbon with no periodic three-dimensional carbon atom arrangements, and non-graphitizing, i.e. it is not transformed to graphite when heated under inert gas at about 3000 °C. Its optical properties, composition, structure and microtextures resemble industrial cokes such as those obtained by pyrolysis of caking cokes at 1000 °C (Rouzaud et al., 1988, Jehlička 1994). The structure of Mítov solid bitumen is very similar to industrial cokes, for which the carbonization occurred by quasi-pure thermal treatment without noticeable anisotropic pressure (Rouzaud et al., 1988).

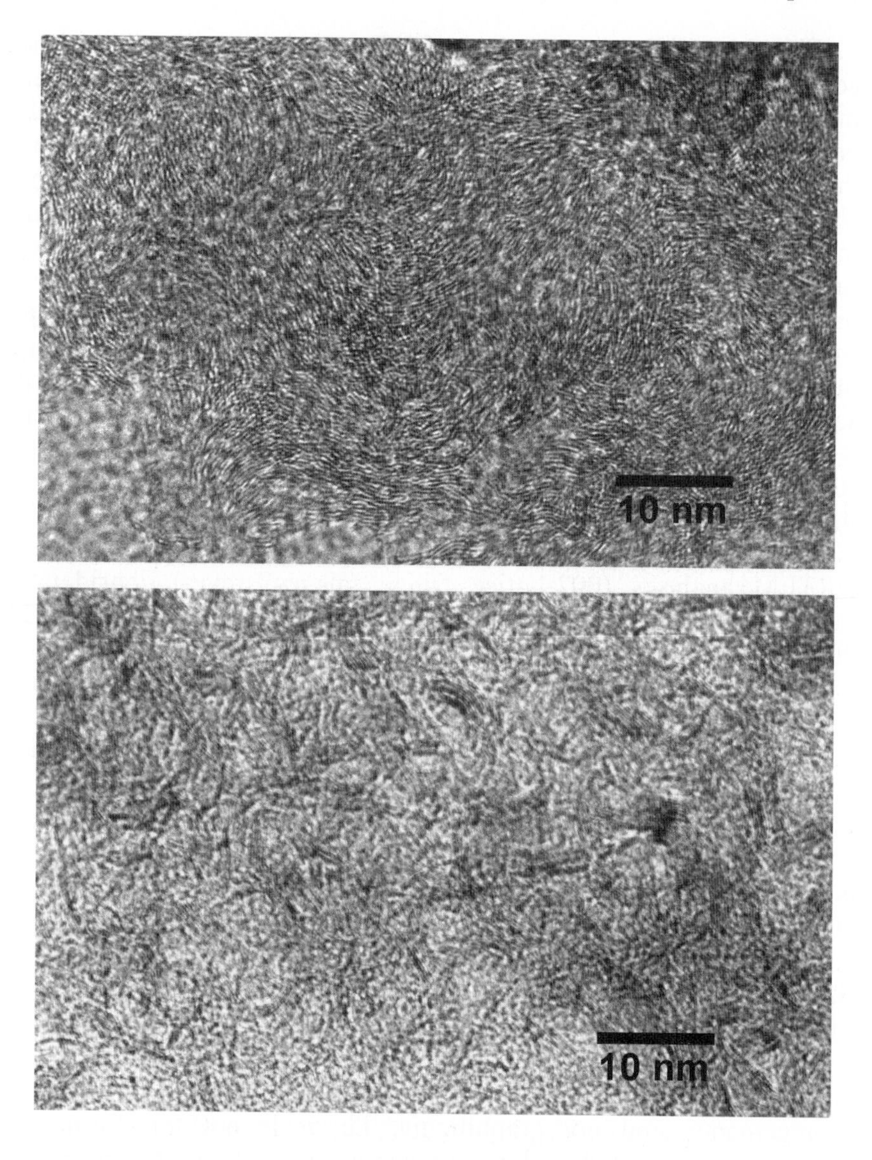

Figure 4. Transmission electron microscope images of solid bitumen from Mítov showing the 002-lattice fringes in (a) a turbostratic arrangement of basic structural units with no preferred orientations and in (b) microporous to mesoporous carbonaceous matter. Modified after Jehlička et al. (2003)

The solid bitumen structure and microtexture is of course different from this of kerogen in the rocks of the *TBU* that experienced regional metamorphism. The observed structural disorder of the solid bitumen at Mitov is explained by the low-pressure conditions of pyrometamorphism. The microtextures of dispersed kerogen that evolved under regional metamorphic conditions (see, Fig. 2) are markedly different from the only moderately preferred orientations of structural domains in solid bitumen that experienced thermal metamorphism with no noticeable lithostatic pressure or strain. Such very limited degrees of preferred orientations could develop in a semi-plastic stage in the first stages of pyro-metamorphism (Jehlička et al., 1997, Jehlička and Rouzaud, 2000). Non-hydrostatic pressures, especially shear stress, are known to affect graphitization of natural anthracite coals (Bonijoly et al., 1982) and kerogen (Jehlička and Rouzaud, 1990), and in heat-treatment experiments (Bustin et al., 1995). In contrast to the kerogen transformations in the *TBU* unit, the preferential orientations of solid bitumen were insufficient to lead to major structural improvement (Jehlička and Rouzaud, 1990).

Contact metamorphic phenomena between coals or bitumens and volcanic rocks, called pyrometamorphism, are well known from geological formations of different ages. This metamorphism is quite different from regional metamorphism: higher temperatures are usually reached with a much higher heating rates but the pressure is generally low (Kisch, 1966). Natural "cokes" produced by pyrometamorphism of coals or solid bitumens are highly carbonised, contain remnants of plastic deformation features with very small devolatilization pores (Vleeskens et al., 1994), optical anisotropy, and fine- to medium-sized mosaic structures. Gize (1993) gives a review of such carbonaceous materials transformed by heating or hydrothermal processes.

6. FULLERENE C_{60} IN SOLID BITUMEN AT MÍTOV

Fullerene C_{60} was detected in toluene extracts of two samples of solid bitumens (labeled SB1 and SB2) by HPLC at concentrations of 0.2 ppm and 0.3 ppm (Jehlička et al., 2000, 2003). Only C_{60} was identified according to its retention time in comparison with the synthetic standard. The EIMS measurements confirmed the presence of C_{60} by M^+ at m/z = 720 and M^{++} at m/z = 360 Da (Jehlička et al., 2003) (Fig. 5)

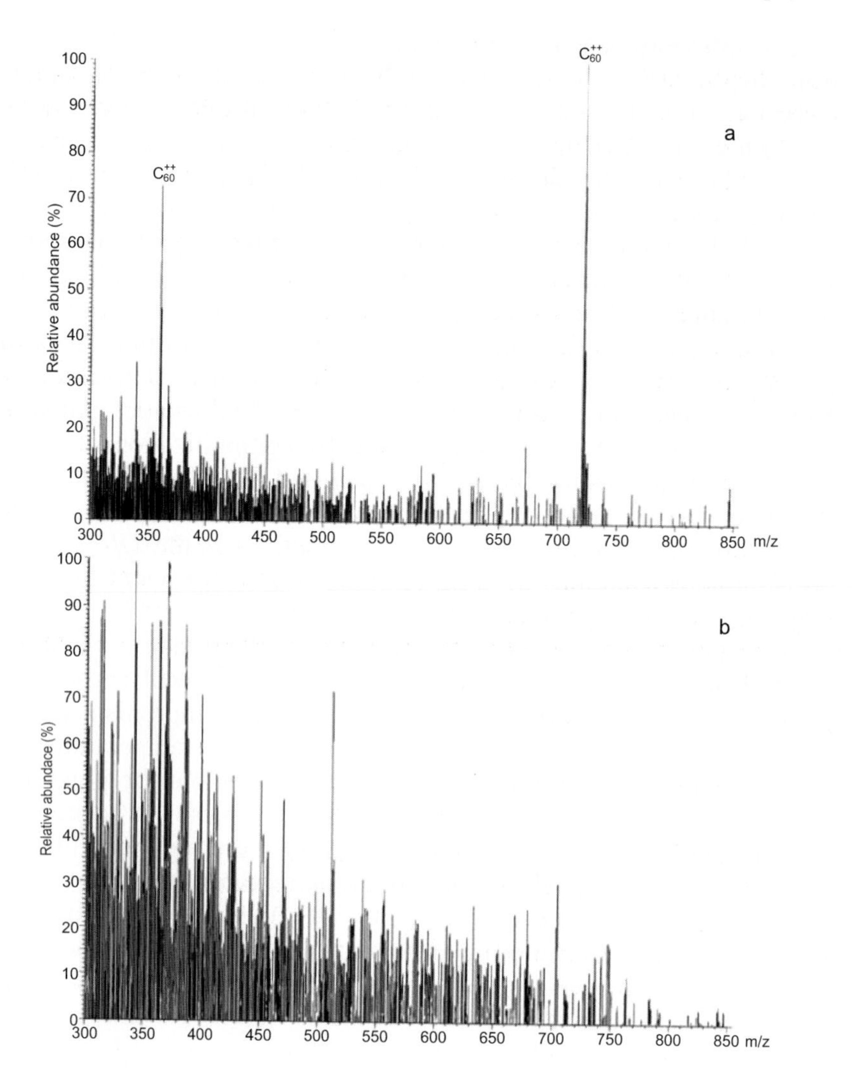

Figure 5. Electron impact mass spectra of toluene-extracted sample SB2 (a), positive detection of C_{60}, and SB4 (b), negative result demonstrating the absence of fullerene in this sample.

The elemental composition of the $M^{+\cdot}$ peak was determined by high-resolution EIMS. The isotopic composition of C_{60} in sample SB 1 was determined in the high-resolution mode as follows. For the m/z = 721 peak adjacent to the $^{12}C_{60}$ peak the accurate mass was measured as 721.0051 (Jehlička et al., 2003). The calculation of the molecular formulae yielded two compositions: $^{13}C^{12}C_{59}$ (721.0034) and $^{1}H^{12}C_{60}$ (721.0079), which correspond to the even-electron ion $^{13}C^{12}C_{59}{}^{+}$ or to the odd-electron ion $^{1}H^{12}C_{60}{}^{+}$, respectively. Although both ions have been simultaneously observed (Bordenave-Montesquieu et al., 2001), our experimental conditions (EI at 70 eV; T = 300 °C) would favor the formation of the even-electron ion (Jin et al., 1994). Therefore, we believe that the observed peak is the ^{13}C peak of fullerene. Finally, the isotopic composition for another peak (m/z = 722) was measured (722.0061; see Jehlička et al., 2003) and calculated as 722.0067, which is consistent with $^{13}C_2{}^{12}C_{58}{}^{+}$. It suggests a ^{13}C isotope distribution in the Mítov samples that is similar to those previously found in samples from the Sudbury Impact Structure (Becker et al., 1994) and shungite (Parthasarathy et al., 1998).

No C_{70} or higher fullerenes were detected but the data clearly establish the presence of natural C_{60} in the hard solid bitumens at Mítov.

6.1 How Widespread C_{60}?

Ten other samples of Mítov solid bitumen and poorly graphitized residual matter from the regional metamorphic Neo-proterozoic shales and schists contain no C_{60} or C_{70} at concentrations higher than 0.01 ppm based on comparison with the elution time of dissolved C_{60} standard. The EIMS spectra for these samples do not display peaks at m/z 720 or 840 Da, confirming the absence of C_{60} and C_{70}. In total, forty samples of solid bitumen and black shales from Mítov, Carboniferous bituminous coals from Kladno area and graphite of Paleozoic and Precambrian ages (Vrbno, Lom, Český Krumlov), formed during regional metamorphism in the Bohemian massif, were tested for the presence of fullerenes (Jehlička et al., 2000, Jehlička, unpubl. data). HPLC and EIMS analyses unambiguously established the presence of C_{60} in only the two different samples of solid bitumen from Mítov. What could be the reason for the apparent paucity of natural fullerene? In an attempt to find answers, we will explore the conditions of fullerene formation and preservation at Mítov.

7. FULLERENE FORMATION AT MÍTOV

How to form fullerenes composed only of carbon in naturally occurring rocks? There is no geological or geochemical evidence that the rocks in the *TBU* (Fig. 1) and particularly those in the Mítov area could be associated with a giant meteoroid impact or global wildfires. We consider that some other process of high-temperature transformation of precursor organic matter was responsible for C_{60} formation. Specifically, we suggest that fullerene formation at Mítov was related to high-temperature pyrometamorphism of precursor bitumen. In order to make our argument we will first review some possible fullerene-forming laboratory processes focusing on pyrolytic transformations and combustion of organic matter.

7.1 Laboratory Fullerene Formation by Pyrolysis and Combustion

It was demonstrated that fullerene could be formed by laser ablation or pyrolysis of suitable carbonaceous precursors. For example, fullerenes were produced in the gas phase that was formed by laser ablation used at low power densities of the pyrolyzed algal product coorongite (Rose et al., 1993). Coorongite is a deposit of *Elaeophyton* algae in saltwater bodies near Coorong (Australia). Other authors have described experimental pyrolytic transformation of aromatic hydrocarbons and the formation of fullerene, such as the generation of C_{60} and C_{70} during pyrolysis of naphthalene in an Ar atmosphere (Taylor et al., 1993). Cyclopentafused polycyclic aromatic hydrocarbons (CP-PAHs) are important precursors for fullerene formation such as during flash-vacuum-thermolysis and other laboratory conditions, as well as in industrial furnaces (Crowley et al., 1996; Jenneskens et al., 1996; Ledesma et al., 1999; Wornat et al., 1999). Under high-temperature conditions (900-1100 °C) many CP-PAH combustion products were reported to have formed in the gas phase for which Jenneskens and Sarobe (1999) proposed various, possible transformation mechanisms. Additionally, cyclotrimerization of *4,5-dihydrobenz[j]acephenanthrylene* $C_{20}H_{14}$ can produce $C_{60}H_{30}$ by heating the monomer with elemental sulfur at lower temperatures (300 °C). The trimer could be considered a C_{60} precursor (Sarobe et al., 1999). Thus, PAHs and especially CP-PAHs can be precursors for fullerene during specific laboratory experiments (Jenneskens and Sarobe, 1999; Heymann et al., 2003a, 2003b).

Fullerenes are high-temperature transformation products of petroleum asphaltenes (Camacho-Bragado et al., 2002). Fullerene-like particles were found in petroleum soot (Grieco et al., 2000; Lee et al., 2002). Fullerene-like structures and fullerene precursors were also detected by HRTEM in different types of carbon blacks (Harris and Tsang, 1997; Cataldo and Pontier-Johnson, 2002). Unfortunately, independent confirmation of fullerenes in these studies is still missing.

The common traits among these aforementioned studies are the rather specific nature of the precursors and high (900-1200 °C) transformation temperatures. Fullerene formation can be very rapid (milliseconds to seconds) (Jenneskens et al., 1996; Jenneskens and Sarobe, 1999) or slow (minutes to hours) (Grieco et al., 2000; Camacho-Bragado et al., 2002; Lee et al., 2002) but such times are very short when compared to the duration of typical geological processes such as pyrometamorphism and regional metamorphisms but comparable to for example the duration of a lightning strike.

7.2 Bitumen as Fullerene Precursor

The stable carbon isotopes data showed that bitumen precursor materials had an algal origin (Pouba and Kříbek, 1986; Jehlička, 1994) and in fact primitive algal remnants were found in graywackes and silicified stromatolites in the area (Vavrdová, 1994, 2000; Konzalová, 2000). This algal-derived organic matter would be an oxygen-poor kerogen precursor. Under favorable conditions during diagenesis, this type of kerogen could produce a large volume of hydrocarbons (Durand and Nicaise, 1985).

Bituminous compounds and mixtures of hydrocarbons, including PAHs, are frequent products of thermal or hydrothermal alteration of carbon-rich sediments. The in situ pyrolysis of oil shales, the generation of volatile hydrocarbons and their accumulation in a micro-reservoir was documented in the Cretaceous Rundle shale (Australia) (Saxby and Stephenson, 1987). Formation and migration of biogenic hydrocarbons were also documented in sedimentary and metasedimentary rocks as old as the Archean (Buick et al., 1998). During hydrothermal generation of petroleum in the Guaymas Basin and in Cretaceous sedimentary systems, produced bitumen is characterized by high contents of PAHs, hopanes, hopenes, and sterenes, and a high content of heteroatomic compounds and sulfur compared to typical reservoir petroleum (Simoneit, 1985, 1993). Similar compounds might have been fullerene precursors in the organic matter derived from the decay and transformation of an algal-derived biomass during diagenesis in the Neo-proterozoic black shales.

Solid bitumen - a*nthraxolite* - occurs within hydrothermal veins at Příbram in a hydrothermal U and Pb-Zn deposit (Kříbek et al., 1999). These veins crosscut the Neo-proterozoic black shales and graywackes in the outer contact of the Central Bohemian plutonic complex. Solid anthraxolite, with and without U-mineralization, was accumulated in carbonate-rich or uraninite-rich hydrothermal veins. This anthraxolite consists of a mixture of partly soluble aliphatic hydrocarbons. It has a relatively high atomic H/C ratio of 0.8, low aromaticity and a very low degree of structural ordering (Jehlička, 1995). These properties point to mild temperatures during accumulation of the original hydrocarbons that during the late carbonate-uraninite stage of hydrothermal vein formation condensed at temperatures below 140 °C (Žák and Dobeš, 1991; Kříbek et al., 1999). Aromatic compounds identified in extracts of this anthraxolite include alkyl benzenes, naphthalene, alkyl naphthalenes, methyl indanes, and dimethyl indane (Kříbek, 1989). Anthraxolite is considered an accumulated and later polymerised product that originated from the same *TBU* black shale formation. This relationship is thought to have existed also in the Mítov area. The black shales of the area are considered to be a source of liquid, perhaps even gaseous, hydrocarbons that could be precursors of anthraxolite in the hydrothermal veins at Příbram and also the solid bitumens at Mítov.

Off-line laboratory pyrolysis of mildly metamorphosed *TBU* black shales produced a mixture of predominantly aromatic compounds, including alkyl benzenes, naphthalene, alkyl naphthalenes, cresols, alkyl phenols, and dimethyl polysulphide, and sulfur (Kříbek et al., 1999). These hydrocarbons that are similar to those identified in anthraxolite extracts could have played an important role as fullerene precursors. It is not surprising that the anthraxolite at Příbram does not contain C_{60} fullerene because the thermal conditions at this location were too low (250 °C) to permit transformations of potential fullerene precursors, such as CP-PAHs, to transform by dehydrogenation to fullerene. Bituminous mixtures similar to those at Příbram were probably present at Mítov before they were pyrolysed to hard bitumens during repeated high-temperature interactions with basalt flows (Kříbek et al., 1993).

We propose that high-temperatures perhaps on the order of 900 °C induced by extruding basaltic lavas in wet, organic carbon-rich sediments produced fullerenes through transformation of organic matter and the formation of PAH precursors as part of the bitumen mixture. The basaltic magma temperatures were estimated at 1000-1200 °C on the basis of geochemical analyses (Waldhauserová, 1997). PAH-trimerization and dehydrogenation with sulfur might be of interest for the patchy occurrences of rare fullerenes in Mítov. These fullerenes could form only in zones wherein PAH precursors and sulfur co-occurred in the proper proportions. From the data obtained so far, we cannot unequivocally decide on the exact pyrolysis mechanism (e.g. Jehlička et al., 2003; Heymann et al., 2003b) for the natural fullerenes in bitumens.

8. HOW TO CONSERVE PRISTINE FULLERENE AT MÍTOV?

One of major problems that arise when studying natural fullerenes is the limited stability of these carbon molecules. For example it was observed that C_{60} is lost when exposed to UV radiation and ozone (Taylor et al., 1991) that could lead to the formation of higher fullerene oxides (Heymann and Chibante, 1993). There are several mechanisms under various conditions of photodecomposition of fullerene that would lead to $C_{60}O$ (Juha et al., 2000). Chibante et al. (1993) described the destruction of C_{60} and C_{70} when heated in air while Chibante and Heymann (1993) investigated the modification of C_{60} samples, including in crystalline form, in various O_3/O_2 mixtures. Thus, the probability of conserving fullerenes in rocks throughout their geological history is low. The ages of the sedimentary and metamorphic rocks in the *TBU* Neo-proterozoic of the Bohemian massif are on the order of 10^8 years. The fullerene-containing rock at Mítov suffered almost no weathering coming from a recently exposed quarry but still there have to be very efficient conditions to ensure the long-term conservation of natural fullerene in the environmental conditions of metasediments and extruded volcanic lavas. We suggest that structural and microtextural features in the solid bitumens were of primary importance. The Mítov solid bitumen resembles synthetic glass-like carbon. Glass-like carbon is a non-graphitizing carbon (Rousseaux and Tchoubar, 1977) that is chemically inert, has high hardness, low gas permeability, and with a microporous to mesoporous texture of closed pores (Yoshida et al., 1991).

We submit that fullerene molecules at Mítov could survive encapsulated in these closed micropores and mesopores in the solid bitumens wherein they were protected against oxidation. We note that the analyzed samples were generally of limited volume, rarely reaching 1 cm^3. Samples in which fullerenes were identified originated from isolated fillings in unweathered basalt. Solid bitumen from Mítov efficiently isolated fullerenes from the changes in their environment, although we do not claim that the concentrations of fullerene that were detected represent the original abundances or are in fact residual fullerenes.

9. CONCLUSIONS

Rare natural C_{60} fullerene, but no C_{70}, occurs in the solid bitumen associated with pillow lavas at Mítov in the Neo-proterozoic (565 Ma) rocks of the Teplá-Barrandian Unit in the Bohemian Massif. Their formation is linked to pyrolysis of biogenic (algal) PAHs precursors that were the derivatives originating from the kerogen evolution in geologically associated black shales. The high temperatures for pyrolysis of bitumen to fullerene-bearing solid bitumen were provided by contemporaneous basaltic extrusions that crosscut the black shale formation. Solid bitumen transformation produced a glassy-carbon with fullerene in closed micropores to mesopores during pyrometamorphism between 700-900 °C. Fullerene could survive within these pores of the glassy carbon. A similar survival scenario might explain the scant fullerene in shungite samples from Karelia.

Acknowledgements: This work was partly supported by a grant of the Grant Agency of the Czech Republic (205/03/1468), CEZ:J13/98:113100005. We are grateful for this support.

10. REFERENCES

Becker, L., Bada, J.L., Winans, R.E., Hunt, J.E., Bunch, T.E. and French B.M. (1994) Fullerenes in the 1.85-billion-year-old Sudbury impact structure. *Science*, 265, 642-644.

Becker, L., Poreda, R.J. and Bada, J.L. (1996) Extraterrestrial helium trapped in fullerenes in the Sudbury impact structure. *Science*, 272, 249-252.

Becker, L., Poreda, R.J., Hunt, J.E., Bunch, T.E. and Rampino, M. (2001) Impact event at the Permian-Triassic boundary: Evidence from extraterrestrial noble gases in fullerenes. *Science*, 291, 1530-1533.

Bénny-Bassez, C. and Rouzaud, J.N. (1985) Characterization of carbonaceous materials by correlated electron and optical microscopy and Raman microspectrometry. *Scanning Electron Microscopy*; 1, 119-132.

Beyssac, O., Rouzaud, J.N. and Goffé, B. (2002) Graphitization in a high-pressure, low-temperature metamorphic gradient: a Raman microspectroscopy and HRTEM study. *Contrib. Mineral. Petrol.*, 143, 19-31.

Bonijoly, M., Oberlin, M. and Oberlin, A. (1982) A possible mechanism for natural graphite formation. *Int. J. Coal Geol.*, 1, 283-312.

Bordenave-Montesquieu, D., Moretto-Capelle, P., Bordenave-Montesquieu, A. and Rentenier, A. (2001) Scaling of C_{60} ionization and fragmentation with the energy deposited in collisions with H^+, H_2^+, H_3^+ and He^+ ions (2-130 keV). *J. Phys. B*, 34, 137-146.

Buick, R., Rasmussen, B. and Krapez, B. (1998) Archean oil: evidence for extensive hydrocarbon generation and migration. *AAPG Bull.*, 82, 50-69.

Buseck, P.R. (2002) Geological fullerenes: review and analysis. *Earth Planet. Sci. Lett.*, 203, 781-792.

Buseck, P.R., Tsipursky, S.J. and Hettich, R. (1992) Fullerenes from the geological environment. *Science*, 257, 215-217.

Buseck, P.R., Galdobina, L.P., Kovalevski, V.V., Rozhkova, N.N., Valley, J.W. and Zaidenberg, A.Z. (1997) Shungites: The C-rich rocks of Karelia, Russia. *Can. Mineral.*, 35, 1363-1374.

Bustin, R.M., Rouzaud, J.N. and Ross, J.V. (1995) Natural graphitization of anthracite - experimental considerations. *Carbon*, 33, 679-691.

Cataldo, F. and Pontier-Johnson, M.A. (2002) Recent discoveries in carbon black formation and morphology and their implications on the structure of interstellar carbon dust. *Fullerenes, Nanotubes, and Carbon Nanostructures*, 10, 1-14.

Camacho-Bragado, G.A., Santiago, P., Marin-Almazo, M., Espinosa, M., Romero, E.T., Murgich, J., Lugo, V.R., Lozada-Cassou, M. and Jose-Yacaman, M. (2002) Fullerenic structures derived from oil asphaltenes. *Carbon*, 40, 2761-2766.

Cháb, J. and Suk, M. (1977) Regional metamorphism in Bohemia and Moravia (*in Czech*). *Knih. Ústř. Úst. Geol.*, 50, 1-156.

Chibante, L.P.F. and Heymann, D. (1993) On the geochemistry of fullerenes - stability of C_{60} in ambient air and the role of ozone. *Geochim. Cosmochim. Acta*, 57, 1879-1881.

Chibante, L.P.F., Pan, C.Y., Pierson, M.L., Haufler, R.E. and Heymann, D. (1993) Rate of decomposition of C_{60} and C_{70} heated in air and the attempted characterization of the products. *Carbon*, 31, 185-193.

Chijiwa, T., Arai, T., Sugai, T., Shinohara, H., Kumazawa, M., Takano, M. and Kawakami, S. (1999) Fullerenes found in the Permo-Triassic mass extinction period. *Geophys. Res. Lett.*, 26, 767-770.

Crowley, C., Taylor, R., Kroto, H.W., Walton, D.R.M., Cheng, P.C. and Scott, L.T. (1996) Pyrolytic production of fullerenes. *Synthetic Metals*., 77, 17-22.

Daly, T.K., Buseck, P.R., Williams, P. and Lewis, C.F. (1993) Fullerenes from a fulgurite. *Science*, 259, 1599-1601.

Durand, B. and Nicaise, G. (1980) Procedures of kerogen isolation. *In Kerogen*, B. Durand, Ed., 35-53, Editions Technip, Paris, France.

Ebbesen, T.W., Hiura, H., Hedenquist, J.W., de Ronde, C.E.J., Andersen, A., Often, M. and Melezhik, V.A. (1995) Origins of fullerenes in rocks. *Science*, 268, 1634-1635.

Fiala, F. (1977) The Upper Proterozoic volcanism of the Barrandian area and the problems of spilites. *Sbor. Geol. Věd.*, 30: 1-247.

Galdobina, L.P. (1993) Shungite rocks. *In Precambrian Industrial Minerals of Karelia*, V. Schciptsov, Ed., 45-53, Karelia Publishing, Petrozavodsk, Russia.

Gize, A.P. (1993) The analysis of organic matter in ore deposits. *In Bitumens in ore deposits*, J. Parnell, P. Kucha and P. Landais, Eds., 28-52, Springer, Berlin Germany.

Grieco, W.J., Howard, J.B., Rainey, L.C. and Vander Sande, J.B. (2000) Fullerenic carbon in combustion-generated soot. *Carbon*, 38, 597-614.

Gu, Y., Wilson, M.A., Fisher, K.J., Dance, I.G., Willet, G.D., Ren, D. and Volkova, I.B. (1995) Fullerenes and shungite. *Carbon*, 33, 862-863.

Harris, P.J.F. and Tsang, S.C. (1997) High-resolution electron microscopy studies of non-graphitizing carbons. *Phil. Mag. A*, 76, 667-677.

Heymann, D. (1995) Search for ancient fullerenes in anthraxolite, shungite, and thucholite. *Carbon*, 33, 237-239.

Heymann, D. and Chibante, L.P.F. (1993) Photo-transformations of C_{60}, C_{70}, $C_{60}O$ and $C_{60}O_2$. *Chem. Phys. Lett.*, 207, 339-341.

Heymann, D., Chibante, L.P.F., Brooks, R.R., Wolbach, W.S. and Smalley, R.E. (1994a) Fullerenes in the Cretaceous-Tertiary boundary layer. *Science*, 265, 645-647.

Heymann, D., Wolbach, W.S., Chibante, L.P.F., Brooks, R.R. and Smalley, R.E. (1994b) Search for extractable fullerenes in clays from the Cretaceous/Tertiary boundary of the Woodside Creek and Flaxbourne River sites, New Zealand. *Geochim. Cosmochim. Acta*, 58, 3531-3534.

Heymann, D., Wolbach, W.S., Chibante, L.P.F. and Smalley, R.E. (1994c) Search for extractable fullerenes in clays from the KT boundary of the Woodside Creek and Flaxbourne River sites, New Zealand (abstract). *In New Developments Regarding the KT Event and Other Catastrophes in Earth History, LPI Contribution*, # 825, 47-48, Lunar and Planetary Institute, Houston, Texas, USA.

Heymann, D., Korochantsev, A., Nazarov, M.A. and Smit, J. (1996) Search for fullerenes C_{60} and C_{70} in Cretaceous-Tertiary boundary sediments from Turkmenistan, Kazakhstan, Georgia, Austria, and Denmark. *Cretaceous Research*, 17, 367-380.

Heymann, D., Yancey, T.E., Wolbach, W.S., Thiemens, M.H., Johnson, E.A., Roach, D. and Moecker, S. (1998) Geochemical markers of the Cretaceous-Tertiary Boundary Event at Brazos River, Texas, USA. *Geochim. Cosmochim. Acta*, 62, 173-181.

Heymann, D., Jenneskens, L.W., Jehlička, J., Koper, C. and Vlietstra, E.J. (2003a) Terrestrial and extraterrestrial fullerenes. *Fullerenes, Nanotubes, and Carbon Nanostructures*, 11, 333-370.

Heymann, D., Jenneskens, L.W., Jehlička, J., Koper, C. and Vlietstra, E.J. (2003b) Biogenic fullerenes? *Int. J. Astrobiology*, 2, 179-183.

Itaya, T. (1981) Carbonaceous material in pelitic schists of the Sanbagawa metamorphic belt in central Shikoku, Japan. *Lithos*, 14, 215-224.

Jacob, H. (1993) Nomenclature, classification, characterization and genesis of natural solid bitumen. *In Bitumens in ore deposits*, J. Parnell, P. Kucha and P. Landais, Eds., 11-27, Springer, Berlin, Germany.

Jehlička, J. (1984) Organic geochemistry of metamorphic pelitic rocks of the Barrandian Upper Proterozoic (*in Czech*). MSc Thesis, 71p., Faculty of Sciences, Charles University,. Prague, Czech Republic.

Jehlička, J. (1994) Étude structurale de matières organiques soumises à des processus métamorphiques (*in French*). PhD Thesis, *Éditions BRGM*, 238, 1-230, Orléans, France.

Jehlička, J. (1995) Composition and structure of solid bitumens from the Příbram hydrothermal deposit. *In Mineral Deposits: From their origin to their environmental impacts*, J. Pašava, B. Kříbek and K. Žák, Eds., 753-756, A.A. Balkema, Rotterdam, the Netherlands.

Jehlička, J. and Bény, C. (1992) Application of Raman microspectrometry in the study of structural changes in Precambrian kerogens during regional metamorphism. *Org. Geochem.*, 18, 211-213.

Jehlička, J. and Rouzaud, J.-N. (1990) Organic geochemistry of Precambrian shales and schists (Bohemian massif, Central Europe). *Org. Geochem.*, 16, 865-870.

Jehlička, J. and Rouzaud, J.-N. (1993a) Glass-like carbon: new type of natural carbonaceous matter from Precambrian rocks. *Carbon*, 30, 1133-1134.

Jehlička, J. and Rouzaud, J.-N. (1993b) Transmission electron microscopy of carbonaceous matter in Precambrian shungite from Karelia. *In Bitumens in ore deposits*, J. Parnell, P. Kucha and P. Landais, Eds., 53-60, Springer, Berlin, Germany.

Jehlička, J. and Rouzaud, J.-N. (2000) Structural and microtextural features of solid bitumens from pillow lavas from Mítov Barrandian Neoproterozoic, Bohemian Massif. *Věst. ČGÚ*, 75, 297-306.

Jehlička, J., Křibek, B. and Weidenhoffer, Z. (1984) Distribution of hydrocarbons and fatty acids in Proterozoic black shales (*in Czech*). *In Correlations of Proterozoic and Paleozoic stratiform deposits*, Z. Pouba, Ed., *UNESCO/IGCP*, 91, 123-138, Prague, Czech Republic.

Jehlička, J., Bény, C. and Rouzaud, J.-N. (1997) Raman microspectrometry of accumulated non-graphitized solid bitumens. *J. Raman Spectr.*, 28, 717-724.

Jehlička, J., Ozawa, M., Slanina, Z. and Osawa, E. (2000) Fullerenes in solid bitumens from pillow lavas of Precambrian age Mítov, Bohemian Massif. *Fullerene Sci. Techn.*, 18, 449-452.

Jehlička, J., Svatoš, A., Frank, O. and Uhlík, F. (2003) Evidence for fullerenes in solid bitumen from pillow lavas of Proterozoic age from Mítov (Bohemian massif, Czech Republic). *Geochim. Cosmochim. Acta*, 67, 1495-1506.

Jehlička, J., Frank, O., PokornýJ. and Rouzaud, J.-N. (2005a) Evaluation of Raman spectroscopy to detect fullerenes in geological materials. *Spectrochim. Acta A* 61, 2364-2367.

Jehlička, J., Frank, O., Hamplová, V., Pokorná, Z., Juha, L., Boháček, Z. and Weishauptová, Z. (2005b) Low extraction recovery of fullerene from carbonaceous geological materials spiked with C_{60}. *Carbon* 43,1909-1917.

Jenneskens, L.W. and Sarobe, M. (1999) Prevalent cyclopentafused polycyclic aromatic hydrocarbon CP-PAH combustion effluents. Build-up and conversion under high temperature conditions in the gas phase. *Polycyclic Arom. Comp.*, 14, 169-178.

Jenneskens, L.W., Sarobe, M. and Zwikker, J.W. (1996) Thermal generation and interconversion of multi cyclopenta-fused polycyclic aromatic hydrocarbons. *Pure Appl. Chem.*, 68, 219-224.

Jin, C.M., Hettich, R., Compton, R., Joyce, D., Blencoe, J. and Burch, T. (1994) Direct solid-phase hydrogenation of fullerenes. *J. Phys. Chem.*, 98, 4215-4217.

Juha, L, Farniková, M., Hamplová, V., Kodymová, J., Mullerová, A., Krása, J., Láska, L., Špalek, O., Kubát, P., Stibor, I., Koudoumas, E. and Couris, S. (2000) The role of the oxygen molecule in the photolysis of fullerenes. *Fullerene Sci. Techn.*, 8, 289-318.

Kisch, H.J. (1966) Carbonisation of semi-anthracitic vitrinite by an analcime basanite sill. *Econ. Geol.*, 61, 1043-1063.

Konkov, O.I., Terukov, E.I. and Pfaunder, N. (1994) Fullerenes in shungite. *Phys. Solid. State*, 36, 1685-1686.

Konzalová, M. (2000) Organic-walled microbiota from the greywackes and other siliciclastic sediments of the Barrandian Neoproterozoic Bohemian Massif, Czech Republic. *Věst. ČGÚ*, 75, 319-331.

Kovalevski, V.V., Buseck, P.R. and Cowley, J.M.(2001) Comparison of carbon in shungite rocks to other natural carbons: An X-ray and TEM study. *Carbon*, 39, 243-256.

Křibek, B. (1989) The role of organic-matter in the metallogeny of the Bohemian Massif. *Econ. Geol.*, 84, 1525-1540.

Křibek, B. and Jehlička, J. (1986) The role of Proterozoic and Paleozoic carbon-rich formations in the metallogeny of the Bohemian Massif. *In Proceedings of the International Conference on Metallogeny of the Precambrian, UNESCO IGCP*, 91, 35-47, Geol. Survey, Prague, Czech Republic.

Křibek, B., Holubář, V., Parnell, J., Pouba, Z. and Hladíková, J. (1993) Interpretation of thermal mesophase in vanadiferous bitumens from Upper Proterozoic lava flows (Mítov, Czechoslovakia). *In Bitumens in ore deposits*, J. Parnell, P. Kucha and P. Landais, Eds., 61-80, Springer, Berlin, Germany.

Křibek, B., Hrabal, J., Landais, P. and Hladíková, J. (1994) The association of poorly ordered graphite, coke and bitumens in greenschist facies rocks of the Ponikla group, Lugicum, Czech Republic - the result of graphitization of various types of carbonaceous matter. J. Metamorphic Geol., 12, 493-503.

Křibek, B., Žák, K., Spangenberg, J., Jehlička, J., Prokeš, S. and Komínek, J. (1999) Bitumens in the Late Variscan hydrothermal vein-type uranium deposit of Příbram, Czech Republic: sources, radiation-induced alteration and relation to mineralization. *Econ. Geol.*, 94, 1093-1114.

Křibek, B., Pouba, Z., Skoček, V. and Waldhauserová, J. (2000) Neoproterozoic as a part of the Cadomian orogenetic belt: A review and correlation aspects. *Věst. ČGÚ*, 75, 175-196.

Ledesma, E.B., Kalish, M.A., Wornat, M.J., Nelson, P.F. and Mackie, J.C. (1999) Observation of cyclopenta-fused and ethynyl-substituted PAH during the fuel-rich combustion of primary tar from a bituminous coal. *Energy Fuels*, 13, 1167-1172.

Lee, T.H., Yao, N., Chen, T.J. and Hsu, W.K. (2002) Fullerene-like carbon particles in petrol soot. *Carbon*, 40, 2275-2279.

Mašek, J. (2000) Stratigraphy of the Proterozoic of the Barrandian area. *Věst ČGÚ*, 75, 197-200.

Masterov, V.F., Chudnovski, F.A., Kozyrev, S.V., Zaidenberg, A.Z., Rozhkova, N.N., Podosenova, N.G. and Stefanovich, G.B. (1994) Microwave absorption in fullerene-containing shungites. *Mol. Mat.*, 4, 213-216.

Melezhik, V.A., Fallick, A.E., Filippov, M.M. and Larsen, O. (1999) Karelian shungite - an indication of 2.0-Ga-old metamorphosed oil-shale and generations of petroleum: geology, lithology and geochemistry. *Earth Sci. Rev.*, 47, 1-40.

Mossman, D., Eigendorf, G., Tokaryk, D., Gauthier-Lafaye, F., Guckert, K.D., Melezhik, V. and Farrow, C.E.G. (2003) Testing for fullerenes in geologic materials: Oklo carbonaceous substances, Karelian shungites, Sudbury Black Tuff. *Geology*, 31, 255-258.

Nakamizo, M., Kammereck, R. and Walker, Jr., P.L. (1974) Laser Raman studies on carbons. *Carbon*, 12, 259-267.

Parthasarathy, G., Srinivasan, R., Vairamani, M., Ravikumar, K. and Kunwar, A.C. (1998) Occurrence of natural fullerenes in low grade metamorphosed Proterozoic shungite from Karelia, Russia. *Geochim. Cosmochim. Acta*, 62, 3541-3544.

Pouba, Z. (1988) Metallogeny of the Neoproterozoic in the Barradian and Železné Hory areas. *In Precambrian in younger fold belts*, V. Zoubek, J. Cogné , D. Kozoukcharov and H.-G. Krautner, Eds., 99-107, Wiley, & Sons, Chichester, New York, USA.

Pouba, Z. and Křibek, B. (1986) Organic-matter and the concentration of metals in precambrian stratiform deposits of the Bohemian massif. *Precambrian Res.*, 33, 225-237.

Pouba, Z., Křibek, B. and Pudilová, M. (2000) Stromatolite-like cherts in the Barrandian Upper Proterozoic: A review. *Věst. ČGÚ*, 75, 285-296.

Rantitsch, G., Grogger, W., Teichert, C., Ebner, F., Hofer, C., Maurer, E.-M., Schaffer, B. and Toth, M. (2004) Conversion of carbonaceous material to graphite within the Greywacke Zone of the Eastern Alps. *Int. J. Earth Sci.*, 93, 959-973.

Rietmeijer, F.J.M., Rotundi, A. and Heymann, D. (2004) C_{60} and giant fullerenes in soot condensed in vapors with variable C/H_2 ratio. *Fullerenes, Nanotubes, and Carbon Nanostructures*, 12, 659-680.

Rietmeijer, F.J.M., Borg, J. and Rotundi, A. (2005) Revisiting C_{60} fullerene in carbonaceous chondrites and interplanetary dust particles: HRTEM and Raman spectroscopy. *Lunar Planet. Sci.*, 36, abstract #1225, Lunar and Planetary Institute, Houston, Texas, USA, CD-ROM.

Rose, H.R., Dance, I.G., Fisher, K.J., Smith, D.R., Willett, G.D. and Wilson, M.A. (1993) Calcium inside C_{60} and C_{70} - from coorongite, a precursor of torbanite. *Chem. Comm.*, 11, 941-942.

Rousseaux, F. and Tchoubar, D. (1977) Structural evolution of glassy carbon as a result of thermal treatment between 1000 and 2700°C - II Tridimensional configuration of a glassy carbon. *Carbon*, 15, 63-68.

Rouzaud, J.N., Vogt, D. and Oberlin, A. (1988) Coke properties and their microtexture 1. Microtextural analysis - A guide for cokemaking. *Fuel Processing Technology*, 20, 143-154.

Sarobe, M., Fokkens, R.H., Cleij, T.J., Jenneskens, L.W., Nibbering, N.M.M., Stas, W. and Versluis, C. (1999) S-8-mediated cyclotrimerization of 4,5-dihydro-benz[l]acephen-anthrylene: trinaphtho-decacyclene $C_{60}H_{30}$ isomers and their propensity towards cyclode-hydrogenation. *Chem. Phys. Lett.*, 313, 31-39.

Saxby, J.D. and Stephenson, L.C. (1987) Effect of an igneous intrusion on oil shale at Rundle (Australia). *Chem. Geol.*, 63, 1-16.

Simoneit, B.R.T. (1985) Hydrothermal petroleum - genesis, migration, and deposition in Guaymas basin, Gulf of California. *Can. J. Earth Sci.*, 22, 1919-1929.

Simoneit, B.R.T. (1993) Aqueous high-temperature and high pressure organic geochemistry of hydrothermal vent systems. *Geochim. Cosmochim. Acta*, 57, 3231-3243.

Taylor, R. and Abdul-Sada, A.K. (2000) There are no fullerenes in the K-T boundary layer. *Fullerene Sci. Techn.*, 8, 47-54.

Taylor, R., Parsons, J.P., Avent, A.G., Rannard, S.P., Dennis, T.J., Hare, J.P., Kroto, H.W. and Walton, D.R.M. (1991) Degradation of C_{60} by light. *Nature*, 357, 277-279.

Taylor. R., Langley. G.J., Kroto. H.W. and Walton, D.R.M. (1993) Formation of C_{60} by pyrolysis of naphthalene. *Nature*, 366, 728-730.

Vavrdová, M. (1994) Silicified microbiota from the Bohemian Late Proterozoic. *J. Czech. Geol. Soc.*, 39, 183-193.

Vavrdová, M. (2000) Microfossils in carbonaceous cherts from Barrandian Neoproterozoic. *Věst. ČGÚ*, 75, 351-358.

Vleeskens, J., Kwiecinska, B, Roos, M. and Hamburg, G. (1994) Coke forms in nature and in power utilities - interpretation with SEM. *Fuel*, 73, 816-822.

Waldhauserová, J. (1997) Geochemistry of volcanites metavolcanites in the western part of the TBU Precambrian and their original geotectonic setting. *In Geological model of the western Bohemia related to the KTB borehole in Germany*, S. Vrána and V. Štědrá, Eds,. *Czech. Geol. Survey, Praha, Sbor. Geol. Věd.*, 47, 85-90.

Wopenka, B. and Pasteris, J.D. (1993) Structural characterization of kerogens to granulite-facies graphite - applicability of Raman microprobe spectroscopy. *Am. Mineral.*, 78, 533-557.

Wornat, M.J., Vriesendorp, F.J.J., Lafleur, A.L., Plummer, E.F., Necula, A. and Scott, L.T. (1999) The identification of new ethynyl-substituted and cyclopenta-fused polycyclic aromatic hydrocarbons in the products of anthracene pyrolysis. *Polycyclic Arom. Comp.*, 13, 221-240.

Yardley, B.W.D. (1989) *An Introduction to Metamorphic Petrology*, 264p., Longman Scientific & Technical, Essex, United Kingdom.

Yoshida, A., Kaburagi, Y. and Hishiyama, Y. (1991) Microtexture and magneto resistance of glass-like carbons. *Carbon*, 29, 1107-1111.

Žák, K. and Dobeš, P. (1991) Stable isotopes and fluid inclusions in hydrothermal ore deposits. *Rozpravy ČSAV, řada Mat.Přír.Věd*, 13, 1-109.

Zoubek, V., Holubec, J., Fiala, F., Pouba, Z. and Chaloupský, J. (1988) Central Bohemian Region. *In Precambrian in younger fold belts*, V. Zoubek, J. Cogné, D. Kozoukcharov and H.-G Krautner, Eds., 76-115, Wiley, & Sons, Chichester, New York, USA.

Chapter 11

FULLERENE SYNTHESIS BY ALTERATION OF COAL AND SHALE BY SIMULATED LIGHTNING

OTA FRANK
Institute of Geochemistry, Mineralogy and Mineral Resources, Faculty of Science, Charles University in Prague, Albertov 6, 12843 Prague 2, Czech Republic.

JAN JEHLIČKA
Institute of Geochemistry, Mineralogy and Mineral Resources, Faculty of Science, Charles University in Prague, Albertov 6, 12843 Prague 2, Czech Republic.

VĚRA HAMPLOVÁ
Academy of Sciences, Institute of Physics, Na Slovance 2, 18221 Prague 8, Czech Republic.

ALEŠ SVATOŠ
Max Planck Institute for Chemical Ecology, Mass Spectrometry Group, Hans-Knöll-Strasse 8, 07745 Jena, Germany.

Abstract: A graptolitic shale, a metamorphosed black schist and a bituminous coal with different organic carbon content but without detectable amounts of fullerenes were subjected to high-energy electric impulses of current amplitudes comparable to those of natural lightning strikes that could lead to the formation of fulgurites. The search for fullerenes concentrated on rock surfaces altered during the impulse experiment using Fourier transform infrared spectroscopy (FTIR). Toluene-extractable materials from the altered zones were investigated by high-performance liquid chromatography (HPLC) and electron-impact ionization mass spectroscopy (EIMS). Powders from altered surfaces were also analyzed by laser desorption time-of-flight mass spectroscopy. Two of four characteristic C_{60} FTIR peaks were only observed in one piece of black schist. As no C_{60} fullerene was detected by HPLC and EIMS in this modified sample, we find no evidence that electric impulse-

Frans J.M. Rietmeijer (ed.), Natural Fullerenes and Related Structures of Elemental Carbon, 241–255.

induced fullerenes were present in any samples. We discuss how the physicochemical conditions in our lightning-strike simulation experiments could have contributed to failure to produce fullerenes in rocks that are rich in organic carbon. Our results underscore the apparent paucity of natural fullerenes induced by lightning strikes.

Key words: Arc discharge; bituminous coal; Cretaceous-Tertiary boundary (KTB); electron impact ionization mass spectroscopy (EIMS); Fourier transform infrared spectroscopy (FTIR); fulgurite; graptolitic black shale; high performance liquid chromatography (HPLC); laser desorption time-of-flight (LD-TOF) mass spectroscopy; lightning; metamorphosed black shale; pulse experiments; shungite

1. INTRODUCTION

One of the classical ways to synthetically produce fullerenes is by arc discharge between graphite electrodes as was introduced by Krätschmer et al. (1990). Since then, electrodes made from coal (Li et al., 2002) have been successfully used to produce fullerenes in direct-current (DC) arc experiments. Using a current magnetic field (JxB) arc reactor, fullerenes were produced from charcoal, synthetic rubber, sumi (Chinese ink stick) and carbon black (Mieno, 2000). Such intense conditions as occur in the arcing experiments occur rarely in natural environments. Lightning is one possible natural process that could produce fullerenes in a fulgurite, a glassy, often tubular or rod-like structure, or crust commonly produced by fusion of loose soil, rarely a compact rock, which is produced when a lightning strike hits the earth's surface.

The first, and so far the only, reported finding of fullerenes in a fulgurite comes from Sheep Mountain (Colorado) (Daly et al., 1993). The time-of-flight mass spectrometry analyses, using a laser that desorbed a pulverized sample from a substrate, found peaks of fullerenes C_{60}^+ and C_{70}^+ in a spectrum that also included peaks of C_{46}^+ to C_{58}^+ and C_{64}^+ to C_{68}^+ with C_{2n}^+ increments. To confirm these findings and to remove any doubt about the laser desorption ionization method, subsequent mass spectrometry with electron ionization was used in the area of m/z 720; C_{60} was detected and divided into three peaks corresponding to $^{12}C_{60}^+$, $^{12}C_{59}{}^{13}C^+$, and $^{12}C_{58}{}^{13}C_2^+$ (Daly et al., 1993). The suggested carbon source for fullerenes in this fulgurite sample was from surrounding pine needles and cones on the soil. Based on this finding, Ebessen et al. (1995) tried to explain other natural fullerene occurrences as being the results of localized events such as lightning, including the fullerenes from Shunga that were previously reported by Buseck et al. (1992).

Natural fullerenes C_{60} and C_{70} were found in the globally deposited soot layers at the geologic Cretaceous-Tertiary Boundary (KTB) (Heymann et al., 1994, 1996, 1998) where they were produced by sooting wildfires triggered by the Chicxulub impact event (e.g. Heymann et al., 1994). Becker et al. (2000) proposed an extraterrestrial origin for fullerenes at the Cretaceous-Tertiary boundary. In order to elucidate the possible role of lightning in fullerene-forming processes in Nature, or more specifically in the KTB sediments, Heymann (1998) made a search for fullerenes in char from a Norway spruce hit by lightning. No C_{60} was found in a concentration higher than 1 ppb in the sample that was analyzed using high-performance liquid chromatography (HPLC) of toluene extract (Heymann, 1998).

In the present experiment, samples of black shale, metamorphosed black shale and bituminous coal were subjected to electric pulses with current amplitudes up to 130 kA. Toluene extracts of powders from the altered zones of these samples were separated by HPLC and subsequently analyzed by electron ionization mass spectrometry (EIMS). In addition, powder from the altered surface was analyzed directly by laser desorption, time-of-flight mass spectrometry and by Fourier transform infrared (FTIR) microspectroscopy.

We report that the physical regime produced in our experiments was apparently insufficient to produce measurable quantities of C_{60} and other fullerene molecules. We discuss that the very physicochemical conditions in our experimental setup could have contributed to our failure to produce measurable quantities of fullerene in natural samples of carbon-rich rocks.

2. SELECTED SAMPLES AND PULSE EXPERIMENTS

The samples that were selected for this experiment are

1. Silurian graptolite (a fossil marine organism)-bearing black shale from Kosov. It was collected in the Kosov quarry (30 km southwest from Prague, Czech Republic). It belongs to Upper Silurian Přídolí Formation (for a detailed description see, Čáp et al., 2000).
2. Metamorphosed black shale containing hard solid bitumen of Neoproterozoic age that was collected in Zbečno quarry (30 km

west of Prague). It belongs to Neoproterozoic of the Teplá-Barrandian unit (Kříbek et al., 2000), and
3. Carboniferous bituminous coal from Kladno. This sample was collected in Kladno coalmine (located 20 km northwest from Prague) in the Carboniferous limnic basin (Pešek, 2004).

The organic carbon content of the samples 1-3 is 2.5%, 0.7% and 79%, respectively. They represent geologic materials with different mechanical properties. They all are weakly, electrically conducting.

Platelets of approximately 10 x 10 centimeters and 1 to 2 centimeters thick were cut from each sample to be used in our experiments. Pulse experiments were conducted in the Laboratory for Very High Voltages in Prague-Běchovice. This laboratory is equipped for testing of resistance of various devices such as fuses or transformers. An impulse current generator was used for our experiment. Samples were placed between electrodes that were cleaned with chloroform and toluene after each run. The sample is placed between the electrodes (Fig. 1).

The electrodes holding the sample were placed inside a closed, stainless steel box. The impulse current was measured using induction-less shunt (1.53 mΩ). The measured signal was transferred to a HiAS 743 measuring system that was developed by Haefely Trench. Samples were stepwise exposed to impulses at higher amplitudes of the passing current; some samples were destroyed when applying the highest possible applied current magnitude. All experiments were carried out in air.

3. EXPERIMENTAL PROCEDURES AND TECHNIQUES

3.1 Extraction Procedures

To prepare the samples for fullerene analyses, 3 grams of each sample was immersed in toluene (30 ml) when pulverized by grinding in an agate mortar. The suspension was sonicated in an ultrasonic bath at room temperature for two hours. The extracts were filtered (0.45 μm PTFE filter), and a fraction of the extract was 10-times condensed in a rotary vacuum evaporator.

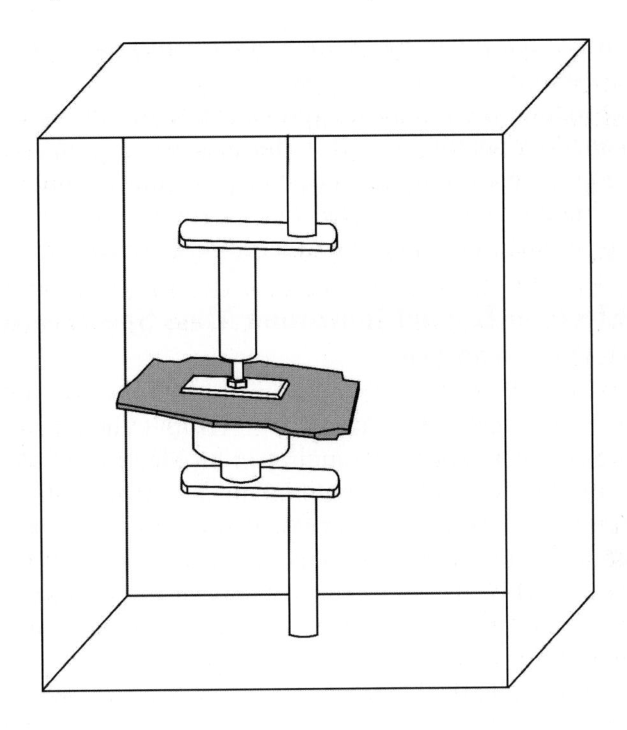

Figure 1. Simplified scheme showing the placement of a sample slab (gray) held between the electrodes of the impulse-current generator

Both original and condensed solutions were subjected to HPLC analysis. A Cosmosil Buckyprep Analytical Column (250 x 4.6 mm) was used for the separation with toluene as the mobile phase. Spectra were taken on a photo-diode-array detector. Each injected volume was 20 μl. As compared to spectra of a C_{60} standard, the limit of detection of C_{60} is approximately 1 ng, which gives a detection limit for the C_{60} content in a sample of approximately 0.01 ppm (sample mass is 3 g, volume of the final extract 0.5 ml; injected volume 20 μl).

Fractions with retention time above seven minutes were collected for EIMS analysis. For selected samples (Kosov1, Zbecno2, and Kladno2), an aliquot of the extract was set aside, and C_{60} standard

(99.5% purity, provided by Xillion, Germany) was added to this aliquot in a concentration of 200 ppb.

We note that all the glassware and the agate mortar used was either new or cleaned by heating at 300 °C for at least 12 hours to ensure no fullerene contamination would occur during the chemical analyses. Instrument blanks were conducted by injecting pure solvent into the system prior to injection of each extract to be analyzed.

3.2 Electron-Impact Ionization Mass Spectroscopy (EIMS) Analyses

Mass spectra were taken using a ZAB-Q (Vacuum Generators, UK) double-sector instrument running in an electron-impact mode at an energy of 70 eV. Sample material in a toluene solution (10 μl) was injected into a fused silica tube sealed at one end and the solvent was evaporated to dryness under vacuum. The probe was inserted into a direct-insert port of the spectrometer and heated to 300 °C with the inlet heated at 330 °C. The resulting spectra are an average of at least fifteen consecutive scans. The high-resolution measurements were performed under identical conditions using a *perfluorohydrocarbon* mixture for calibration.

3.3 Laser Desorption Time-of-Flight Mass Spectroscopy (LD-TOF-MS) Analyses

The TOF spectra were taken on Tof Spec 2E (Micromass) spectrometer running in a linear mode with positive ions. The operating voltage was 20 kV; the pulse voltage was 1.8 kV. A N_2 laser was used for ionization with 50% 'laser course' and 50% 'laser fine'. For these measurements powder from the altered surface was dispersed in a small amount of toluene to make a paste; no matrix was used for our fullerene analyses.

3.4 Fourier Transform Infrared Spectroscope Analyses

Reflectance spectra of the altered surfaces were made in the near and middle infrared spectral regions using a FTIR spectrometer Magna-IR 760 E.S.P. (Nicolet Instrument Corporation), with low-temperature heated tungsten lamp emitting radiation in the range to 3.5 μm. The selected MCT/A detector allows the detection of radiation in the 0.9 μm to 5 μm spectral range.

4. RESULTS

4.1 Electric Pulses

Graptolitic shales reacted to the pulse by breaking-up preferably along cleavage planes at all pulse magnitudes. Shale break-up occurred also in the area between the electrodes, i.e. perpendicular to the cleavage planes and perpendicular to the surface of the electrodes (see, Fig. 1). The pulses caused thermal alteration on all fracture planes. Macroscopically this induced alteration had the appearance of a grayish coating approximately 0.5 mm thick. The samples from Zbečno developed perpendicular fractures only in the area between the electrodes when the pulse magnitude was 88.4 kA. For the 125 kA pulse, break-up occurred also in directions parallel to the electrodes' surfaces. The fracture planes in this sample also showed a grayish thermal alteration. Bituminous coal from Kladno fractured with pulse magnitudes higher than 60 kA. In this sample, the fracture planes did not show the macroscopic evidence for thermal alteration. The profile of the electric pulse for each sample as a function of time is shown in Fig. 2 and the measured pulse values are summarized in Table 1.

After each run, a strong smell of ozone could be detected when the apparatus, containing the sample, was opened.

Table 1. Locality and rock type of the carbonaceous samples selected for the electric-pulse experiments listed #1 through #9, and the measured peak intensity (Ipk) of the current impulse in each experiment that caused surface fracturing. The right hand-column identifies the processed samples selected for fullerene analyses

Locality	Rock Type	#	I_{pk}	Analyzed Samples
KOSOV	Graptolitic shale	1	52.9	Kosov1
		2	14.1	Kosov2
		3	16.8	Kosov3, Kosov4
ZBECNO	Metamorphosed black shale	4	88.4	Zbecno1, Zbecno2
		5	121.7*	Zbecno3, Zbecno4
KLADNO	Bituminous coal	6	17.9	Kladno1
		7	61.4*	Kladno1
		8	60.8	Kladno2
		9	137.7	Kladno3

Note: * designates the estimated value when the pulse magnitude was out of the sensitivity range setting of the instrument

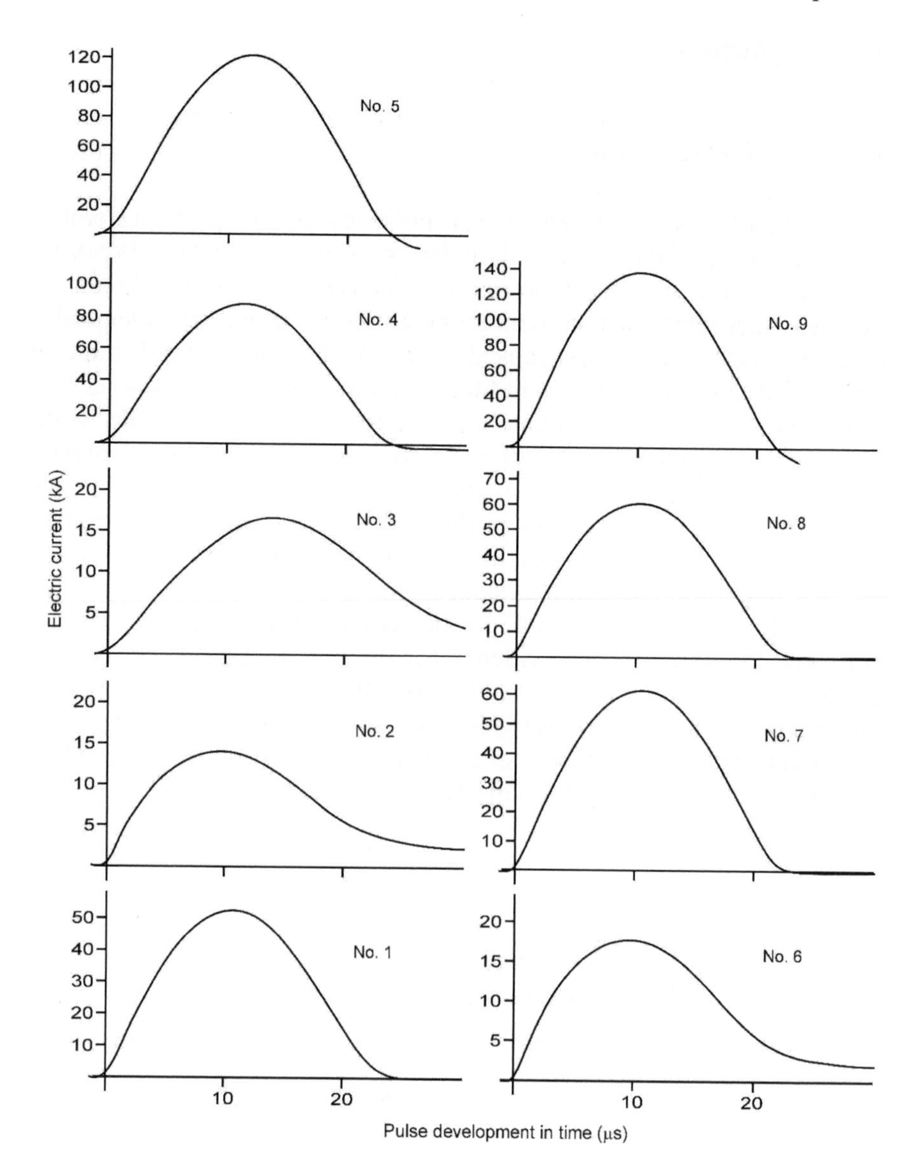

Figure 2. Induced pulse (kA) profiles as a function of time for the experiments #1 through #9 of the Kosov, Zbecno and Kladno samples that are summarized in Table 1. Modified after Frank et al. (2003)

4.2 Fullerene Analyses

No C_{60} was detected in concentrations higher than 10 ppb in original samples or in the concentrated toluene extracts of the samples that were subjected to the electric pulses (Table 1). The retention spectrum of sample Zbecno2 shows only an unresolved, complex mixture of smaller organic molecules (e.g. PAHs) in the region between 3 and 4 minutes (Fig. 3, solid line). For comparison, the retention spectrum of the same extract with an added C_{60} standard in 200 ppb concentration shows the dominant C_{60} peak at the retention time 8.1 minutes (Fig. 3, dashed line).

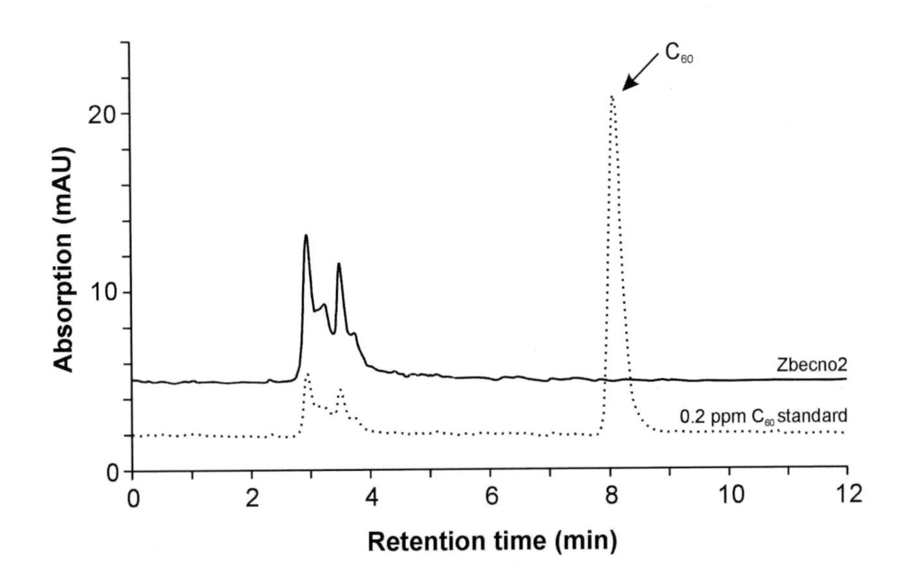

Figure 3. The HPLC chromatograms of the toluene extract Zbecno2 (solid line) and of the same sample with an added synthetic C_{60} standard in concentration of 200 ppb (dotted line). The peaks in both chromatograms prior to 4 minutes represent a mixture of small hydrocarbons; the peak at 8.1 minutes in the chromatogram of the extract with added synthetic C_{60} standard represents C_{60}. Modified after Frank et al. (2003)

In the case that C_{60} would be present in concentrations less than 10 ppb, fractions with retention times longer than seven minutes were collected for each sample. For each sample at least 20 experimental runs were done to obtain enough solution for EIMS analysis. The EIMS performance was first tested on extracts that were not previously separated by HPLC. No fullerene peaks expected at 360 amu (C_{60}^{++}) or 720 amu (C_{60}^{+}) or 840 amu (C_{70}^{+}) were observable in the resulting spectra due to a strong background caused by other organic substances, probably hydrocarbons, in the extracts of the natural samples.

Therefore, aliquots of the fractions that were used for HPLC were subsequently analyzed. This method should provide enough resolution for detecting fullerenes in a trace concentration of 10 ppt. No fullerenes were detected in any of the fractions of the samples that were treated by electric pulses. The LD-TOF-MS analyses of powders from the thermally altered surfaces did not detect fullerenes either. Similar to EIMS spectra of the extracts that were not previously separated by HPLC, the LD-TOF-MS spectra also showed a strong background caused by other organic substances present in the samples.

The C_{60} fullerene molecule has four infrared-active mode frequencies at 526.5 cm^{-1}, 575.8 cm^{-1}, 1182.9 cm^{-1} and 1429.2 cm^{-1} (Wang et al., 1993). Due to the technical constraints of our particular spectrometer with microscope configuration, it was only possible to obtain reflection spectra at wave numbers >600 cm^{-1}. Thus the possibility to identify fullerenes directly on the altered surfaces was limited to only two lines. Only the sample Zbecno2, a metamorphosed black shale, shows the 1182.9 cm^{-1} and 1429.2 cm^{-1} spectral lines in the IR spectrum but with a very strong background and reflection peaks from the sample matrix (Fig. 4). None of the other spectra, including those from the unaltered and reference samples, show these two spectral lines.

5. DISCUSSION

High-energy electric pulses up to 138 kA and 20 μsec under ambient atmospheric conditions failed to induce C_{60} fullerene formation in several natural carbonaceous rock samples in any measurable quantities. Only the IR spectrum of one electric pulse-altered surface of metamorphosed black shale from Zbečno shows two vibration lines that could perhaps be due to C_{60} but no fullerene was detected in this sample by HPLC and EIMS analyses.

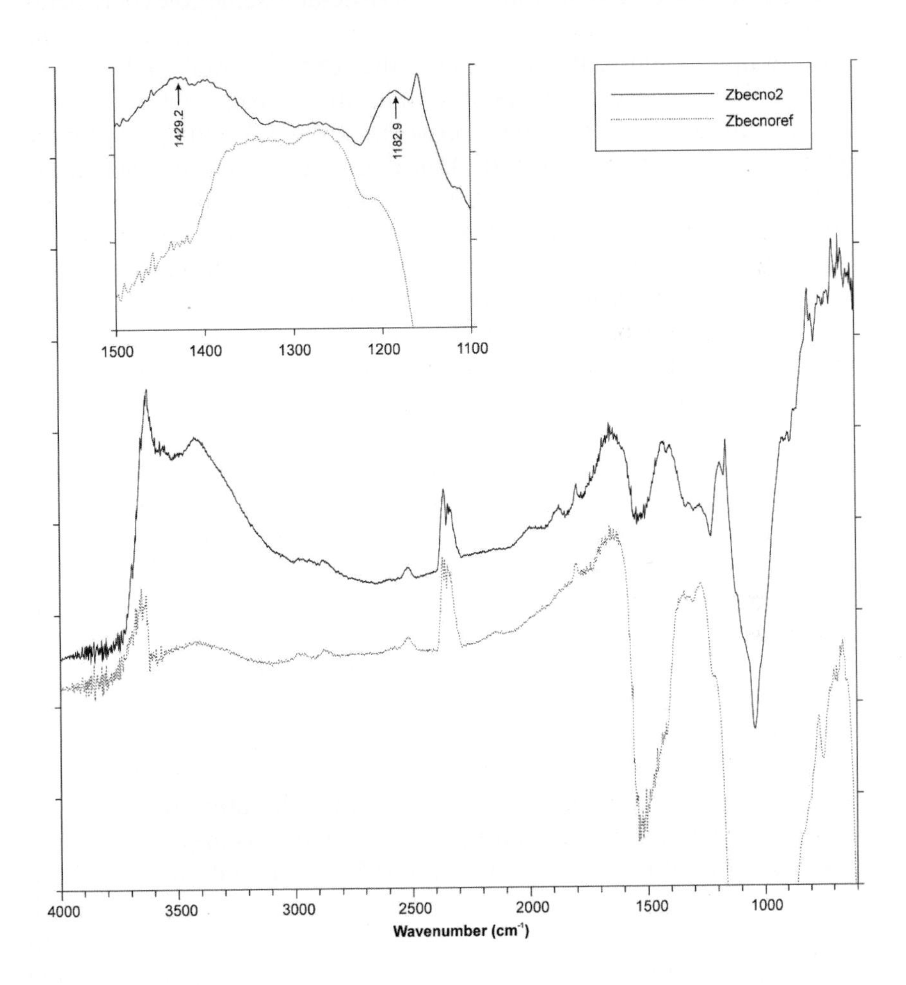

Figure 4. Fourier transform infrared spectra of the altered surface layer of Zbecno2 (solid line) and the unaltered surface of this metamorphosed shale sample (Zbecnoref). The inset is the enlarged spectral region between 1130 cm^{-1} and 1470 cm^{-1}. Modified after Frank et al. (2003)

Also noting that the interpretation of only the 1182.9 cm^{-1} and 1429.2 cm^{-1} IR spectral lines is somewhat ambiguous to determine C_{60}, we submit that no fullerene was formed in this sample. The lack of C_{60} in the pulse-modified samples is somewhat surprising because the carbon content of the selected samples ranging from 0.7% up to 79% should have been sufficiently high for fullerene formation. Thus, we have to look for possible reasons to explain the absence of fullerene in the very conditions of our artificial discharge experiments.

In terms of current magnitude, the experimental pulses were comparable to energies in natural lightning strikes wherein typical peak currents are in the range between 10-20 kA with a maximum around 110 kA (Uman, 1969). Based on properties of the current generator we used for the experiment, the peak temperatures in our pulses reached 20,000-25,000 K (Totev, 2004, personal comm.). However, the applied voltage and duration of our pulses was lower and thermal modification could not propagate deeper into the sample than the macroscopically altered zone. In our experiments no sample melting, let alone sample evaporation, occurred. The duration of the return stroke of a lightning strike is up to milliseconds and the heat input of the return stroke can raise the channel temperature to as much as 30,000 K (Uman, 1969), which would be more than enough to fuse and vaporize quantities of target material. Channel pressures can reach tens to up to hundreds of atmospheres followed by the return to normal atmospheric pressure within microseconds. The pulse duration in our experiments was approximately 30 μsec (Fig. 2). For comparison, gas (He) temperatures in DC arc discharge experiments may reach 3,000-5,000°C during fullerene formation in a condensing carbon vapor via the assembly from two-ring carbon clusters in the cooling gas (Alekseyev and Dyuzhev, 2003).

In our experiments only superficial thermal alteration was observed as grayish coatings at the slab surface but with no apparent thermal effects deeper into the slab below the 0.5-mm alteration zone. From this point of view, even if fullerene had formed, the majority would have been formed at the slab surface. We noticed a strong ozone odor upon opening the experimental chamber. We hypothesize that fullerenes at the surface, which might have formed during the electric pulse experiment, could have been exposed directly to interactions with ozone produced in the experiment. If so, such interactions would lead to fullerene decomposition reactions such as described by Chibante and Heymann (1993) and Juha et al. (2000). Langer et al.

(2004), have documented a similar scenario for arc-discharge production of fullerenes from graphite and anthracite (a natural high-rank coal) electrodes. They observed that even traces of O_2 present in the reaction chamber were enough to oxidize all fullerene molecules produced in the condensation experiment. A lightning strike clearly produced fullerenes in a fulgurite at Sheep Mountain (Colorado) (Daly et al., 1993). Our simulation experiments of lightning strike-induced fullerene formation in carbon-rich rock samples failed for reasons already explained. In natural fulgurites the original soil or rock is molten which then has the potential that fullerenes once they had formed could become enclosed in the quenched glass. In this manner the natural fullerenes would be protected from weathering. The Sheep Mountain sample (Daly et al., 1993) appeared to be fresh, possibly indicating it had not yet experienced significant weathering, which could have destroyed fullerene by interactions with the atmosphere.

We report a negative result but we note that Daly et al. (1993) found similar negative results for fulgurites formed in quartz sand and glacial till. These deposits as well as the ash-flow tuff of Sheep Mountain have low carbon contents. Fullerene formation under such conditions would require a rich external source of carbon such as pinecones and needles on Sheep Mountain. For our selection of rock samples with organic carbon contents up to 79%, the carbon supply was not likely a limiting factor.

Our failure to produce C_{60} in a coal sample suggests constraints other than a source of carbon and the exact ambient atmospheric conditions on lightning strike-induced fullerene formation. For example, Daly et al. (1993) described the existence of small glass tubes that pass through the fullerite-bearing rock interior that would testify to extreme heating along well-defined pathways during the lightning strike. Such features were not produced in our experiments despite the fact that our induced current energies were of an order of magnitude consistent with those described by Uman (1969) in natural lightning.

Our experiments show that the pulse durations that were used were insufficient to induce kinetically controlled melting of the samples, thermal decomposition of the organic matter (hydrocarbons, kerogen) disseminated in the samples, and the generation of a carbon vapor. It appears that extreme heating rates are required to create the appropriate physiochemical regime for lightning strike-induced fullerene formation to occur.

6. CONCLUSIONS

The conditions during electric pulses generated in the laboratory could not cause the formation of fullerene in natural, carbonaceous geologic rock samples. The results indicate that fullerene formation caused by electric discharge in natural rocks (fulgurites) is limited to the long-lasting pulses of natural lightning. Along with conclusions from other studies (Heymann, 1998; Daly et al., 1993), the role of multiple lightning strikes as the cause of natural fullerenes in terrestrial surface rocks such as in the lower-Proterozoic shungite (Ebbesen et al., 1995) remains uncertain.

It will require further laboratory experiments using more powerful electric pulses than were used in this study on carbonaceous materials before quantitative statements can be made on the efficiency of C_{60} and other fullerenes formation by multiple lightning strikes. Continued experiments will also have to address the physicochemical conditions of fullerene formation during a lightning strike and, for example, avoid the copious production of ozone. Such information will be required to assess whether widespread or only localized natural fullerene formation induced by lightning is possible at the Earth's surface.

Acknowledgements: The authors appreciate the review by Dieter Heymann. We wish to thank J. Bolech, I. Totev and J. Novotny in Laboratory for Very High Voltages in Prague-Běchovice, who assisted during the experiments. This work was partly supported by the Grant Agency of the Charles University project Nr. 212/1999 B GEO, Ministry of Education of Czech Republic project Nr. 2307/2002 and CEZ:J13/98:113100005.

7. REFERENCES

Alekseyev, N.I. and Dyuzhev, G.A. (2003) Fullerene formation in an arc discharge. *Carbon*, 41, 1343-1348.

Becker, L., Poreda, R.J. and Bunch, T.E. (2000) Fullerenes: An extraterrestrial carbon carrier phase for noble gases. *Proc. Natl. Acad. Sci.*, 97, 2979-2983.

Buseck, P.R. (2002) Geological fullerenes: review and analysis. *Earth. Planet. Sci. Lett.*, 203, 781-792.

Buseck, P.R., Tsipursky, S.J. and Hettich, R. (1992) Fullerenes from the geological environment. *Science*, 257, 215-217.

Čáp, P., Vacek, F. and Vorel, T. (2003) Microfacies analysis of Silurian and Devonian type sections (Barrandian, Czech Republic). *Czech Geol. Survey Spec. Papers*, 15, 1-40.

Chibante, L.P.F. and Heymann, D. (1993) On the geochemistry of fullerenes - stability of C_{60} in ambient air and the role of ozone. *Geochim. Cosmochim. Acta*, 57, 1879-1881.

Daly, T.K., Buseck, P.R., Williams, P. and Lewis, C.F. (1993) Fullerenes from a fulgurite. *Science*, 259, 1599-1601.

Ebbesen, T.W., Hiura, H., Hedenquist, J.W., de Ronde, C.E.J., Andersen, A., Often, M. and Melezhik, V.A. (1995) Origins of fullerenes in rocks; response by P.R. Buseck and S. Tsipursky. *Science*, 268, 1634-1635.

Frank, O., Jehlička, J. and Hamplová, V. (2003) Search for fullerenes in geological carbonaceous samples altered by experimental lightning. *Fullerenes, Nanotubes, and Carbon Nanostructures,* 11, 257-267.

Heymann, D. (1998) Search for C_{60} fullerene in char produced on a Norway spruce by lightning. *Fullerene Sci. Techn.*, 6, 1079-1086.

Heymann, D., Chibante, L.P.F., Brooks, R.R., Wolbach, W.S. and Smalley, R.E. (1994) Fullerenes in the Cretaceous-Tertiary boundary layer. *Science*, 265, 645-647.

Heymann, D., Korochantsev, A., Nazarov, M.A. and Smit, J. (1996) Search for fullerenes C_{60} and C_{70} in Cretaceous-Tertiary boundary sediments from Turkmenistan, Kazakhstan, Georgia, Austria, and Denmark. *Cretaceous Res.*, 17, 367-380.

Heymann, D., Yancey, T.E., Wolbach, W.S., Thiemens, M.H., Johnson, E.A., Roach, D. and Moecker, S. (1998) Geochemical markers of the Cretaceous-Tertiary boundary event at Brazos River, Texas, USA. *Geochim. Cosmochim. Acta*, 62, 173-181.

Juha, L., Farníková, M., Hamplová, V., Kodymová, J., Mullerová, A., Krása, J., Láska, L., Špalek, O., Kubát, P., Stibor, I., Koudoumas, E. and Couris, S. (2000) The role of the oxygen molecule in the photolysis of fullerenes. *Fullerene Sci. Techn.*, 8, 289-318.

Krätschmer, W., Lamb, L.D., Fostiropoulos, K. and Huffman, D.R. (1990) Solid C_{60}: a new form of carbon. *Nature*, 347, 354-358.

Křibek, B., Pouba, Z., Skoček, V. and Waldhauserová, J. (2000) Neoproterozoic as a part of the Cadomian orogenetic belt: A review and correlation aspects. *Věstník ČGÚ*, 75, 175-196.

Langer, J.J., Golczak, S., Zabinski, S. and Gibinski, T. (2004) Fullerenes and carbon nanotubes formed in an electric arc at and above atmospheric pressure. *Fullerenes, Nanotubes, and Carbon Nanostructures*, 12, 593-602.

Li, Y.F., Qiu, J.S., Zhou, Y. and Mieno, T. (2002) Laser desorption time-of-flight mass spectrometry and gas chromatography-mass spectrometry analysis of extracts from coal-derived fullerene containing soots. *Chinese J. Anal. Chem.*, 30, 769-773.

Mieno, T. (2000) Production of fullerenes from plant materials and used carbon materials by means of a chip-injection-type JxB arc reactor. *Fullerene Sci. Techn.*, 8, 179-186.

Pešek, J. (2004) Late Paleozoic limnic basins and coal deposits of the Czech Republic. *Folia musei rerum naturalium Bohemiae occidentalis Geologica. Editio Specialit Volumen*, 1, 1-188.

Uman, M.A. (1969) Lightning. *Adv. Phys. Monograph Ser.*, 1-264, McGraw-Hill, New York. USA.

Wang, K.A., Rao, A.M., Eklund, P.C., Dresselhaus, M.S. and Dresselhaus, G. (1993) Observation of higher-order infrared modes in solid C-60 films. *Phys. Rev. B*, 48, 11375-11380.

Chapter 12

FULLERENE IN SOME COAL DEPOSITS IN CHINA

P.H. FANG
F.S. Lab, 156 Common St., Belmont, Massachusetts, USA and Department of Physics, Yunnan University, Kunming, Yunnan, China.

FAWEN CHEN
Yunnan Institute of Technology and Information, Kunming, Yunnan, China.

RUZHAO TAO
Yipinglang Coal Mine Company, Lufeng, Yunnan, China.

BINGHOU JI
Department of Physics, Inner Mongolia University, Huhehaote, Inner Mongolia, China.

CHONGJUN MU
Department of Physics, Yunnan University, Kunming, Yunnan, China.

ERGANG CHEN
Department of Physics, Yunnan University, Kunming, Yunnan, China.

YUANJIAN HE
Huaping Coal Mine Company, Lijiang, Yunnan, China.

Abstract: Highly variable concentrations of natural C_{60} are reported from several coals from mines in the Yunnan-Guizhou provinces and the Inner Mongolia-Shanxi-Xinjiang provinces of China. The fullerene-extraction procedures from coals

Frans J.M. Rietmeijer (ed.), Natural Fullerenes and Related Structures of Elemental Carbon, 257–266.

and C_{60} identification by Fourier Transform Infrared spectroscopy are described.

Key words: C_{60}; coal deposits; China; fullerene extraction; fullerene identification; Fourier Transform Infrared Spectroscopy (FTRIR)

1. INTRODUCTION

Artificial C_{60} fullerene was first made in 1985 by Kroto et al. (1985), while Buseck et al. (1992) reported the first finding of natural fullerene in a sample of shungite, a carbon-rich Precambrian rock from Russia, in 1992. In 1990, a method was found to produce synthetic C_{60} fullerene in macroscopic quantities (Krätschmer et al., 1990) and, as a result, fullerene research and applications have expanded ever since. There are only a few reported discoveries of natural fullerene in various geological settings across the globe (Heymann et al., 2003). Probably the slow development in natural fullerene research is due to a paucity of knowledge on the geological conditions and the mechanisms required for the formation of natural fullerenes that were found in lightning-strike produced fulgurite (Daly et al., 1993), at the geological Cretaceous-Tertiary (KT) and Permian-Triassic boundaries associated with meteor impacts and their associated global wildfire deposits, and associated with Neo-Proterozoic volcanic rocks (see, Heymann et al., 2003).

The aim of the present work is to describe the search for natural fullerenes in coal deposits from two different regions in China. At the outset, we note that an extraterrestrial origin of fullerenes in coal is excluded but fullerene formation induced by impacting meteorites remains as an option. As an initial effort, we have selected a limited number of mines to study from among many more coalmines in these geological regions. Fullerene was found in ca. 10% of the mines we have studied. However, we hesitate to rule out that coals in the remaining 90% of the mines would have no fullerene. There are multiple factors that could cause a negative result, among them the method of fullerene extraction and the limitation of the fullerene detection method we employed. Fullerene as referred to in this paper will be exclusively C_{60}. Usually, it is accompanied by a lesser amount of C_{70} but in the samples described here, which were analyzed using Fourier Transform Infrared Spectrometry (FTIR), C_{70} was not examined.

2. PREVIOUS REPORTS OF FULLERENE

The exact physicochemical conditions of each occurrence of natural fullerene are not always well understood, but it appears there is no unique set of conditions required for the formation of C_{60}, C_{70} and other fullerenes in the natural environments. It seems plausible though that carbon-rich rocks or deposits would be the most likely environments wherein, under the right conditions, fullerenes might form. Intuitively, such environments would also be conducive to fullerene preservation on a geological timescale. From the Yipinglang coalmine (Yunnan Province, SW. China), with coals of Jurassic age, two types of coals were selected for a search of fullerenes, (1) a bright and hard coal (ca. 75% C with Si as a major impurity) and (2) coal with ca. 80% carbon content that formed meters-long, millimeter to centimeter thick, layers (Fang and Wong, 1997). High-performance liquid chromatography confirmed C_{60} and C_{70} in both coals with concentrations of 2.6×10^{-5} and 2.6×10^{-4}, respectively. The C_{60} to C_{70} ratio in the first coal was 5.7; in the second coal this ratio was 3.1 (Fang and Wong, 1997). The results of this study indicated that natural fullerenes in coal have (1) variable concentrations as a function of coal type and/or occurrence, and (2) variable C_{60} to C_{70} ratios with some coals virtually without C_{70} (Fang and Wong, 1997).

One coal sample from the Yipinglang Mine has an extraordinarily high C_{60} content of ca. 0.1 percent by weight (Fang et al., 1996; Fang and Wong, 1997). An examination of other samples from this mine found a C_{60} fullerene content of only 30 parts per million (Osawa et al., 2001). Similarly large variations were found in many C_{60}-bearing coal samples, which serves as a point of caution for the high variability of fullerene contents in coal from the same location in the same mine. These results were from the very first group of five coal samples each weighing 100 grams wherein the fullerene yields varied from null to ca.10^{-4} grams. The significance, if any, of these variations remains an area of active research. In the same coalmine, Li et al. (2000) and Liang et al. (2002) also reported the finding of fullerenes.

3. COAL MINES

The precursor materials of coal were obviously carbon rich and once formed subsurface coal beds would be conducive to the preservation of fullerenes. The coalmines from which the samples were selected are located in two regions of China, viz. the

Yunnan-Guizhou and the Inner Mongolia-Shanxi-Xinjiang Provinces, which will be discussed separately.

3.1 Yunnan-Guizhou Provinces

From a geological point of view, Yunnan and Guizhou make up the so-called Yun-Gui Plateau with its high mountains. The coals in these two provinces were principally formed during the Jurassic period of the Mesozoic era. A sketch of these provinces delineated by the dashed line shows the locations of the major geological faults (Fig. 1).

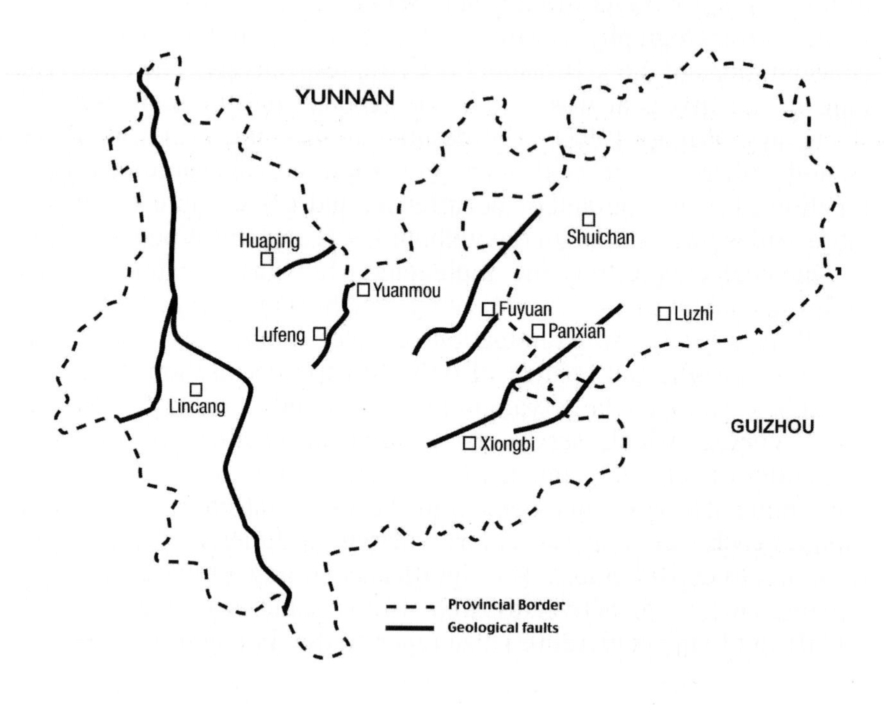

Figure 1. Geographical map of Yunnan and Guizhou provinces with the locations of major geological fault lines

The coalmines are generally located along these major faults that are reverse faults that exerted strong compressive stresses on the coal strata. As we submit, such stress energies could have promoted C_{60} formation. We conjecture that other possible fullerene occurrences in the area could be associated with violent earthquakes, volcanic eruptions, major forest fires, or lightning storms, but in this paper we will make no attempts to correlate such natural events with specific, coal occurrences. The coals in the Yun-Gui region are mainly bituminous. The coalmines in which fullerene was found are represented by open squares in Fig. 1, which does not show the numerous other coalmines in these provinces that were either (1) not examined for the presence of fullerenes or (2) where the initial analyses gave a null result.

3.2 Inner Mongolia-Shanxi-Xinjiang Provinces

The fullerene search in Inner Mongolia, Shanxi, and Xinjiang coals will be an extension of our fullerene search in Yunnan and Guizhou coals. Coal formation in these regions occurred during the Permian. This coal is thus older than those in the Yunnan-Guizhou Provinces. Intense intra-continental rifting and extensional tectonics in the Inner Mongolia-Shanxi-Xinjiang Provinces occurred during the early Cenozoic era (Zhang et al., 2003). Coal formation followed by mechanical stresses could have led to fullerene formation. The presence of tectonic stress fields is reminiscent of the situation in the Yunnan and Guizhou Provinces. A search for natural fullerene in coals of this region is reasonable, assuming the prevailing tectonic regimes either triggered, sustained, or both, fullerene formation.

An initial search was made in two coalmines in Inner Mongolia. These mines are several hundred kilometers from the Daqing Mountains. The Zhungeer Mine with an area of 65 km x 20 km has a coal reserve of about 0.3 billion tons. Coal is mined to a depth of 50 to 100 meters. The Qingshuihe Mine is a small open pit mine. Its coal reserves are quite depleted after a long history of coal exploitation. An initial analysis showed C_{60} at less than one part per billion in a coal from this mine (pers. comm. D. Heymann, Rice University, Houston).

4. ENVIRONMENTAL CONDITIONS OF COAL-FULLERENE FORMATION

In laboratory experiments formation of C_{60} and C_{70} typically requires rapid cooling from very high flash-heating temperatures (>1,700 °C). Such conditions could have existed in some environments of natural fullerene formation (see, Heymann et al., 2003) but they would have been unlikely in geological terrains during tectonic stress fields. The highly variable C_{60} to C_{70} ratios in the Chinese coals might contain a clue to their environment. For example, Wang et al. (1998) submitted that absence of C_{70} in a 60-Ma old fossilized dinosaur egg that contained C_{60} could be the result of a reaction of C_{70} with highly concentrated atmospheric (volcanic) SO_2 when the egg was exposed at the surface. This reaction does not appear to affect C_{60} (Wang et al., 1998). We note that the dinosaur eggshell contained a trace of $C_{60}O$ (Wang et al., 1998). Sulfur-containing organic and mineral matter are common to almost all coals on the Northern Hemisphere that during coal maturation could be oxidized to sulfur dioxide. Thus, C_{70} formation might either have been inhibited or destroyed during coal maturation op to temperatures of ca. 600 °C inferred for the environmental conditions that could have existed in the eggshell (Wang et al., 1998).

5. FULLERENE EXTRACTION FROM COAL

Fullerene extraction from natural coals has to consider that all natural coals invariably contain major amounts of mineral impurities such as silicates, oxides, sulfides, nitrides, and hydrocarbons. Also, the fullerene content is typically less than one part per million to one part per billion. The coal samples collected from the coalmines are usually lumps of hundreds to thousands of grams. A first step is to break these lumps into small fragments of about 10 g by weight. For coal purification, a common practice is to wash the crushed material with a salt-water solution using zinc chloride (Waugh and Bowling, 1984; Wilson et al., 1992). Since this coarse procedure could wash away fine, fullerene-containing coal particles, this process of coal washing was not applied. The individual steps used are as follows:

1. A 100 g fragment of a selected coal sample was ground into fine powders of 100 microns or smaller using a steel sieve. The powder

was dispersed in 500 ml of toluene solvent and agitated by a magnetic stirrer.

2. Mechanical stirring was halted after ca. 2 hours when the liquid became sluggish, which could trap the fullerene in the solution. The rate of the formation of sluggishness was dependent on the different coal samples.
3. The solution from step 2 was filtered through a mesh to remove toluene-wetted coal and toluene-insoluble materials, including minerals and ashes. The result was a liquid with about 60% of the original volume that includes toluene and coal. In some samples the color of the solution was reddish brown that is a possible signature of a fullerene presence.
4. The liquid of step 3 possibly contains fullerene and various hydrocarbons dissolved in toluene. To extract the fullerene, a fullerene non-soluble liquid such as methanol was added to precipitate the fullerene in the solution. For effectiveness, the excessive toluene in the original liquid of step 3 was removed by evaporation. About 100 cc of methanol, which does not dissolve fullerene, was added to this dried residue.
5. After sonification for half an hour followed by centrifugation, black precipitates of fullerene were obtained. The remaining liquid with hydrocarbons and oils was discarded. This toluene-methanol procedure was repeated to obtain higher fullerene purity. This purified fullerene sample was usually of minute quantity and a direct transfer of the solid fullerene material for measurement or storage is difficult. Re-dissolving the solid sample in toluene facilitates the transfer. The samples are found to be stable under ambient light and temperature condition for a period of days.

6. C_{60} IDENTIFICATION

The C_{60} identification was made by infrared (IR) spectroscopy, a preference that was primarily based on the available equipment. To prepare a specimen for IR analysis a drop of the toluene solution containing fullerene is placed on a polished silicon substrate and slowly heated at 50 °C for several hours to remove toluene.

The reflection spectra were taken using a Perkin Elmer Fourier Transform Infrared (FTIR) spectrometer. A FTIR spectrum of fullerene extracted from coal of the Huaping mine (Fig. 2) shows the clearly marked and numbered, five known absorption peaks of C_{60} (see for comparison Fig. 14 in Dresselhaus et al., 1993). The remaining, unidentified peaks (Fig. 2) are chiefly from the silicon substrate and from residual toluene

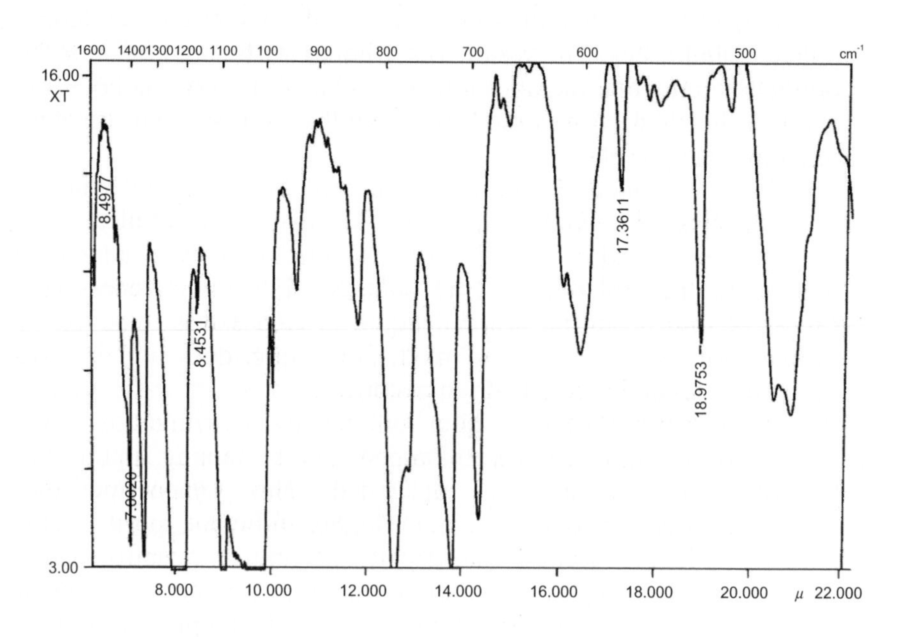

Figure 2. Fourier Transform Infrared Spectrum (FTIR) of fullerene in a coal sample from the Huaping coalmine (see, Fig. 1). Due to the instrumental configuration of the spectrometer, the absorption peaks appear as valleys in this figure

7. COAL AS A SOURCE FOR FULLERENE PRODUCTION

Qiu et al. (2002, 2004) have reported the production of fullerenes and fullerenic carbon nanotubes by the electric arc discharge or plasma condensation processes. Using coal as the carbon source to replace more expensive graphite source has an apparent economic advantage, but the resultant yield should be factored in. The reported yields were a factor of two to five times smaller than for graphite (Qiu et al., 2002, 2004). Other relevant considerations are:

1. Impurity content of coal (Qiu et al., 2002), including hydrocarbons (Fang and Wong, 1997),
2. Intrinsic factors such as the fact that coal is a molecular solid with weak bonds while graphite is a crystalline solid, and therefore a facilitation to form the fullerene could be effected, and
3. Coals could contain fullerene in various stages of incomplete formation (Amato, 1992), which is consistent with experimental observations of fullerene formation (Burden and Hutchison, 1998).

8. CONCLUSIONS

This paper reported on a search for natural fullerene in coals from China. The preliminary results suggest the possible widespread occurrence of natural C_{60} in coals. The preliminary results also indicate considerable small-scale variations in C_{60} concentrations within the coalmine. The search for high volumes of natural fullerene will be a demanding effort. Still, we can posit that natural C_{60} is not necessarily a rare oddity. In the future we plan exploratory work in Shanxi and Xinjiang coals. We note that finding evidence for quasi-fullerenes would provide an important link between the matured natural fullerenes and juvenile natural fullerenes.

Acknowledgements: We wish to express our appreciation to the authorities of Yunnan University, who for many years provided laboratory space. We acknowledge the support of Zhaoxi He and Zhiping Zang of Lufeng County of Yipinglang coal mine, and Zhixing He of Lijiang Municipality where the Huaping coalmine is located. Generous supply of coal samples from other locations shown in Fig. 1 is appreciated. In 1996, E. Osawa (NanoCarbon Research Institute, Japan) confirmed the presence of fullerene in several Yipinglang coal samples. T. Lowe (MER Corp., Tucson, Arizona) advised on fullerene purification. A.J. Fang and M.B. Fang were instrumental in the preparation of this paper.

9. REFERENCES

Amato, I. (1992) A first sighting of Buckyballs in the wild. *Science*, 257, 167.

Burden, A.P. and Hutchinson, L.H. (1998) In situ fullerene formation - the evidence presented. *Carbon*, 36, 1167-1173.

Buseck, P.R., Tsipursky, S.J. and Hettich, R. (1992) Fullerenes from the Geological Environment. *Science*, 257, 215-217.

Daly, T.K., Buseck, P.R., Williams, P. and Lewis, C.F. (1993) Fullerenes from a fulgurite. *Science*, 259, 1599-1601.

Dresselhaus, M.S., Dresselhaus, G. and Eklund, P.C. (1993) Fullerenes. *J. Mat. Res.*, 8, 2054-2097.

Fang, P.H. and Wong, R. (1997) Evidence for fullerene in a coal of Yunnan, Southwestern China. *Innov. Mat. Res.*, 1, 130-132.

Fang, P.H., Zhou, X., Tao, R., Wang, Q., Mu, C. and Wu, X. (1996) Fullerenes discovered in coal mines in Yunnan, China. *Innov. Mat. Res.*, 1, 129-134.

Heymann, D., Jenneskens, L.W., Jehlička, J., Koper, C. and Vlietstra, E. (2003) Terrestrial and extraterrestrial fullerenes. *Fullerenes, Nanotubes, and Carbon Nanostructures*, 11, 333-370.

Krätschmer, W., Lamb, L.D., Fostiropoulos, K. and Huffman, D.R. (1990) Solid C_{60}: a new form of carbon. *Nature*, 347, 354-358.

Kroto, H.W., Heath, J.R., O'Brien, S.C., Curl, R.F. and Smalley, R.E. (1985) C_{60}: Buckminsterfullerene. *Nature*, 318, 162-163.

Li Yanfang, Liang Handong, Zuo Danying, Yu Chunhai, and Tao Ruzao (2000) Review of technique of extracting C_{60}/C_{70} from geological samples (*in Chinese*). *J. China University Mining and Technology*, 10, 157-160.

Liang Handong, Li Yanfang, Liu Dunyi, Tao Ruzao and Lin Yuchen (2002) Primary investigation of the possible occurrence of geological fullerene (C_{60}) in the coal seams and their wall rocks located in Lufeng of southwestern China. *J. Acta Petrologica Sinica*, 18, 419-423.

Osawa, E., Ozawa, M., Chijiwa, K., Hoyanagi, K., Tanaka, K. and Kusunoki, M. (2001) Survey of natural fullerenes in Southwestern China. *In Nanonetwork Materials: Fullerenes, Nanotubes, and Related Systems*. S. Saito, T. Ando, Y. Iwasa, K. Kikuchi, M. Kobayashi and Y. Saito, Eds., *Am. Inst. Phys. Conf. Proc.*, 590, 421-424, American Institute of Physics Press, Woodbury, New York, USA.

Qiu, J.S., Zhang, F., Zhou, Y., Han, H.M., Hu, D.S., Tsang, S.C. and Harris, P.J.F. (2002) Carbon nanomaterials from eleven coking coals. *Fuel*, 81, 1509-1514.

Qiu, J., Li, Y., Wang, Y. and Li, W. (2004) Production of carbon nanotubes from coal. *Fuel Processing Techn.*, 85, 1663-1670.

Wang, Z., Li, X., Wang, W., Xu, X., Tang, Z., Huang, R. and Zheng, L. (1998) Fullerenes in the fossil of dinosaur egg. *Full. Sci. Techn.*, 6, 715-720.

Waugh, A.B. and Bowling, K. (1984) Removal of mineral matter from bituminous coals by aqueous chemical leaching. *Fuel Proc. Tech.*, 9, 217-233.

Wilson, M.A., Pang, L.S.K., Willett, G.D., Fisher, K.J. and Dance, I.G. (1992). Fullerenes – Preparation, properties, and carbon chemistry. *Carbon*, 30, 675-693.

Zhang, Y., Ma, Y., Yang, N., Shi, W. and Dong, S. (2003) Cenozoic extensional stress evaluation in north China. *J. Geodynamics*, 36, 591-613.

Chapter 13

BIOGENIC FULLERENES

DIETER HEYMANN
Rice University. Department of Earth Science, MS 126, Houston, Texas, 77251-1892, USA.

Abstract: This paper presents the hypothesis that algal remains were precursors of the fullerenes in the hard carbonaceous rock Shungite (Russia), in carbonaceous matter from the hard rocks of the Black Member of the Onaping Formation at Sudbury (Canada), and in solid, hard bitumens from Mitov (Czech Republic). The paper argues further that PAHs from the biogenic matter were transformed, perhaps by cyclomerization followed by zip-up, during the geologic metamorphic stages experienced by these rocks.

Key words: Algae; cyclotrimerization; fullerenes; polycyclic aromatic hydrocarbons (PAHs); zip-up

1. INTRODUCTION

The findings of natural fullerenes in shungite, a carbonaceous rock, from Karelia in Russia (Buseck et al., 1992), in carbonaceous matter from the Black Member of the Onaping Formation (*BMOF*) at the impact crater of Sudbury in Canada (Becker et al., 1994), and in solid bitumens from Mitov in the Czech Republic (Jehlička et al., 2000) are perhaps the most mesmerizing among the discoveries of naturally occurring fullerenes. They are all from "hard rock" occurrences that have experienced substantial geologic metamorphism by high temperatures and lithostatic pressures. The multi-ring Sudbury impact feature, which has an estimated original diameter of 180 to 250 km (Dressler et al., 1987), was formed by a meteorite impact that had occurred at 1.85 Ga during the Precambrian era. Becker et al. (1996) have argued that the fullerenes in the *BMOF* must have come from

Frans J.M. Rietmeijer (ed.), Natural Fullerenes and Related Structures of Elemental Carbon, 267–277.

the, obviously large, impacting meteorite because they found helium atoms inside the C_{60} cages with a bulk non-terrestrial $^{3}He/^{4}He$ ratio. The fundamental problem with this interpretation is that the known greenschist facies metamorphism (ca. 400 °C) that the *BMOF* rocks have experienced should have totally driven this gas out of the cages when one considers

1. The outgassing results of the He-bearing fullerenes reported by Becker et al. (1996), and
2. Typical time-temperature relationships of greenschist facies metamorphism.

There is no evidence for meteorite-impact cratering that could have introduced fullerenes to the Shunga and Mitov sites. Considering all available evidence fullerenes at the former site could have a lightning origin (Buseck, 2002).

Heymann et al. (1994) suggested that the fullerenes discovered in the clays at the Cretaceous-Tertiary boundary had been derived from 'global wildfires' triggered by the ca. 200 km in diameter Chicxulub meteorite impact structure at 65 Ma (Anders et al., 1986). However, insufficient land-based biota was available at the Earth's surface for 'global wildfires' at the time of the formation of shungite of Lower Proterozoic age (2.1-2.0 Ga) and the *BMOF* sediments during the Precambrian era. There is no evidence for 'global wildfires' at the Mitov site where the contemporaneous rocks of Neoproterozoic age (ca. 700 to 540 Ma) adjacent to pillow lavas with fullerene-bearing bitumens do not contain wildfire soot or fullerenes.

2. ALGAE: PRECURSORS OF SOME CARBONACEOUS MATTER

The elemental carbon content of the Black Member of the Onaping Formation is approximately 1 wt% and the estimated total mass of this element in the *BMOF* is on the order of 10^{17} g. The $^{13}C/^{12}C$ ratio (below -30 per mille in all measured samples) shows that carbon in the Black Member of the Onaping Formation is incontestably biogenic (Heymann et al., 1999). The most likely biological source was blue-green algae that bloomed in a warm lake following the impact wherein algae co-accumulated with the sediments that became the Black Member rocks (Heymann et al., 1999). Unfortunately, the greenschist facies metamorphism that subsequently transformed the kerogen to carbonaceous matter also destroyed all algal evidence with the possible exception of a single occurrence of apparent fossil algae described by Avermann (1994). Opinions on an algal parentage of the carbonaceous matter at Shunga are divided (Melezhik et al., 1999; Kovalevski et al., 2001). There is a general consensus for an algal biogenic source of carbon in the black shales, schists, and silicified stromatolites in the Mitov area (Pouba and Kříbek, 1986). For many years the idea that the fullerenes had somehow formed inside the rock by solid state processing of precursor materials under geologic conditions was not seriously considered because of the strongly expressed opinion of experts that this was highly improbable as expressed for example by Amato (1992) in a *Science* editorial.

Osawa (1999) proposed that fullerenes might form from some unspecified biologic remains in solid charcoal, coal, and shungite but he offered no detailed speculations about their possible evolution to fullerenes. Heymann et al. (2003) presented a hypothesis of 'biogenic fullerenes'. It postulated in essence that the natural decomposition of algae had generated organic compounds, most likely polycyclic aromatic hydrocarbons (PAHs), which had become transformed to fullerenes during the regional metamorphism of their parent geologic formations.

3. FULLERENE FORMATION BY "ZIP-UP"

Fullerenes had actually already been produced in laboratories from biological materials and from PAHs. Rose et al. (1993, 1994) had synthesized fullerenes from the rubbery algal product coorongite pyrolized by laser ablation at low power densities. Fisher et al. (1996) had obtained fullerenes from pyrolized algal-derived oil shales, also by laser ablation. Coorongite is usually obtained from the alga *Botryococcus braunii*. Its hydrocarbons are principally n-alkanes, n-alkadienes, and n-alkatrienes (Cane and Albion, 1973; Gatellier et al., 1993). When coorongite was heated to 400 °C under helium flow it did not form C_{60}, but the $C_{aromatic}/C_{total}$ ratio increased greatly. The chemical vapor deposit of a toluene extract from the diterpenoid compound camphor yielded C_{60} when heated to 600-700 °C (Mukhopadhyay et al., 1994). Gas phase pyrolysis of the (non)-alternant PAHs naphthalene ($C_{10}H_8$), corannulene ($C_{20}H_{10}$) and benzo[k]fluoranthene ($C_{20}H_{12}$) also produced fullerenes (Taylor et al., 1993; Crowley et al., 1996). The significance of the PAH pyrolysis is that ring rearrangements must have occurred during the pyrolysis process to produce the required 12 pentagonal carbon rings of the fullerene molecule from precursors that contained only hexagonal carbon rings (Sarobe et al., 1999a).

Sarobe et al. (1999b) have experimentally demonstrated that three molecules of 4,5 dihydrobenz[j]acephenanthrylene ($C_{20}H_{14}$) can be readily cyclotrimerized to the very insoluble non-alternant PAH $C_{60}H_{30}$ by heating the monomer with elemental sulfur which removes hydrogen. The product for these conditions was a mixture of two $C_{60}H_{30}$ isomers, viz.

1. Isomer I: tribenzo[*l,l',l''*]benzo[1,2-e,3,4-e':5,6-e''] (Fig. 1a), and
2. Isomer II: tribenzo[*l,l',l''*]benzo[1,2-e,3,4-e':6,5-e''] acephenanthrylene (Fig. 1b).

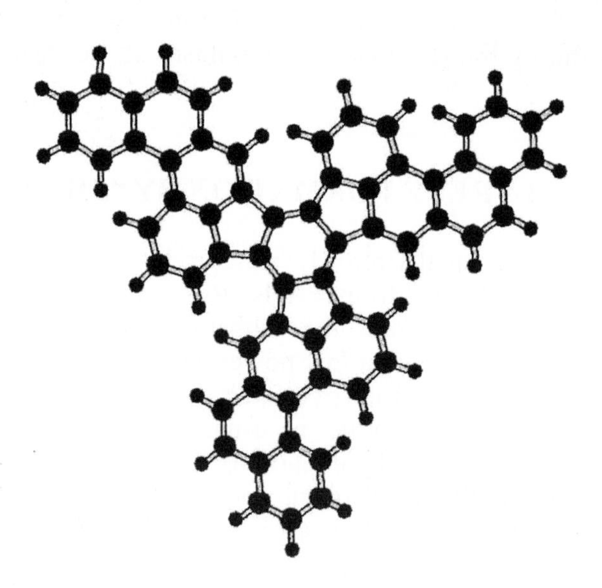

Figure 1a. Isomer I, the $C_{60}H_{30}$ isomer tribenzo[*l,l',l''*]benzo[1,2-e,3,4-e':5,6-e'']-acephenanthrylene. The molecule is not planar but is propeller-shaped. This molecule can be zipped-up to fullerene C_{60}. NOTE: Large black balls are carbon atoms in this and other figures; smaller black balls are hydrogen atoms

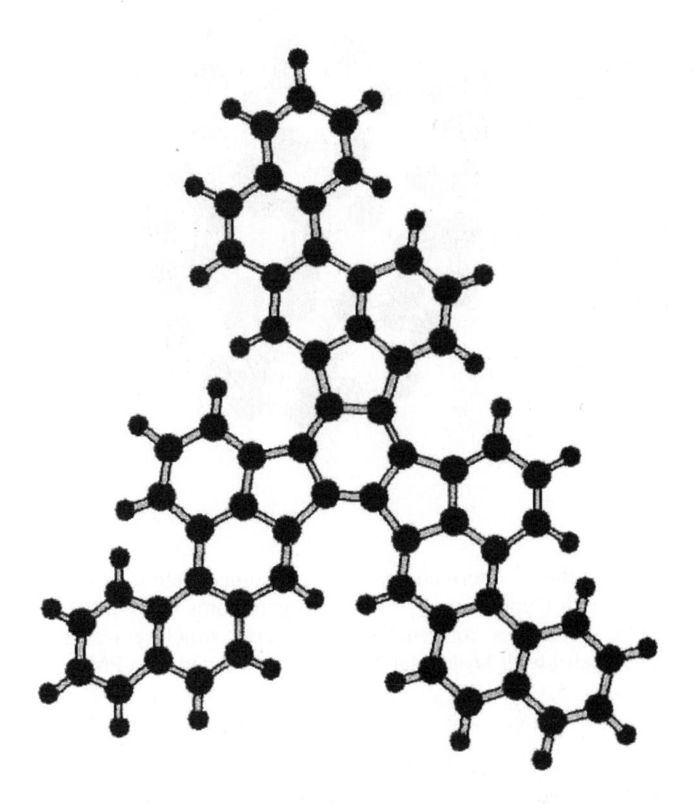

Figure 1b. Isomer II, the $C_{60}H_{30}$ isomer tribenzo[*l,l',l''*]benzo[1,2-e,3,4-e':6,5-e''] acephenanthrylene. Because of its lower symmetry, this precursor cannot be zipped past $C_{60}H_{10}$

Theoretically, isomer I, which has the proper carbon skeleton of the C_{60} Schlegel diagram, can form C_{60} by dehydrogenation and 'zip-up', meaning the formation of new C-C bonds (Fig. 1c).

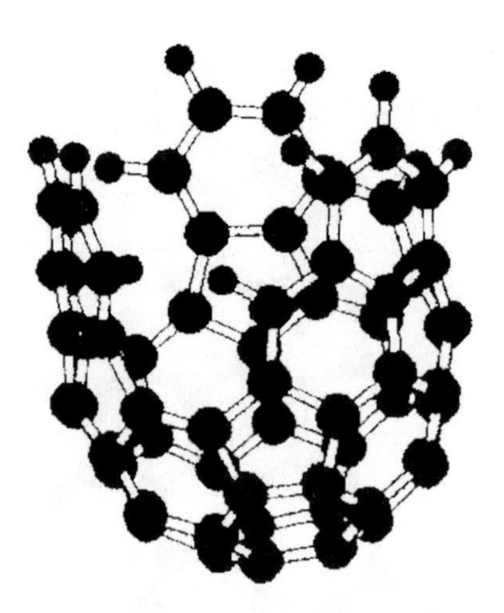

Figure 1c. $C_{60}H_{12}$; the 7th intermediary between isomer I and C_{60} proper. The zip-ups were calculated with Hyperchem 7. Two hydrogen atoms were removed in each step and a new C-C bond was constructed. Every new structure was first treated for equilibrium geometry with Molecular Mechanics followed by a PM3 calculation

Isomer II does not have the proper carbon skeleton, hence its dehydrogenation and zip-up leads to $C_{60}H_{10}$ without rearrangement of its basic carbon skeleton. Indeed, when the independently synthesized isomer I was investigated by MALDI TOF-MS it was zipped-up by 15 successive cyclodehydrogenations at relatively low power densities of ca. 60 $mWcm^{-2}$ without fragmentation and buildup (Boorum et al., 2001; Gómez-Lor et al., 2002).

C_{70} can also be synthesized by 'zip-up on paper'. Several cyclotrimerized PAHs are possible precursors; $C_{70}H_{34}$ is shown here as an example (Fig. 2a). This $C_{70}H_{34}$ is not the only potential precursor that can be zipped up to C_{70} as demonstrated by Heymann et al. (2003) who discussed the zip-up of $C_{70}H_{26}$ (Fig. 2b) and $C_{70}H_{40}$ precursors, both non-cyclotrimerized PAHs, to C_{70} fullerene. Cyclopentafusing of PAHs was observed in combustion (Lafleur et al., 1996a, 1996b, 1998; Marsh et al., 2000), in anthracene pyrolysis (Wornat et al., 1999), and in brown coal pyrolysis (Wornat et al., 1998).

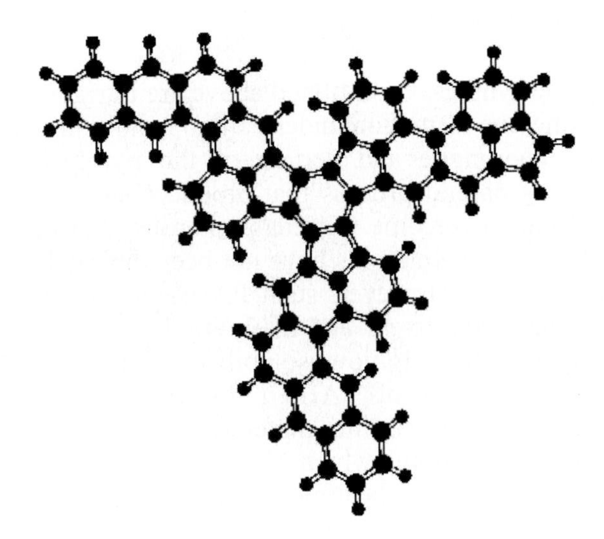

Figure 2a. $C_{70}H_{34}$, a precursor that can be zipped-up to the fullerene C_{70}. This molecule is also propeller-shaped (compare Fig. 1a)

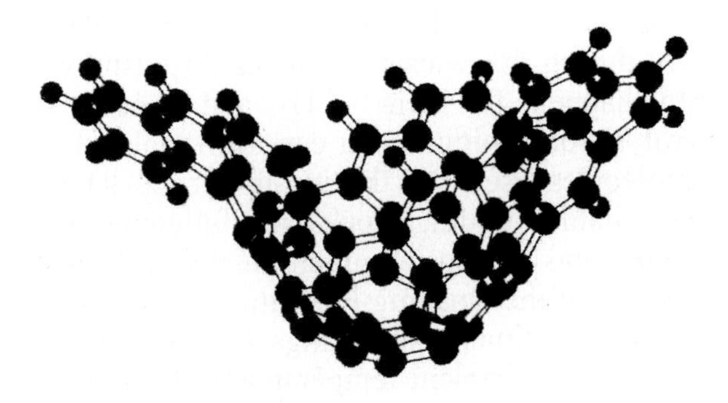

Figure 2b. $C_{70}H_{26}$, one of the intermediates between $C_{70}H_{34}$ and C_{70}

4. DISCUSSION

The fullerene syntheses mentioned above are intriguing but they do not involve fullerene formation under natural conditions. Although the studies using coorongite and certain of the pyrolysis studies are encouraging, any natural process that produces such fullerenes from algal remains must overcome formidable hurdles. For openers, PAHs with sixty or more carbon atoms have not been definitely identified in nature (Fetzer, 2000). However, such PAHs, or their O-, S-, N-, or halogen-carrying derivatives, are likely to have escaped detection because of their extremely low solubility in the solvents used to prepare rock extractions of PAHs (Yoshimura et al., 2001). An alternate and more likely starting material consists of molecules with less than sixty carbon atoms which became fused during early stages of their metamorphisms by heating above about 100 °C with molten elemental sulfur which is commonly present in rocks, especially in sedimentary rocks that contain the mineral pyrrhotite (Fe_7S_8) that produces sulfur upon weathering and oxidation. Unfortunately, the overwhelming majority of molecules generated by the diagenesis of algae contain only hexagonal carbon rings. PAHs with the favorable acenaphtylenic units such as $C_{20}H_{14}$ are very rare in, or absent from sediments, crude oil and oil shales (cf. Table 109.2 of the Standard Reference Materials Program of the National Institute of Standards and Technology, Gaithersburg, Maryland, USA).

Living algae have hopanoids, steroids, and sterols whose structures have 'external' pentagonal carbon rings (Miller, 1962; Simoneit, 1990; Kannenberg and Poralla, 1999) but it is far from clear how these molecules evolve during diagenesis and metamorphism, especially as to what would happen to their $-CH_3$ and longer side-chains. Apparently, pyrolysis of lignitic matter can form molecules analogous to $C_{20}H_{14}$ or molecules used for the zip-up of C_{70} (Lafleur et al., 1996a). Perhaps suitable building blocks and fullerene precursors did form from algal remains during the greenschist facies metamorphism of the Black Member of the Onaping Formation at the Sudbury impact structure and the metasediments at Shunga (the locality of shungite) and during the high emplacement temperatures of bitumens at Mitov estimated to have been in the range 800-1000 °C (Jehlička, pers. comm.).

5. CONCLUSIONS

The hypothesis of 'biogenic fullerenes' is still a long way from being proven or disproven. One salient fact needs to be remembered. Namely that the reported concentrations of 200 ppb C_{60} in the Mitov bitumens (Jehlička et al., 2000) is some five to six orders of magnitude smaller than fullerene concentrations typically obtained in soot deposits from controlled hydrocarbon burning, although we cannot be sure that this amount of natural fullerene at Mitov represents the original abundance or that it could be the relics remaining after geological evolution of the host rocks. At least two aspects of the hypothesis have remained essentially unstudied. First, the pathways and products of photochemical processes involving algal remains in the upper water columns are largely unknown. Second, a possible role of unpaired electrons, which are so abundantly present in kerogens and carbonaceous matter (Lewis et al., 1982), in the processes of zipping-up algal PAHs into fullerenes or in the production of PAHs precursors is unknown.

Acknowledgements: I thank Dr. Jan Jehlička for providing abundant information on the Mitov bitumens and their host rocks prior to publication.

6. REFERENCES

Amato, I. (1992) A first sighting of buckyballs in the wild. *Science*, 257, 167.

Anders, E., Wolbach, W.S. and Lewis, R.S. (1986) Cretaceous extinctions and wildfires. *Science*, 234, 261-264.

Avermann, M.E. (1994) Origin of the polymict, allochtonous breccias of the Onaping Formation, Sudbury Structure, Ontario, Canada. *In Large Meteorite Impacts and Planetary Evolution*, B.O. Dressler, R.A.F. Grieve and V.L. Sharpton, Eds., Geol. Soc. Amer. Spec. Paper, 293, 264-267.

Becker, L., Bada, J.L., Winans, R.E., Bunch, T.E. and French, B.M. (1994) Fullerenes in the 1.85-billion-year-old Sudbury impact structure. *Science*, 265, 642-645.

Becker, L., Poreda, R.J. and Bada, J.L. (1996) Extraterrestrial helium trapped in fullerenes in the Sudbury Impact Structure. *Science*, 272, 249-252.

Boorum, M.M., Vasil'ev, Y.V., Drewello, T. and Scott, L.T. (2001) Groundwork for rational synthesis of C_{60}: Cyclodehydrogenation of a $C_{60}H_{30}$ polyarene. *Science*, 294, 828-831.

Buseck, P.R. (2002) Geological fullerenes: review and analysis. *Earth Planet. Sci. Lett.*, 203, 781-792.

Buseck, P.R., Tsipurski, S.J. and Hettich, R. (1992) Fullerenes from the geological environment. *Nature*, 247, 215-217.

Cane, R.F. and Albion, P.R. (1973) The organic geochemistry of torbanite precursors. *Geochim. Cosmochim. Acta*, 37, 1543-1549.

Crowley, C., Taylor, R., Kroto, H.W., Walton, D.R.M., Cheng, P.C. and Scott, L.T. (1996) Pyrolytic production of fullerenes. *Synth. Metals*, 77, 17-22.

Dressler, B.O., Morrison, G.G., Peredery, W.V. and Rao, B.V. (1987) The Sudbury Structure, Ontario, Canada-A review. *In Research in Terrestrial Impact Structures*, J. Pohl, Ed., 39-68, Viehweg & Sohn, Braunschweig, Germany.

Fetzer, J.C., Ed. (2000) Large (C>24) Polycyclic Aromatic Hydrocarbons, 288p., Wiley Interscience, New York, New York, USA.

Fischer, K., Largeau, C. and Derenne, S. (1996) Can oil shales be used to produce fullerenes? *Org. Geochem.*, 24, 715-723.

Gatellier, J.P.L.A., de Leeuw, J.W., Sinninghe Damsté, J.S., Derenne, S., Largeau, C. and Metzger, P. (1993) A comparative study of macromolecular substances of a Coorongite and cell walls of the extant alga Botryococcus braunii. *Geochim. Cosmochim. Acta*, 57, 2053-2068.

Gómez-Lor, B., Koper, C., Fokkens, R.H., Vlietstra, E.J., Cleij, T.J, Jenneskens, L.W., Nibbering, N.M.M. and Echevarren, A.M. (2002) Zipping up the 'crushed fullerene' $C_{60}H_{30}$: C_{60} by fifteen intramolecular H_2 losses. J. *Chem. Soc., Chem. Comm.*, 2002, 370-371.

Heymann, D., Chibante, L.P.F., Brooks, R.R., Wolbach, W.S. and Smalley, R.E. (1994) Fullerenes in the K/T boundary layer. *Science*, 265, 645-647.

Heymann, D., Dressler, B.O., Knell, J., Thiemens, M.H., Buseck, P.R., Dunbar, R.B. and Mucciarone, D. (1999) Origin of carbonaceous matter, fullerenes, and elemental sulfur in rocks of the Whitewater Group, Sudbury Impact Structure, Ontario, Canada. *In Large Meteorite Impacts and Planetary Evolution II*, B.O. Dressler and V.L. Sharpton, Eds., Geol. Soc. Amer., Spec. Paper, 339, 345-360.

Heymann, D., Jenneskens, L.W., Jehlička, J., Koper, C. and Vlietstra, E.J. (2003) Biogenic Fullerenes? *Int. J. Astrobiol.*, 2, 179-183.

Jehlička, J., Ozawa, M., Slanina, Z. and Osawa, E. (2000) Fullerenes in solid bitumens from pillow lavas of Precambrian age (Mitov, Bohemian massif). *Fullerene Sci. Techn.*, 8, 449-452.

Kannenberg, E.L. and Poralla, K. (1999) Hopanoid biosynthesis and function in bacteria. *Naturwissenschaften*, 86,168-176.

Kovalevski, V.V., Buseck, P.R. and Cowley, J.M. (2001) Comparison of carbon in shungite rocks to other natural carbons: An X-ray and TEM study. *Carbon*, 39, 243-256.

Lafleur, A.L., Howard, J.B., Plummer, E., Taghizadeh, K., Necula, A., Scott, L.T., Swallow, K.C. (1998) Identification of some novel cyclopenta-fused polycyclic aromatic hydrocarbons in ethylene flames. *Polycyclic Aromatic Compounds*, 12, 223-237.

Lafleur, A.L., Howard, J.B., Taghizadeh, K., Plummer, E., Scott, L.T., Necula, A. and Swallow, K.C. (1996a) Identification of $C_{20}H_{10}$ Dicyclopentapyrenes in flames: Correlation with corannulene and fullerenes formation. *J. Phys. Chem.*, 100, 17421-17428.

Lafleur, A.L., Taghizadeh, K., Howard, J.B., Anacleto, J.F. and Quilliam, M.A. (1996b) Characterization of flame-generated C^{10} to C_{160} polycyclic aromatic hydrocarbons by atmospheric-pressure chemical ionization mass spectrometry with liquid introduction via nebulizer interface. *J. Am. Soc. Mass Spectrometry*, 7, 276-286.

Lewis, R.S., Ebihara, M. and Anders, E. (1982) Unpaired electrons: An association with primordial gases in meteorites (abstract). *Meteoritics*, 17, 244-245.

Marsh, N.D., Wornat, M.J., Scott, L.T., Necula, A., Lafleur, A.L. and Plummer, E.F. (2000) The identification of cyclopenta-fused and ethynyl-substituted polycyclic aromatic hydrocarbons in benzene droplet combustion products. *Polycyclic Aromatic Compounds*, 13, 379-402.

Melezhik, V.A., Fallick, A.E., Filippov, M.M. and Larsen, O. (1999) Karelian shungite - an indication of 2.0-Ga-old metamorphosed oil-shale and generation of petroleum: geology, lithology and geochemistry. *Earth Sci. Rev.*, 47, 1-40.

Miller, J.D.A. (1962) Fats and Steroids. *In Physiology and Biochemistry of Algae*, R.A. Levin, Ed., 357-370, Academic Press, New York, New York, USA.

Mukhopadhyay, K., Krishna, K.M. and Sharon, M. (1994) Fullerenes from camphor: A natural source. *Phys. Rev. Lett.*, 72, 3182-3185.

Osawa, E. (1999) Natural fullerenes-Will they offer a hint to the selective synthesis of fullerenes? *Fullerene Sci. Techn.*, 7, 637-652.

Pouba, Z. and Kříbek, B. (1986) Organic matter and the concentration of metals in Precambrian stratiform deposits of the Bohemian Massif. *Precambrian Res.*, 33, 225-237.

Rose, H.R., Dance, I.G., Fischer, K.J., Smith, D.R., Willett, G.D. and Wilson, M.A. (1993) Calcium inside C_{60} and C_{70}-from coorongite, a precursor of torbanite. *J. Chem. Soc., Chem. Commun.*, 1993, 941-942.

Rose, H.R., Dance, I.G., Fischer, K.J., Smith, D.R., Willett, G.D. and Wilson, M.A. (1994) From green algae to calcium inside buckyballs. *Org. Mass Spectrom.*, 29, 470-474.

Sarobe, M., Fokkens, R.H., Cleij, T.J., Jenneskens, L.W., Nibbering, N.M.M., Stas, W. and Versluis, C. (1999a) S_8-mediated cycvlotrimerization of 4,5-dihydrobenz[l] acephenanthrylene : trinaphtodecacyclene ($C_{60}H_{30}$) isomers and their propensity towards cyclodehydrogenation. *Chem. Phys. Lett.*, 313, 31-39.

Sarobe, M., Kwint, H.C., Fleer, T., Havenith, R.W.A., Jenneskens, L.W., Vlietstra, E.J., van Lenthe J.H. and Wesseling, M. (1999b) Flash vacuum thermolysis of acenaphtol[1,2-a]acenaphtylene, fluoranthene, benzo[k]-and benzo[j]fluoranthene-Homolytic scission of carbon-carbon single bonds of internally fused cyclopenta moieties at T>1100 degrees C. *Eur. J. Org. Chem.*, 1999, 1191-1200.

Simoneit, B.R.T. (1990) Petroleum generation, an easy and widespread process in hydrothermal systems; an overview. *Appl. Geochem.*, 5, 3-15.

Taylor, R., Langley, G.J., Kroto, H.W. and Walton, D.R.M. (1993) Formation of C_{60} by pyrolysis of naphthalene. *Nature*, 366, 728-731.

Wornat, M.J., Vernaglia, B.A., Lafleur, A.L., Plummer, E.F., Taghizadeh, K., Nelson, P.F., Li, C.Z., Necula, A. and Scott, L.T. (1998) Cyclopenta-fused polycyclic aromatic hydrocarbons from brown coal pyrolysis. *Twenty-seventh Internl. Symp. Combustion*, 1677-1686, The Combustion Institute, Pittsburgh, Pennsylvania, USA.

Wornat, M.J., Vriesendorp, J.J., Lafleur, A.L., Plummer, E.F., Necula, A. and Scott, L.T. (1999) The identification of new ethynyl-substituted and cyclopenta-fused polycyclic aromatic hydrocarbons in the products of anthracene pyrolysis. *Polycyclic Aromatic Compounds*, 13, 221-240.

Yoshimura, K., Przybilla, L., Ito, S., Brand, J.D., Wehmeir, M., Räder, H.J. and Müllen, K. (2001) Characterization of large synthetic polycyclic aromatic hydrocarbons by MALDI- and LD-TOF mass spectrometry. *Macromol. Chem. Phys.*, 202, 215-222.

Chapter 14

FUTURE PROCEDURES FOR ISOLATION OF HIGHER FULLERENES IN NATURAL AND SYNTHETIC SOOT

LUANN BECKER
Department of Geological Sciences, Institute of Crustal Studies, University of California, Santa Barbara, California 93106, USA.

ROBERT J. POREDA
Department of Earth and Environmental Sciences, University of Rochester, Rochester, New York 14627, USA.

JOSEPH A. NUTH
Goddard Space Flight Center, Greenbelt, Maryland 20771, USA.

FRANK T. FERGUSON
Chemistry Department, The Catholic University of America, Washington D.C. 20064, USA.

FENG LIANG
Rice Chemistry and Biochemistry Department, Rice University, Houston, Texas, USA.

W. EDWARD BILLUPS
Rice Chemistry and Biochemistry Department, Rice University, Houston, Texas, USA.

Abstract: We describe the extraction methodologies for the isolation of fullerenes in synthetic "Graphitic Smokes" soot. These same methods were used to isolate natural fullerenes in some carbonaceous chondrite meteorites and in meteor impact-related sedimentary deposits previously described in the literature. In addition, a new functionalizatoin methodology that significantly enhances the yield of fullerenes extracted in synthetic fullerene material (up to 20%

Frans J.M. Rietmeijer (ed.), Natural Fullerenes and Related Structures of Elemental Carbon, 279–295.

fullerene) will be discussed. Both laser desorption-mass spectrometry and high-resolution transmission electron microscopy were used to characterize fullerenes extracted from the synthetic soot. These extraction methods promise to reveal new insights on the encapsulaton of noble gases in a variety of fullerene and fullerene-related structures up to C_{1000}.

Key words: Carbonaceous chondrites; carbon vapor condensation; graphitic smokes; high-resolution transmission electron microscopy (HRTEM); higher fullerenes; Murchison meteorite; laser desorption-mass spectrometry (LDMS); noble gas-carbon atmospheres; soot

1. SYNTHETIC FULLERENES IN GRAPHITIC SMOKES

First, we review details of laboratory techniques we have adapted to investigate fullerenes as a carrier phase of noble gases in natural, e.g. carbonaceous chondrite meteorites (Becker et al., 2000; Pizzarello et al., 2001), and meteor-impact related sedimentary deposits (Becker et al., 2000, 2001; Poreda and Becker, 2003), and synthetic fullerenes in "Graphitic Smokes" (*GS*) material (Olsen et al., 2000). Olsen et al. (2000) had previously demonstrated that xenon was trapped in this material in higher amounts than found in the acid resistant residues of carbonaceous chondrite meteorites, but the actual synthetic carbon carrier phase of Xe was poorly understood, as is also the case for the meteorites. Both the *GS* and arc-discharge apparatuses, such as the original arc-discharge evaporation apparatus used in the discovery of fullerene (Krätschmer et al., 1990), produce a carbon soot material. Saunders et al. (1996) and Giblin et al. (1997) had already shown that fullerenes (C_{60}, C_{70}, C_{84}) extracted from the arc-evaporated soot are capable of trapping all of the noble gases.

We obtained ten *GS* samples synthesized in a fixed noble gas mixture of 49% Ne, 49% Ar, 1% Xe and 1% Kr that was mixed with helium to balance the total pressure at 200 torr. The matrix of these *GS* experiments consisted of smokes condensed from an atmosphere that contained 25%, 50%, 75% and 100% of this noble gas mixture with proportionally added helium. These samples with a demonstrated high amount of trapped Xe, and most likely also of the other noble gases, were prepared in order to further evaluate the role of fullerene as a noble gas carrier phase in some extraterrestrial environments and whether there would be a preference for a particular fullerene cage size or sizes. To determine the various fullerene distributions and the contents of these samples we used the two-step extraction method for isolating fullerenes from natural samples:

1. Toluene extraction to separate predominately C_{60}, C_{70} up to C_{100}, followed by
2. Extraction with a high boiling point solvent (*1,2,4 trichlorobenzene or 1,2,3,5 tetramethylbenzene*) (*TCB*) to separate fullerenes in the C_{100} to C_{300} range (Becker et al., 2000).

Using this two-step extraction technique we find that the *GS* soot material is dominated by fullerenes C_{60} up to C_{300}, which suggests that the higher fullerenes play an important role in the trapping of noble gases in their environments.

Second, we describe new extraction methodologies that should result and higher yields of pure fullerenes (up to 20%) and better separation of individual fullerene cages. These methodologies will be applied to synthetic *GS* smokes that contain several percent fullerenes that are the carriers of noble gases trapped in this material.

2. TOLUENE EXTRACTS OF C_{60}, C_{70} AND SOME HIGHER FULLERENES

Laser desorption-mass spectrometry (LDMS) was used to identify fullerenes that were extracted from soot produced in the "Graphitic Smokes" apparatus. The LDMS spectrum for sample *GS*#1 (Fig. 1), obtained at 200 torr of He and 0 torr of pre-mixed noble gas mixture, revealed fullerenes that were mostly C_{60}^+ and C_{70}^+ but up to C_{94}^+ with an increase in the abundance of C_{70}^+ in 2 to 1 proportions rather than the C_{60}^+ to C_{70}^+ ratios of ~5 to 1 obtained when using the arc-discharge carbon evaporation apparatus for the laboratory production of fullerenes (e.g. Krätschmer et al., 1990).

The *GS*#2 sample (Fig. 2) obtained at 150 torr of He and 50 torr of the pre-mixed noble gases yielded a much higher abundance of C_{70}^+ that was almost equal compared to C_{60}^+ abundance, and trace amounts of higher fullerenes up to C_{94}^+. This observation is the first indication that increasing the partial pressure of the noble gas mixture enhances the ability for a *specific fullerene cage* to form that is known to encapsulate noble gases; in this case, high C_{70}^+ relative to C_{60}^+. It further suggests that even a slight increase in the noble gas pressures influences the formation (kinetics) for the production of these carbon cages in the "Graphitic Smokes" apparatus.

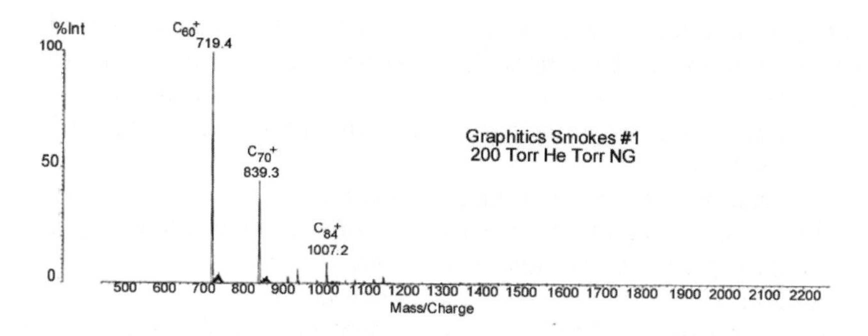

Figure 1. Laser desorption-mass spectrum (% intensity vs. mass/charge ratio) of the *GS*#1 toluene extract shows mostly C_{60}^+ and C_{70}^+ and low abundances of the higher fullerenes up to C_{94}^+. The "Graphitic Smokes" apparatus produced a higher C_{70}^+ abundance relative to C_{60}^+ than in the arc evaporator apparatus

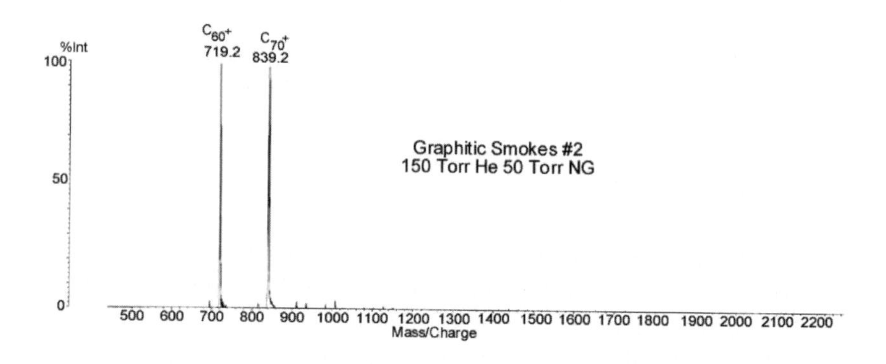

Figure 2. Laser desorption-mass spectrum (% intensity vs. mass/charge ratio) of the *GS*#2 toluene extract with almost equal abundances of C_{60}^+ and C_{70}^+ and only trace amounts of higher fullerenes

The formation of higher fullerenes, up to C_{200}, continued with increasing pressure of the noble gas mixture and even larger fullerene cages forming at enhanced abundances as observed in sample *GS*#4 obtained at 50 torr of He and 150 torr of the noble gas mixture (Fig. 3). This was the first observation of larger than C_{94}^+ fullerene cages up to C_{200}^+ that were in this experimental matrix of the *GS* samples. This trend of forming increasingly larger fullerenes as a function of noble gas-mixture pressure reversed in the toluene extract sample *GS*#6, obtained at 150 torr of He and 50 torr of the noble gas mixture with the formation of mostly C_{60}^+ and C_{70}^+, which is similar to the result obtained for sample *GS*#1.

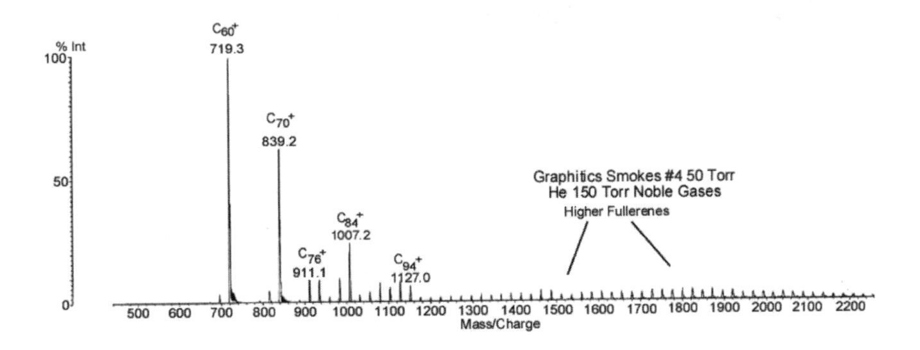

Figure 3. Laser desorption-mass spectrum (% intensity vs. mass/charge ratio) of the *GS*#4 toluene extract. With increasing pressure on the noble gases in the graphitic smokes apparatus several higher fullerenes up to C_{96}^+ were observed

A gradual increase in the abundances of the higher fullerenes from 1% to 4% was observed for the toluene-extracted samples *GS*#7 through *GS*#10 with more-or-less the same 2 to 1 ratios for C_{60}^+ and C_{70}^+. In all samples *GS*#7 through *GS*#10, the abundances of C_{60}^+ and C_{70}^+ in the toluene-extracted soot were decreasing while some larger fullerene cages up to C_{94}^+ were increasing in size but not necessarily in abundance (Fig. 3). This result is likely due to the limited solubility of the higher fullerenes in toluene in the samples *GS*#5 through

GS#10, which is not necessarily related to the yield of higher fullerenes that may be formed during smoke condensation.

As observed in several carbonaceous chondrites (Becker et al., 2000; Pizzarello et al., 2001), it is the higher fullerenes C_{100} to C_{300} that are the dominant carrier phase for the noble gases. Thus, we expect to learn more about what types of extraterrestrial environments are most conducive to fullerene formation and the specific fullerene cages that most efficiently retain the various noble gases.

3. *TCB* EXTRACTS FOR HIGHER FULLERENES $>C_{100}$

The results for the *TCB* extracts were remarkably similar for all *GS* samples with a distribution of higher fullerenes between C_{100}^+ and C_{260}^+ and no obvious variation in the abundances of any particular subset of these fullerene cages (Fig. 4).

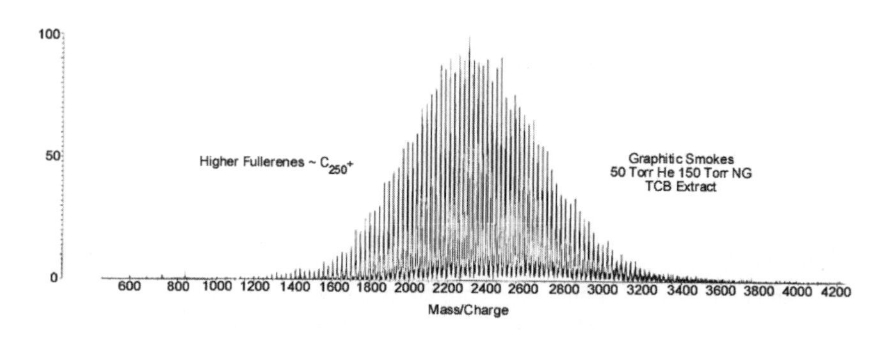

Figure 4. A typical laser desorption-mass spectrum (% intensity vs. mass/charge ratio) of the higher fullerenes in *GS* sample *TCB*#4. Note: The distribution of fullerenes did not change for these extracts with increasing partial pressures of the pre-mixed noble gas mixture

A small amount of C_{60}^+ and C_{70}^+ was present in sample *GS#5* (Fig. 5), but these fullerenes were not observed in any of the other *TCB* extracts. Their presence in sample *GS#5* is attributed to either an inefficient extraction of these fullerenes in the toluene step or perhaps inadvertent contamination from the glassware when switching to the *TCB* solvent extract step.

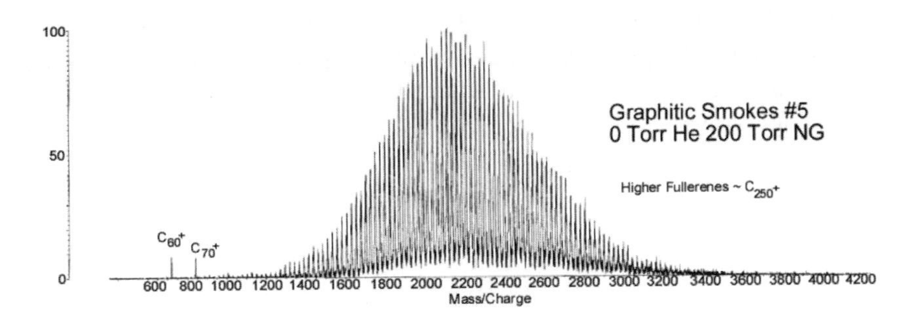

Figure 5. Laser desorption-mass spectrum showing (% intensity vs. mass/charge ratio 600-4200) trace amounts of C_{60}^+ and C_{70}^+ detected in the *GS* sample extract *TCB#5*, which is attributed to either an inefficient toluene extraction or possible, inadvertent, contamination from the glassware

The observed distribution of higher fullerenes smaller than about C_{300}^+ may, in part, be due to the resolution of the LDMS instrument with increasing mass, or because the extraction methodology is not efficiently isolating fullerene cages greater than C_{300}^+. Work is underway to further evaluate the extraction protocol including the introduction of a new method to extract even larger fullerene cages.

4. A NEW EXTRACTION METHODOLOGY – FUNCTIONALIZING FULLERENES

4.1 Methodology

The preliminary LDMS results for the *GS* toluene and *TCB* extracts demonstrated that C_{60} and higher fullerenes are formed during graphitic smoke condensation in the apparatus that was designed to identify the condensed carbon solids that would be likely carriers of the noble gases. The mass for a typical condensed *GS* soot sample is only a few 100 milligrams and the subsequent solvent-extracted yields are quite low, i.e. ca. 100 to 200 μg per condensed sample. Thus, future investigations will either have to process more *GS* material (i.e. grams) to enhance fullerene abundances, or more efficient fullerene extraction procedures must be developed.

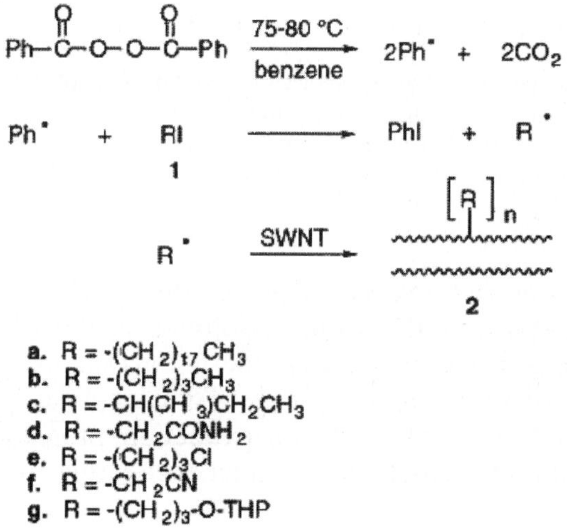

$$Ph{-}C(=O){-}O{-}O{-}C(=O){-}Ph \xrightarrow[\text{benzene}]{75\text{-}80\ °C} 2Ph^{\bullet} + 2CO_2$$

$$Ph^{\bullet} + \underset{\mathbf{1}}{RI} \longrightarrow PhI + R^{\bullet}$$

$$R^{\bullet} \xrightarrow{\text{SWNT}} \mathbf{2}$$

a. $R = -(CH_2)_{17}CH_3$
b. $R = -(CH_2)_3CH_3$
c. $R = -CH(CH_3)CH_2CH_3$
d. $R = -CH_2CONH_2$
e. $R = -(CH_2)_3Cl$
f. $R = -CH_2CN$
g. $R = -(CH_2)_3\text{-O-THP}$

Figure 6. A large variety of functional groups (a-g) may be preferentially added to the larger fullerene cages and nanotubes when benzoyl peroxide is gently heated and decomposes in the presence of alkyl iodides (Ying et al., 2003)

We are exploring some new extraction approaches that will enhance the yield of fullerenes in laboratory-produced analogs and natural fullerenes in terrestrial and extraterrestrial samples. Dr. Billups and colleagues at Rice University have developed a method for functionalizing the larger fullerene cages from the "Rice HiPco synthetic soot material" following the preparation scheme in Fig. 6.

This soot material contains small nanotubes and higher fullerenes (Fig 7). The latter includes about 20% more higher fullerenes from C_{100} to C_{300} than obtained in a typical arc-discharge process.

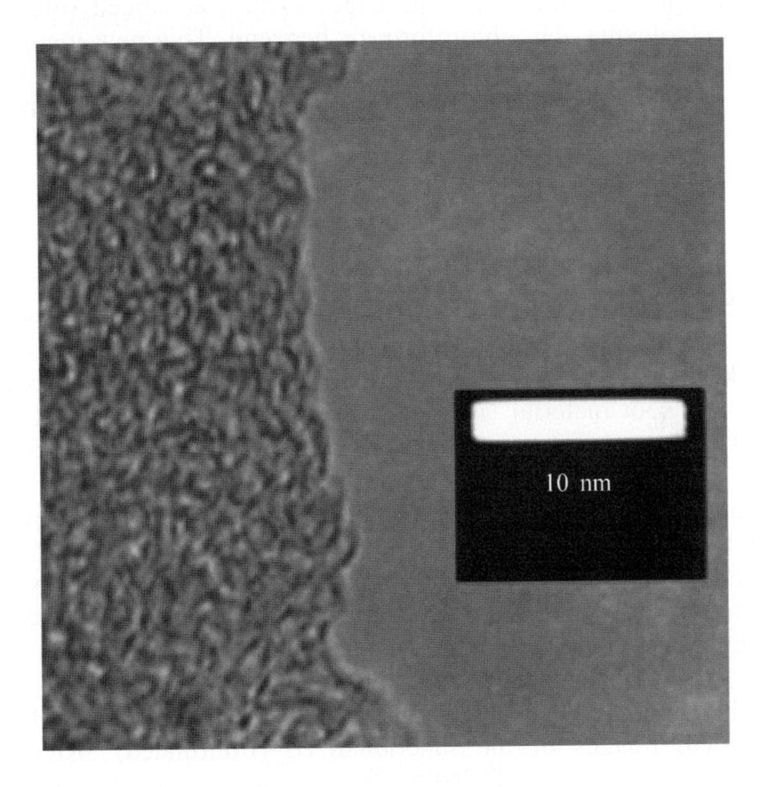

Figure 7. High-Resolution Transmission Electron micrograph of a functionalized (Ying et al., 2003) "Rice HiPCo synthetic fullerene-containing soot material" showing circular rings of higher fullerenes and slightly elongated ring structures and tubular, nanotube structures. Scale bar is 10 nm

Ying et al. (2003) have demonstrated that the larger fullerene cages and some small nanotubes can be more readily extracted with organic solvents when first functionalizing the higher fullerenes present in the "HiPco soot" by the addition of free radicals. They found that benzoyl peroxide is a suitable source for generating phenyl radicals. Benzoyl peroxide is readily available and breaks down under mild temperatures (70-80 ^{o}C) forming CO_2 and phenyl radicals. Since alkyl iodides react in a diffusion-controlled process with radicals, a large variety of functional groups can be added to the fullerene cages when, for example, benzoyl peroxide decomposes in the presence of alkyl iodides. The process results in enhanced yields of isolated, higher fullerenes from the bulk "Rice HiPco synthetic soot material" of up to 1 wt% after purification (Ying et al., 2003).

The functionalized fullerene cages exhibit enhanced solubility in organic solvents, e.g. chloroform or benzene. This new method may eventually replace the high boiling solvent step we have used to extract the higher fullerenes and allow for a simple dissolution step of the functionalized higher fullerenes in a lower boiling point solvent (Ying et al., 2003; Sadana et al., 2005).

4.2 Preliminary Results

The functionalized "Rice HiPco synthetic soot material" was analyzed using LDMS and compared to *TCB* extracts of "Graphitic Smokes" material (Fig. 8). The LDMS mass spectrum for "Rice HiPco synthetic soot material" indicates carbon that is dominated by large-fullerene cages. There is a marked shift in the LDMS spectra toward even larger fullerene cages in the benzene-extracted HiPco material in comparison to the *TCB* extracted *GS* material (Fig. 8). This result may be due to the increasing solubility of the higher fullerenes using the functionalizing method in comparison to our two-step extraction method, or it might be that very large (up to C_{1000}) fullerene cages are preferentially produced in the HiPco apparatus.

The High Resolution Transmission Electron Microscope (HRTEM) dark-field images of the functionalized (fluorinated) "Rice HiPco synthetic soot material" taken at low-magnification (top image) showing the sample supported on the lacey carbon support film and at high magnification (lower image) showing the high density of large fullerenes (Fig. 9).

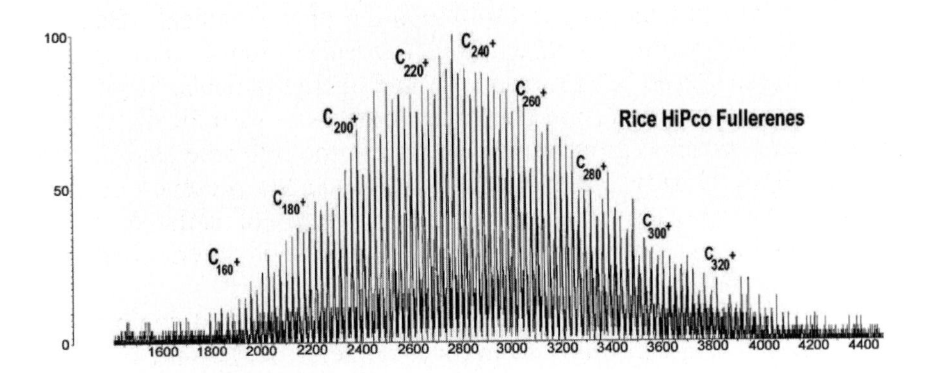

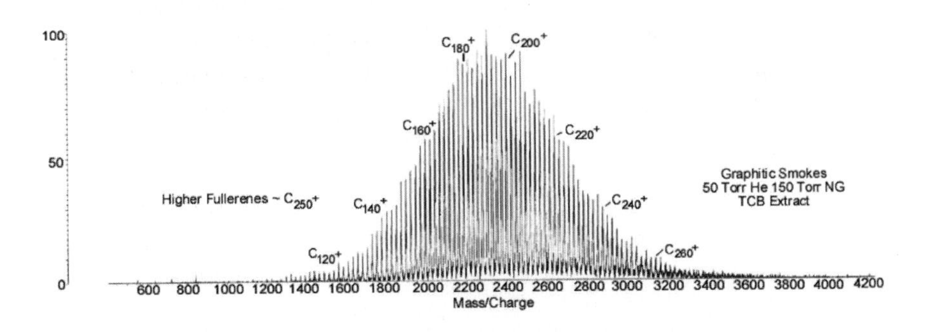

Figure 8. Laser-desorption-mass spectra showing the higher fullerene distributions between C_{160}^+ and C_{320}^+, in the "Rice HiPco" sample functionalized using the scheme in Fig. 6 and extracted with chloroform (% intensity vs. mass/charge ratio) (at the top) and between C_{120}^+ and C_{260}^+, in a *GS* soot that was extracted using *1,2,4 trichlorobenzene* (*TCB*) (% intensity vs. mass/charge ratio) (at the bottom)

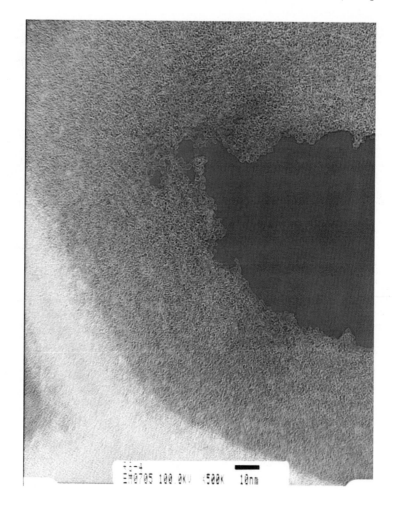

Figure 9a. High-Resolution Transmission Electron, dark-field, micrographs of the functionalized (fluorinated) synthetic fullerene “Rice HiPco material”. Scale bars are 10 nm

Image taken at low magnification showing the sample (gray) placed on top of the lacey carbon film shown white in the lower left-hand corner)

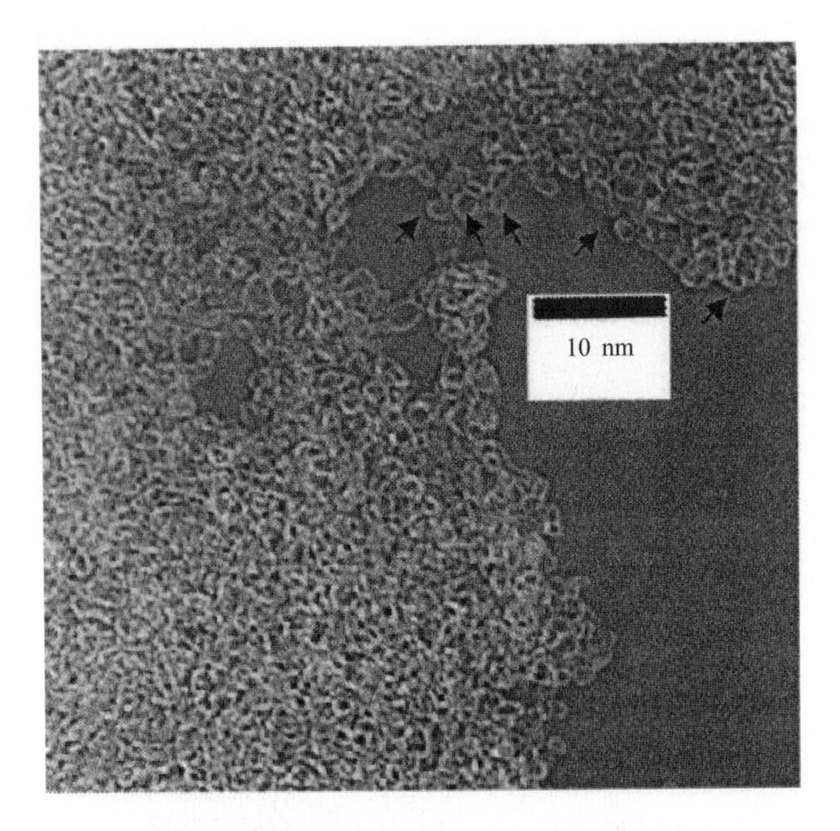

Figure 9b. Image taken at high magnification showing the circular and elongated ring structures for large fullerene cages of different diameter in this "Rice HiPco material"

We applied the procedure outlined in the functionalization scheme (see, Fig. 6) to the Murchison CM carbonaceous chondrite meteorite in an attempt to separate larger fullerene cages and fullerene-related carbon material from is bulk acid residue. The Murchison material was analyzed by LDMS (Fig. 10) and compared to the functionalized synthetic "Rice HiPco" material.

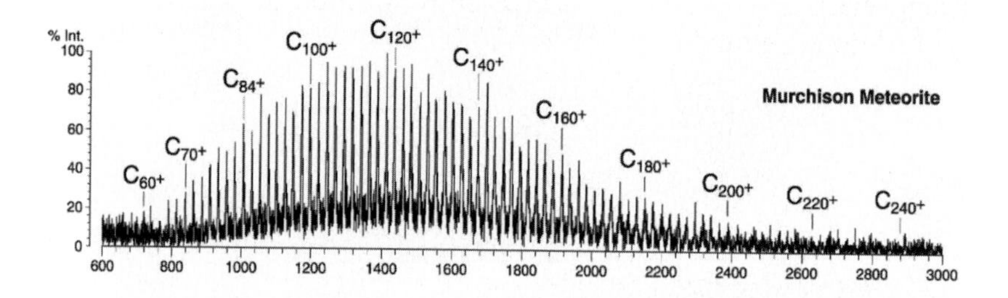

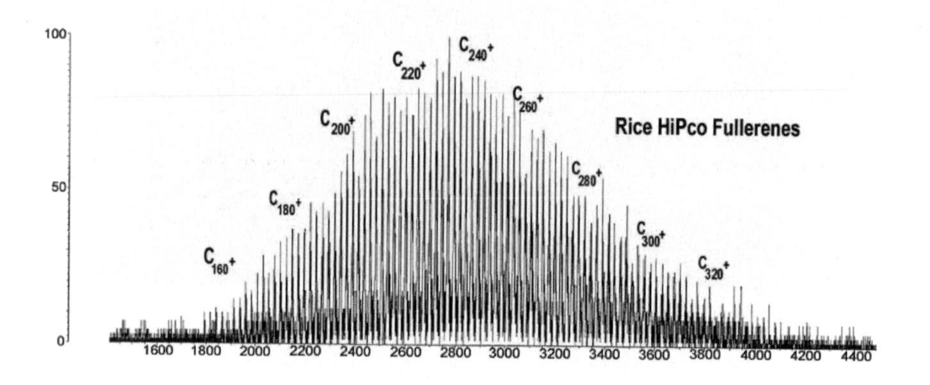

Figure 10. Laser desorption-mass spectra of the Murchison natural fullerenes extracted with *TCB* (top; % intensity vs. mass/charge ratio) is compared to the "Rice HiPco" fullerenes (bottom; % intensity vs. mass/charge ratio) and shows a marked shift of the functionalized "HiPco" fullerenes attributed to an enhanced yield of higher fullerenes in this synthetic material

While the HRTEM image of the functionalized (fluorinated) "Rice HiPco synthetic fullerene material" (see, Fig. 9) shows an abundance of large fullerene-related cages, the preliminary HRTEM results for the functionalized Murchison residue show limited or patchy areas of closed fullerene-related cages (Fig. 11).

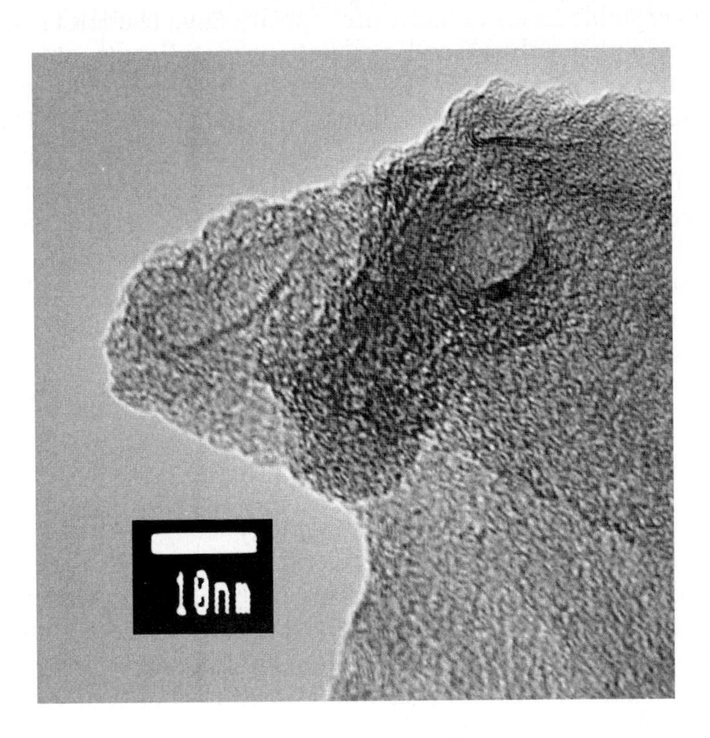

Figure 11. High-Resolution Transmission Electron micrograph of the functionalized Murchison residue showing closed fullerene-related cages, i.e. the ring structures along he edge of the sample. This functionalized acid resistant Murchison residue likely reflects even larger cages then was previously found in extracted Murchison material using *TCB* (Becker et al., 2001). Scale bar is 10 nm

These results applying functionalization of the higher fullerenes to meteorite residues are preliminary and more work is needed to properly assess the functionalization method of natural fullerene-bearing samples.

4.3 Significance of Results

The ability to isolate a specific carbon fraction of the carbonaceous chondrite acid residue that only complexes to a higher fullerene-related phase, is a critical step in unraveling the nature and abundances of carbon phases in general and the nature of the carbon carriers of noble gases in meteorites specifically. The HRTEM images of the two-step solvent and functionalized fullerene extracts have demonstrated that some of the carbon in meteoritic acid residues is *large* and has *structure* rather than *small* and *amorphous*. The initial HRTEM results for the Murchison residue indicate that many of the larger fullerenes could be nested fullerenes, that is, a concentric nanostructure of smaller fullerenes enclosed by larger ones like a Russian doll. The next challenge will be to ascertain if nested fullerenes are closed cages capable of encapsulating noble gases inside of their structure that requires the smallest fullerene in this structure can be no less than C_{60}.

5. CONCLUSIONS

We applied our extraction methodology using organic solvents to synthetic soot to isolate fullerenes up to C_{300}. We explored the formation of higher fullerenes by varying the partial pressures of the noble gas mixture in a 200-torr atmosphere with the helium fill-gas maintained inside the "Graphitic Smokes" apparatus. Increasing the noble gas mixture's partial pressure resulted in enhanced yields, up to 5% of fullerenes C_{60} to C_{94} in toluene-extracted residues. The fullerenes extracted using *1,2,4 trichlorobenzene* (*TCB*) did not vary in abundance or size with increasing partial pressure of the pre-mixed noble gas mixture.

A new functionalized method for the extraction of fullerenes was also applied to the *GS* soot smokes and the Murchison meteorite. This method resulted in enhanced yields of the higher fullerenes up to 20% in the "Graphitic Smokes" materials and even larger fullerene cages up to C_{800} in the "Rice HiPco soot material" in comparison to our solvent extraction method. Future investigations will include studies of individual fullerene cages and the encapsulation of noble gases inside the fullerene structure.

Clearly the higher fullerenes are the predominant carrier of the noble gases in the carbonaceous chondrites. As the preliminary *GS* experiments show, the higher fullerenes are forming with increasing pressure on the noble gases and by implications may lead to a better understanding of how and where fullerenes form in circumstellar and interstellar environments. As fullerenes are accreted into grains and incorporated into asteroids and comets, the distribution of fullerenes will likely vary due to the changing conditions (e.g. temperature) and processing on the parent bodies.

Acknowledgements: We thank Rick Smalley at Rice University in Houston (TX) for the use of the HRTEM facility to image the fullerenes in our study. We also thank Anil Sandana at Rice University for assistance with the functionalization of the Murchison meteorite and the Rice HiPco soot material. We thank the Robert A. Welch Foundation (C-0490) and the National Science Foundation (CHE-0011486).

6. REFERENCES

Becker, L., Poreda, R.J. and Bunch, T.E. (2000) Fullerene: A new extraterrestrial carbon carrier phase for noble gases. *Proc. Natl. Acad. Sci.*, 97, 2979-2983.

Becker, L., Poreda, R.J., Hunt, A.G., Bunch, T.E. and Rampino, M. (2001) Impact event at the Permian-Triassic boundary: Evidence from extraterrestrial noble gases in fullerenes. *Science*, 291, 1530-1533.

Giblin, D.E., Gross, M.L., Saunders, M., Jimenez-Vazquez, H.A. and Cross, R.J. (1997) Incorporation of helium and endohedral complexes of C60 and C70 containing noble-gas atoms: A tandem mass spectrometry study. *J. Am. Chem. Soc.*, 119, 9883-9890.

Krätschmer, W., Lamb, L.D., Fostiropoulos, K. and Huffman, D.R. (1990) Solid C_{60}: A new form of carbon. *Nature*, 347, 354-357.

Olsen, E.K., Swindle, T.D., Nuth, J.A. and Ferguson, F. (2000) Noble gases in graphitic smokes. *Lunar Planet. Sci.*, 31, abstract #1479, Lunar and Planetary Institute, Houston, Texas, USA (CD-ROM).

Pizzarello, S., Hang, Y., Becker, L., Podera, R.J., Nieman, R.A., Cooper, G. and Williams, M. (2001) The organic content of the Tagish Lake meteorite. *Science*; 293, 2236-2239.

Poreda, R.J. and Becker, L. (2003) Fullerenes and interplanetary dust at the Permian-Triassic boundary. *Astrobiology*, 3, 120-136.

Sadana, A.K., Liang, F., Brinson, B., Arepalli, S., Farhat, S., Hauge, R.H., Smalley, R.E. and Billups, W.E. (2005) Functionalization and extraction of large fullerenes and carbon-coated metal formed during the synthesis of single wall carbon nanotubes by laser oven, direct current arc, and high-pressure carbon monoxide production methods. *J. Phys. Chem. B*, 109, 4416-4418.

Saunders, M., Jimenez-Vasquez, H.A., Cross, R.J. and Poreda, R.J. (1993) Stable compounds of helium and neon He@C_{60} and He@C_{70}. *Science*, 259, 1428-1431.

Ying, Y., Saini, R.K., Liang, F., Sadana, A.K and Billups, W.E. (2003) Funtionalization of the carbon nanotubes by free radicals. *Org. Lett.*, 5, 1471-1473.